Contents

KU-767-761

This book has been written for the *AQA GCSE Physics* and *AQA GCSE Combined Science: Trilogy courses,* making them completely co-teachable. Physics only lessons are easily identifiable with their own black-bordered design, and are also formatted in italics in the below contents list for quick access.

Required Practicals

Practical work is a vital part of physics, helping to support and apply your scientific knowledge, and develop your investigative and practical skills. As part of your GCSE physics course, there are 10 required practicals that you must carry out. Questions in your exams could draw on any of the knowledge and skills you have developed in carrying out these practicals.

A·Required practical feature box has been included in this student book for each of your required practicals. Further support is available on Kerboodle.

Required practicals	Topic
1 **Determining specific heat capacity.** Determine the specific heat capacity of a metal block of known mass by measuring the energy transferred to the block and its temperature rise, and using the equation for specific heat capacity.	P2.4
2 **Investigating thermal insulators.** Use different materials and different thicknesses of the same material to insulate identical beakers of hot water, and measure the change in temperature of the water at regular intervals.	P2.1
3 **Investigating resistance.** Set up circuits and investigate the resistance of a wire, and of resistors in series and parallel.	P4.3 P4.6
4 **Investigating electrical components.** Correctly assemble a circuit and investigate the potential difference-current characteristics of circuit components.	P4.4
5 **Calculating densities.** Measure the mass and volume of objects and liquids and calculate their densities using the density equation.	P6.1
6 **Investigate the relationship between force and extension for a spring.** Hang weights of known mass from a spring and, using the correct apparatus, measure the resulting extension. Use the results to plot a force-extension graph.	P10.8
7 **Investigate the relationship between force and acceleration.** Using a newton-metre, investigate the effect on the acceleration of an object of varying the force on it and of varying its mass.	P10.1
8 **Investigating plane waves in a ripple tank and waves in a solid.** Determine which apparatus are the most suitable for measuring the frequency, speed, and wavelength of waves in a ripple tank, and investigate waves on a stretched string.	P12.4
9 **Investigate the reflection and refraction of light.** Use different substances and surfaces to investigate the refraction and relection of light.	P14.2 P14.3
10 **Investigating infrared radiation.** Determine how the properties of a surface affect the amount of infrared radiation absorbed or radiated by the surface.	P13.2

How to use this book

Learning objectives

- Learning objectives at the start of each spread tell you the content that you will cover.
- Any objectives marked with the higher tier icon **H** are only relevant to those who are sitting the higher tier exams.

This book has been written by subject experts to match the new 2016 specifications. It is packed full of features to help you prepare for your course and achieve the very best you can.

Key words are highlighted in the text. You can look them up in the glossary at the back of the book if you are not sure what they mean.

The diagrams in this book are as important for your understanding as the text, so make sure you revise them carefully.

Synoptic link

Synoptic links show how the content of a topic links to other parts of the course. This will support you with the synoptic element of your assessment.

There are also links to the Maths skills for Physics chapter, so you can develop your maths skills whilst you study.

Practical

Practicals are a great way for you to see science in action for yourself. These boxes may be a simple introduction or reminder, or they may be the basis for a practical in the classroom. They will help your understanding of the course.

Required practical

These practicals have important skills that you will need to be confident with for part of your assessment. Your teacher will give you additional information about tackling these practicals.

Anything in the Higher Tier spreads and boxes must be learnt by those sitting the higher tier exam. If you will be sitting foundation tier, you will not be assessed on this content.

Higher

Study tip

Hints giving you advice on things you need to know and remember, and what to watch out for.

Using maths

This feature highlights and explains the key maths skills you need. There are also clear step-by-step worked examples.

Go further!

Go further feature boxes encourage you to think about science you have learnt in a different context and introduce you to science beyond the specification. You do not need to learn any of the content in the Go further boxes.

Summary questions

Each topic has summary questions. These questions give you the chance to test whether you have learnt and understood everything in the topic. The questions start off easier and get harder, so that you can stretch yourself.

The Literacy pen **✏** shows activities or questions that help you develop literacy skills.

Key points

Linking to the Learning objectives, the Key points boxes summarise what you should be able to do at the end of the topic. They can be used to help you with revision.

Any questions marked with the higher tier icon **H** are for students sitting the higher tier exams.

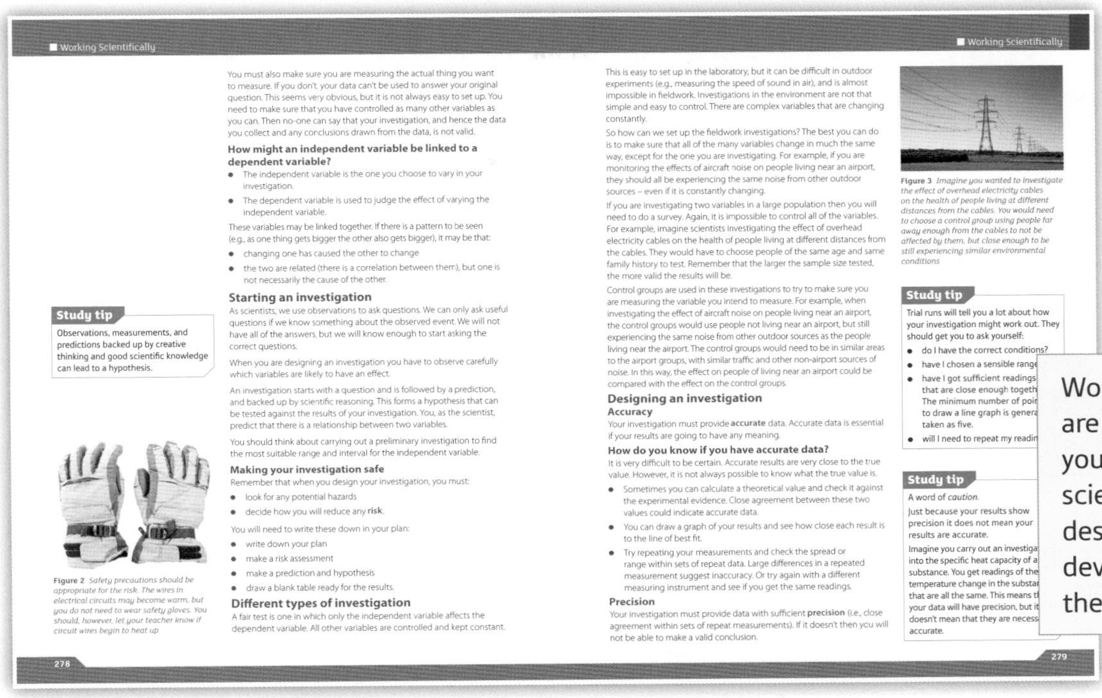

You must also make sure you are measuring the actual thing you want to measure. If you don't, your data can't be used to answer your original question. This seems very obvious, but it is not always easy to set up. You need to make sure that you have controlled as many other variables as you can. Then no-one can say that your investigation, and hence the data you collect and any conclusions drawn from the data, is not valid.

How might an independent variable be linked to a dependent variable?
- The independent variable is the one you choose to vary in your investigation.
- The dependent variable is used to judge the effect of varying the independent variable.

These variables may be linked together. If there is a pattern to be seen (e.g., as one thing gets bigger the other also gets bigger), it may be that:
- changing one has caused the other to change
- the two are related (there is a correlation between them), but one is not necessarily the cause of the other.

Starting an investigation
As scientists, we use observations to ask questions. We can only ask useful questions if we know something about the observed event. We will not have all of the answers, but we will know enough to start asking the correct questions.

When you are designing an investigation you have to observe carefully which variables are likely to have an effect.

An investigation starts with a question and is followed by a prediction, and backed up by scientific reasoning. You, as the scientist, predict that there is a relationship between two variables.

You should think about carrying out a preliminary investigation to find the most suitable range and interval for the independent variable.

Making your investigation safe
Remember that when you design your investigation, you must:
- look for any potential hazards
- decide how you will reduce any **risk**.

You will need to write these down in your plan:
- write down your plan
- make a risk assessment
- make a prediction and hypothesis
- draw a blank table ready for the results.

Different types of investigation
A fair test is one in which only the independent variable affects the dependent variable. All other variables are controlled and kept constant.

Figure 2 *Safety precautions should be appropriate for the risk. The wires in electrical circuits may become warm, but you do not need to wear safety gloves. You should, however, let your teacher know if circuit wires begin to heat up*

Study tip
Observations, measurements, and predictions backed up by creative thinking and good scientific knowledge can lead to a hypothesis.

This is easy to set up in the laboratory, but it can be difficult in outdoor experiments (e.g., measuring the speed of sound in air), and is almost impossible in fieldwork. Investigations in the environment are not that simple and easy to control. There are complex variables that are changing constantly.

So how can we set up the fieldwork investigations? The best you can do is to make sure that all of the many variables change in much the same way, except for the one you are investigating. For example, if you are monitoring the effects of aircraft noise on people living near an airport, they should all be experiencing the same noise from other outdoor sources – even if it is constantly changing.

If you are investigating two variables in a large population then you will need to do a survey. Again, it is impossible to control all of the variables. For example, imagine scientists investigating the effect of overhead electricity cables on the health of people living at different distances from the cables. They would have to choose people of the same age and same family history to test. Remember that the larger the sample size tested, the more valid the results will be.

Control groups are used in these investigations to try to make sure you are measuring the variable you intend to measure. For example, when investigating the effect of aircraft noise on people living near an airport, the control groups would use people not living near an airport, but still experiencing the same noise from other outdoor sources as the people living near the airport. The control groups would need to be in similar areas to the airport groups, with similar traffic and other non-airport sources of noise. In this way, the effect on people of living near an airport could be compared with the effect on the control groups.

Designing an investigation
Accuracy
Your investigation must provide **accurate** data. Accurate data is essential if your results are going to have any meaning.

How do you know if you have accurate data?
It is very difficult to be certain. Accurate results are very close to the true value. However, it is not always possible to know what the true value is.
- Sometimes you can calculate a theoretical value and check it against the experimental evidence. Close agreement between these two values could indicate accurate data.
- You can draw a graph of your results and see how close each result is to the line of best fit.
- Try repeating your measurements and check the spread or range within sets of repeat data. Large differences in a repeated measurement suggest inaccuracy. Or try again with a different measuring instrument and see if you get the same readings.

Precision
Your investigation must provide data with sufficient **precision** (i.e., close agreement within sets of repeat measurements). If it doesn't then you will not be able to make a valid conclusion.

Figure 3 *Imagine you wanted to investigate the effect of overhead electricity cables on the health of people living at different distances from the cables. You would need to choose a control group using people far away enough from the cables to not be affected by them, but close enough to be still experiencing similar environmental conditions*

Study tip
Trial runs will tell you a lot about how your investigation might work out. They should get you to ask yourself:
- do I have the correct conditions?
- have I chosen a sensible range?
- have I got sufficient readings that are close enough together?
- will I need to repeat my readings?

Study tip
A word of caution.
Just because your results show precision it does not mean your results are accurate.
Imagine you carry out an investigation into the specific heat capacity of a substance. You get readings of the temperature change in the substance that are all the same. This means that your data will have precision, but it doesn't mean that they are necessarily accurate.

> Working scientifically skills are an important part of your course. The working scientifically section describes and supports the development of some of the key skills you will need.

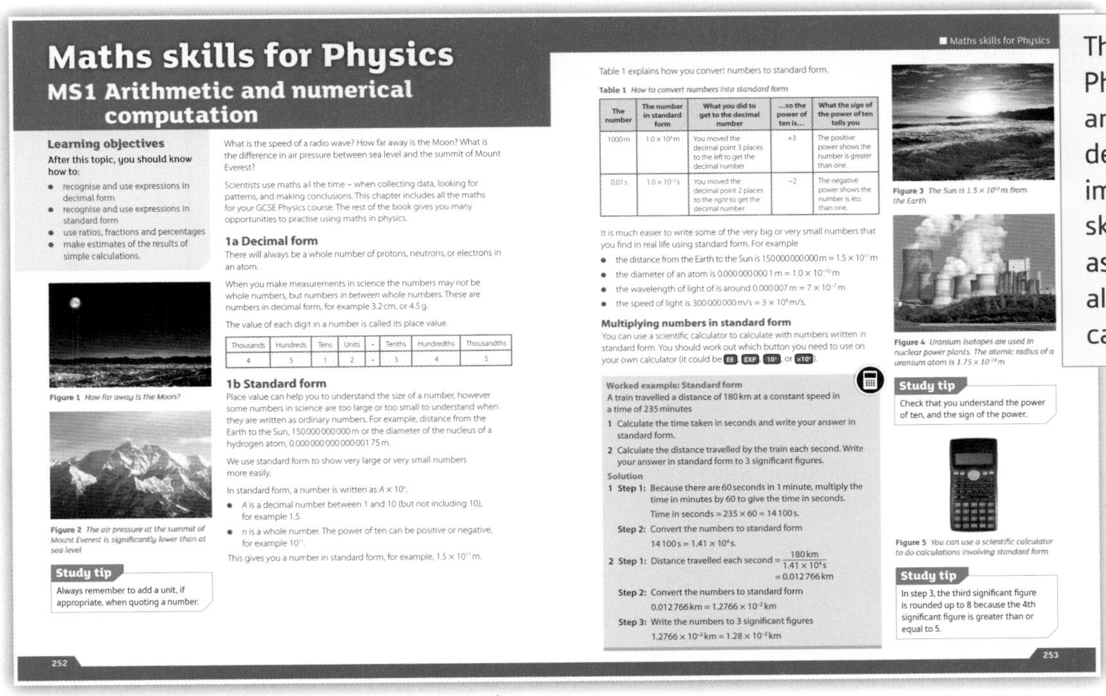

Maths skills for Physics
MS1 Arithmetic and numerical computation

Learning objectives
After this topic, you should know how to:
- recognise and use expressions in decimal form
- recognise and use expressions in standard form
- use ratios, fractions and percentages
- make estimates of the results of simple calculations.

Figure 1 *How far away is the Moon?*

Figure 2 *The air pressure at the summit of Mount Everest is significantly lower than at sea level*

Study tip
Always remember to add a unit, if appropriate, when quoting a number.

What is the speed of a radio wave? How far away is the Moon? What is the difference in air pressure between sea level and the summit of Mount Everest?

Scientists use maths all the time – when collecting data, looking for patterns, and making conclusions. This chapter includes all the maths for your GCSE Physics course. The rest of the book gives you many opportunities to practise using maths in physics.

1a Decimal form
There will always be a whole number of protons, neutrons, or electrons in an atom.

When you make measurements in science the numbers may not be whole numbers, but numbers in between whole numbers. These are numbers in decimal form, for example 3.2 cm, or 4.5 g.

The value of each digit in a number is called its place value.

Thousands	Hundreds	Tens	Units	.	Tenths	Hundredths	Thousandths
4	5	1	2	.	3	4	5

1b Standard form
Place value can help you to understand the size of a number, however some numbers in science are too large or too small to understand when they are written as ordinary numbers. For example, distance from the Earth to the Sun, 150 000 000 000 m or the diameter of the nucleus of a hydrogen atom, 0.000 000 000 000 001 75 m.

We use standard form to show very large or very small numbers more easily.

In standard form, a number is written as $A \times 10^n$.
- A is a decimal number between 1 and 10 (but not including 10), for example 1.5.
- n is a whole number. The power of ten can be positive or negative, for example 10^{11}.

This gives you a number in standard form, for example, 1.5×10^{11} m.

Table 1 explains how you convert numbers to standard form.

Table 1 *How to convert numbers into standard form*

The number	The number in standard form	What you did to get to the decimal number	...so the power of ten is...	What the sign of the power of ten tells you
1000 m	1.0×10^3 m	You moved the decimal point 3 places to the left to get the decimal number	+3	The positive power shows the number is greater than one.
0.01 s	1.0×10^{-2} s	You moved the decimal point 2 places to the right to get the decimal number	−2	The negative power shows the number is less than one.

It is much easier to write some of the very big or very small numbers that you find in real life using standard form. For example
- the distance from the Earth to the Sun is 150 000 000 000 m = 1.5×10^{11} m
- the diameter of an atom is 0.000 000 000 1 m = 1.0×10^{-10} m
- the wavelength of light of is around 0.000 007 m = 7×10^{-7} m
- the speed of light is 300 000 000 m/s = 3×10^8 m/s.

Multiplying numbers in standard form
You can use a scientific calculator to calculate with numbers written in standard form. You should work out which button you need to use on your own calculator (it could be **EE**, **EXP**, **10^x**, or **$\times 10^x$**).

Worked example: Standard form
A train travelled a distance of 180 km at a constant speed in a time of 235 minutes
1 Calculate the time taken in seconds and write your answer in standard form.
2 Calculate the distance travelled by the train each second. Write your answer in standard form to 3 significant figures.
Solution
1 **Step 1:** Because there are 60 seconds in 1 minute, multiply the time in minutes by 60 to give the time in seconds.
   Time in seconds = 235 × 60 = 14 100 s.
   Step 2: Convert the numbers to standard form
   14 100 s = 1.41×10^4 s.
2 **Step 1:** Distance travelled each second = $\dfrac{180 \text{ km}}{1.41 \times 10^4 \text{ s}}$ = 0.012 766 km
   Step 2: Convert the numbers to standard form
   0.012 766 km = 1.2766×10^{-2} km
   Step 3: Write the numbers to 3 significant figures
   1.2766×10^{-2} km = 1.28×10^{-2} km

Figure 3 *The Sun is 1.5×10^{11} m from the Earth*

Figure 4 *Uranium isotopes are used in nuclear power plants. The atomic radius of a uranium atom is 1.75×10^{-10} m*

Study tip
Check that you understand the power of ten, and the sign of the power.

Figure 5 *You can use a scientific calculator to do calculations involving standard form*

Study tip
In step 3, the third significant figure is rounded up to 8 because the 4th significant figure is greater than or equal to 5.

> The Maths skills for Physics chapter describes and supports the development of the important mathematical skills you will need for all aspects of your course. It also has questions so you can test your skills.

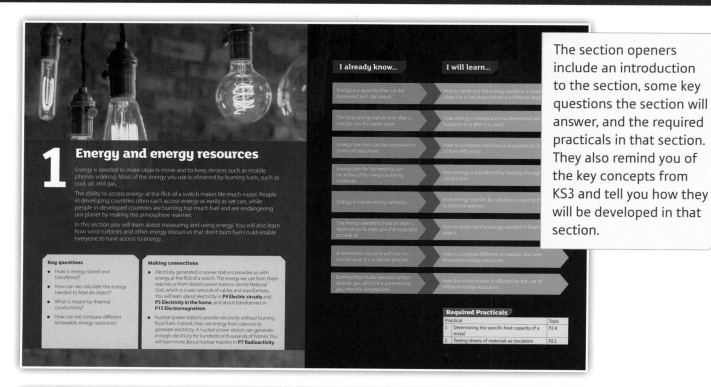

The section openers include an introduction to the section, some key questions the section will answer, and the required practicals in that section. They also remind you of the key concepts from KS3 and tell you how they will be developed in that section.

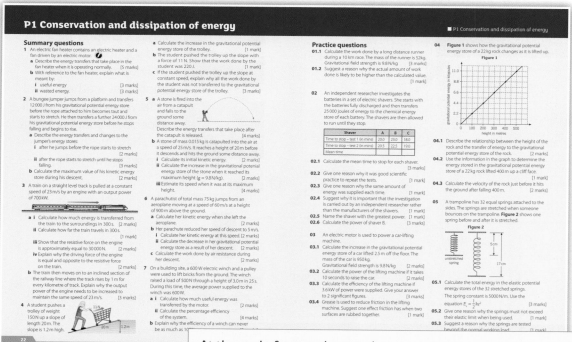

At the end of every chapter there are summary questions and practice questions. The questions test your literacy, maths, and working scientifically skills, as well as your knowledge of the concepts in that chapter. The practice questions can also call on your knowledge from any of the previous chapters to help support the synoptic element of your assessment.

There are also further practice questions at the end of the book to cover all of the content from your course.

Kerboodle

This book is also supported by Kerboodle, offering unrivalled digital support for building your practical, maths and literacy skills.

If your school subscribes to Kerboodle, you will find a wealth of additional resources to help you with your studies and revision:

- animations, videos, and revision podcasts
- webquests
- maths and literacy skills activities and worksheets
- on your marks activities to help you achieve your best
- practicals and follow-up activities
- interactive quizzes that give question-by-question feedback
- self-assessment checklists

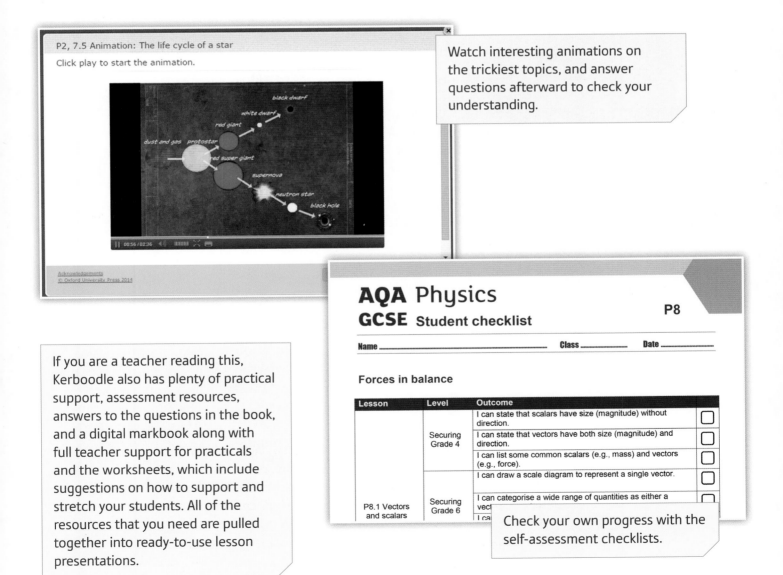

P2, 7.5 Animation: The life cycle of a star

Click play to start the animation.

black dwarf
white dwarf
red giant
dust and gas   protostar
red super giant
supernova
neutron star
black hole

00:56 / 02:36

Acknowledgements
© Oxford University Press 2014

Watch interesting animations on the trickiest topics, and answer questions afterward to check your understanding.

If you are a teacher reading this, Kerboodle also has plenty of practical support, assessment resources, answers to the questions in the book, and a digital markbook along with full teacher support for practicals and the worksheets, which include suggestions on how to support and stretch your students. All of the resources that you need are pulled together into ready-to-use lesson presentations.

AQA Physics
GCSE Student checklist
P8

Name .... Class .... Date ....

Forces in balance

Lesson	Level	Outcome	
	Securing Grade 4	I can state that scalars have size (magnitude) without direction.	☐
		I can state that vectors have both size (magnitude) and direction.	☐
		I can list some common scalars (e.g., mass) and vectors (e.g., force).	☐
		I can draw a scale diagram to represent a single vector.	☐
P8.1 Vectors and scalars	Securing Grade 6	I can categorise a wide range of quantities as either a vect...	☐
		I ca...	

Check your own progress with the self-assessment checklists.

1 Energy and energy resources

Energy is needed to make objects move and to keep devices such as mobile phones working. Most of the energy you use is obtained by burning fuels, such as coal, oil, and gas.

The ability to access energy at the flick of a switch makes life much easier. People in developing countries often can't access energy as easily as we can, while people in developed countries are burning too much fuel and are endangering our planet by making the atmosphere warmer.

In this section you will learn about measuring and using energy. You will also learn how wind turbines and other energy resources that don't burn fuel could enable everyone to have access to energy.

Key questions

- How is energy stored and transferred?

- How can we calculate the energy needed to heat an object?

- What is meant by thermal conductivity?

- How can we compare different renewable energy resources?

Making connections

- Electricity generated in power stations provides us with energy at the flick of a switch. The energy we use from them reaches us from distant power stations via the National Grid, which is a vast network of cables and transformers. You will learn about electricity in **P4 Electric circuits** and **P5 Electricity in the home**, and about transformers in **P15 Electromagnetism**.

- Nuclear power stations provide electricity without burning fossil fuels. Instead, they use energy from uranium to generate electricity. A nuclear power station can generate enough electricity for hundreds of thousands of homes. You will learn more about nuclear reactors in **P7 Radioactivity**.

AQA Physics
Third edition

Jim Breithaupt
Gary Calder
Editor: Lawrie Ryan

GCSE

Message from AQA

This textbook has been approve   means that we have checked th   are satisfied with the overall qu   found on our website.

We approve textbooks because we know how important it is for teachers and students to have the right resources to support their teaching and learning. However, the publisher is ultimately responsible for the editorial control and quality of this book.

Please note that when teaching the *AQA GCSE Physics* or *AQA GCSE Combined Science: Trilogy* course, you must refer to AQA's specification as your definitive source of information. While this book has been written to match the specification, it cannot provide complete coverage of every aspect of the course.

A wide range of other useful resources can be found on the relevant subject pages of our website: www.aqa.org.uk.

OXFORD
UNIVERSITY PRESS

OXFORD
UNIVERSITY PRESS

Great Clarendon Street, Oxford, OX2 6DP, United Kingdom

Oxford University Press is a department of the University of Oxford. It furthers the University's objective of excellence in research, scholarship, and education by publishing worldwide. Oxford is a registered trade mark of Oxford University Press in the UK and in certain other countries

© Jim Breithaupt 2016

The moral rights of the authors have been asserted

First published in 2016

British Library Cataloguing in Publication Data
Data available

978019 835939 5

10 9 8 7 6 5 4 3

MIX
Paper from responsible sources
FSC
www.fsc.org FSC® C007785

Paper used in the production of this book is a natural, recyclable product made from wood grown in sustainable forests. The manufacturing process conforms to the environmental regulations of the country of origin.

Printed in Great Britain by Bell and Bain Ltd., Glasgow

Acknowledgements

The author wishes to acknowledge the support, advice, and contributions he has received from Marie Breithaupt, Gary Calder, and Darren Forbes, and from Emma Craig, Sadie Garratt, and their colleagues at Oxford University Press.

AQA examination questions are reproduced by permission of AQA.

Index compiled by INDEXING SPECIALISTS (UK) Ltd., Indexing house, 306A Portland Road, Hove, East Sussex, BN3 5LP United Kingdom.

COVER: Eric James Azure / Offset

I already know...

Energy is a quantity that can be measured and calculated.

The total energy before and after a change has the same value.

Energy transfers can be compared in terms of usefulness.

Energy transfer by heating can be reduced by using insulating materials.

Energy is transferred by radiation.

The energy needed to heat an object depends on its mass and the material it is made of.

A renewable resource will not run out because it is a natural process.

Burning fossil fuels releases carbon dioxide gas, which is a greenhouse gas, into the atmosphere.

I will learn...

How to work out the energy stored in a moving object or in an object when it is lifted or stretched.

How energy is stored and transferred and what happens to it after it is used.

How to compare machines and appliances in terms of their efficiency.

How energy is transferred by heating through conduction.

How energy transfer by radiation is causing the Earth to become warmer.

How to work out the energy needed to heat an object.

How to compare different renewable and non-renewable energy resources.

How the environment is affected by the use of different energy resources.

Required Practicals

Practical		Topic
1	Determining the specific heat capacity of a metal	P2.4
2	Testing sheets of materials as insulators	P2.1

Learning objectives

After this topic, you should know:

- the ways in which energy can be stored
- how energy can be transferred
- the changes in energy stores that happen when an object falls
- the energy transfers that happen when a falling object hits the ground without bouncing back.

On the move

Cars, buses, planes, and ships all use fuels as chemical energy stores. They carry their own fuel. Electric trains use energy transferred from fuel in power stations. Electricity transfers energy from the power station to the train.

Figure 1 *The French Train à Grande Vitesse electric train can reach speeds of more than 500 km/hour*

Energy can be stored in different ways and is transferred by heating, waves, an electric current, or when a force moves an object. Here are some examples:

- Chemical energy stores include fuels, foods, or the chemicals found in batteries. The energy is transferred during chemical reactions.
- Kinetic energy stores describe the energy an object has because it is moving.
- Gravitational potential energy stores are used to describe the energy stored in an object because of its position, such as an object above the ground.
- Elastic potential energy stores describe the energy stored in a springy object when you stretch or squash it.
- Thermal energy stores describe the energy a substance has because of its temperature.

Energy can be transferred from one store to another. In a torch, the torch's battery pushes a current through the bulb. This makes the torch bulb emit light, and also get hot.

When an electric kettle is used to boil water, the current in the kettle's heating element transfers energy to the thermal energy store of the water and the kettle.

When an object is thrown into the air, the object slows down as it goes up. Here, energy is transferred from the object's kinetic energy store to its gravitational potential energy store.

You can show the energy transfers by using a flow diagram:

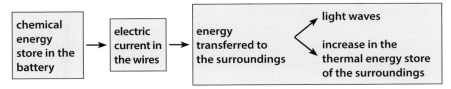

Figure 2 *Changes in energy stores in a torch lamp*

Energy transfers

When an object starts to fall freely, it speeds up as it falls. The force of gravity acting on the object causes energy to be transferred from its gravitational potential energy store to its kinetic energy store.

Look at Figure 3. It shows an object that hits the floor with a thud. All of the energy in its kinetic energy store is transferred by heating to the thermal energy store of the object and the floor, and by sound waves moving away from the point of impact. The amount of energy transferred by sound waves is much smaller than the amount of energy transferred by heating.

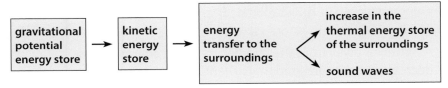

Figure 4 *An energy transfer diagram for an object when it falls and when it hits the ground*

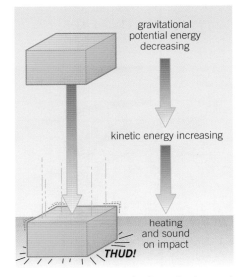

Figure 3 *An energetic drop. On impact, energy is transferred to the thermal energy store of the surroundings by heating and by sound waves*

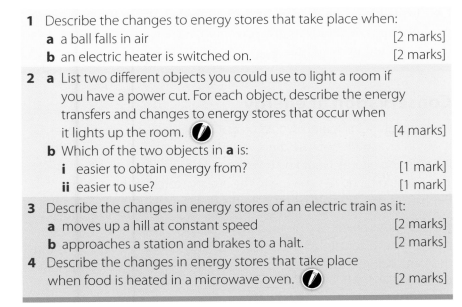

1 Describe the changes to energy stores that take place when:
 a a ball falls in air [2 marks]
 b an electric heater is switched on. [2 marks]

2 **a** List two different objects you could use to light a room if you have a power cut. For each object, describe the energy transfers and changes to energy stores that occur when it lights up the room. ✏ [4 marks]
 b Which of the two objects in **a** is:
  i easier to obtain energy from? [1 mark]
  ii easier to use? [1 mark]

3 Describe the changes in energy stores of an electric train as it:
 a moves up a hill at constant speed [2 marks]
 b approaches a station and brakes to a halt. [2 marks]

4 Describe the changes in energy stores that take place when food is heated in a microwave oven. ✏ [2 marks]

Key points

- Energy can be stored in a variety of different energy stores.
- Energy is transferred by heating, by waves, by an electric current, or by a force when it moves an object.
- When an object falls and gains speed, its store of gravitational potential energy decreases and its kinetic energy store increases.
- When a falling object hits the ground without bouncing back, its kinetic energy store decreases. Some or all of its energy is transferred to the surroundings – the thermal energy store of the surroundings increases, and energy is also transferred by sound waves.

P1.2 Conservation of energy

Learning objectives

After this topic, you should know:

- what conservation of energy is
- why conservation of energy is a very important idea
- what a closed system is
- how to describe the changes to energy stores in a closed system.

Figure 1 *Energy transfers on a roller coaster*

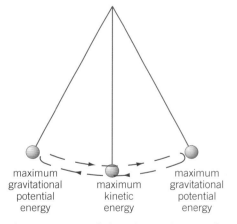

maximum gravitational potential energy   maximum kinetic energy   maximum gravitational potential energy

Figure 2 *A pendulum in motion. As the pendulum swings down and towards the centre, its gravitational potential energy store decreases as its kinetic energy store increases. As the pendulum moves upwards and away from the centre, its gravitational potential energy store increases as its kinetic energy store decreases*

At the funfair

Funfairs are very exciting places because changes to stores of energy happen quickly. As a roller coaster climbs an incline, its gravitational potential energy store increases. This energy is then transferred to other energy stores as the roller coaster races downwards.

As the roller coaster descends, its gravitational potential energy store decreases. Most of this energy is transferred to its kinetic energy store, which therefore increases. However, some energy is transferred to the thermal energy store of the surroundings by air resistance and friction, and some energy is transferred by sound waves.

Investigating pendulums

When changes to energy stores happen, does the total amount of energy stay the same? You can investigate this question with a simple pendulum.

Figure 2 shows a pendulum bob swinging from side to side.

As it moves towards the middle, energy is transferred by the force of gravity from its gravitational potential energy store to its kinetic energy store. So its gravitational potential energy store decreases and its kinetic energy store increases.

As it moves away from the middle, its kinetic energy stores decreases and its gravitational potential energy store increases. If the air resistance on the bob is very small the bob will reach the same height on each side.

- Describe the changes to energy stores that take place in the bob when it goes from one side at maximum height to the other side at maximum height.
- Explain why it is difficult to mark the exact height the pendulum bob rises to. Suggest how you could make your judgement of height more accurate.

Conservation of energy

The pendulum in Figure 2 would probably keep on swinging for ever if it was in a vacuum because there would be no air resistance acting on it, and so no energy would be transferred from any of its energy stores. There would be no net change to the energy stored in the system. Because of this, it would be an example of a closed system.

A system is an object or a group of objects. Scientists have done lots of tests and have concluded that the total energy of a closed system is always the same before and after energy transfers to other energy stores within the closed system.

This important result is known as the principle of conservation of energy. It says that:

energy cannot be created or destroyed.

Energy can be stored in various ways. For example:

- when a rubber band is stretched, its elastic potential energy store increases
- when an object is lifted, its gravitational potential energy store is increased.

Bungee jumping

What energy transfers happen to a bungee jumper after jumping off the platform?

- When the rope is slack, energy is transferred from the gravitational potential energy store to the kinetic energy store as the jumper accelerates towards the ground due to the force of gravity.

- When the rope tightens, it slows the bungee jumper's fall. This is because the force of the rope reduces the speed of the jumper. The jumper's kinetic energy store decreases and the rope's elastic potential energy store increases as the rope stretches. Eventually the jumper comes to a stop – the energy that was originally in the kinetic energy store of the jumper has all been transferred into the elastic potential energy store of the rope.

After reaching the bottom, the rope recoils and pulls the jumper back up. As the jumper rises, the energy in the elastic potential energy store of the rope decreases and the bungee jumper's kinetic energy store increases until the rope becomes slack. After the rope becomes slack, and at the top of the ascent, the bungee jumper's kinetic energy store decreases to zero. The bungee jumper's gravitational potential energy store increases throughout the ascent.

The bungee jumper doesn't return to the original height. This is because some energy was transferred to the thermal energy store of the surroundings by heating as the rope stretched and then shortened again.

1 When a roller coaster gets to the bottom of a descent, describe the energy transfers and changes to energy stores that happen if:
   a the brakes are applied to stop it   [4 marks]
   b it goes up and over a second hill.   [5 marks]

2 **a** A ball dropped onto a trampoline returns to almost the same height after it bounces. Describe the energy transfers and changes to the energy stores of the ball from the point of release to the top of its bounce.   [5 marks]
   b Describe the energy stores of the ball at the point of release compared with its energy stores at the top of its bounce. [1 mark]
   c Describe how you would use the test in **a** to see which of three trampolines is the bounciest.   [5 marks]

3 One exciting fairground ride acts like a giant catapult. The capsule in which you are strapped is fired high into the sky by the rubber bands of the catapult. Explain the changes to the energy stores that take place in the ride as you move upwards.   [4 marks]

Bungee jumping

You can try out the ideas about energy transfers during a bungee jump using the experiment shown in Figure 3.

Figure 3 *Testing a bungee jump*

Safety: Make sure the stand is secure. Protect feet and bench from falling objects.

Key points

- Energy cannot be created or destroyed.
- Conservation of energy applies to all energy changes.
- A closed system is an isolated system in which no energy transfers take place out of or into the energy stores of the system.
- Energy can be transferred between energy stores within a closed system. The total energy of the system is always the same, before and after, any such transfers.

P1.3 Energy and work

Learning objectives

After this topic, you should know:

- what work means in science
- how work and energy are related
- how to calculate the work done by a force
- what happens to work done to overcome friction.

Figure 1 *Working out*

Worked example

A builder pushed a wheelbarrow a distance of 5.0 m across flat ground with a force of 50 N. How much work was done by the builder?

Solution

work done = force applied × distance moved

$= 50\,N \times 5.0\,m$

$= \mathbf{250\,J}$

Figure 2 *Pulling a lorry*

Working out

In a fitness centre or a gym, you have to work hard to keep fit. Lifting weights and pedalling on an exercise bike are just two ways to keep fit. Whichever way you choose to keep fit, you have to apply a force to move something. So the work you do causes a transfer of energy.

When an object is moved by a force, **work** is done on the object by the force. So the force transfers energy to the object. The amount of energy transferred to the object is equal to the work done on it. For example, to raise an object, you need to apply a force to it to overcome the force of gravity on it. If the work you do on the object is 20 J, the energy transferred to it must be 20 J. So its gravitational potential energy store increases by 20 J.

$$\text{energy transferred} = \text{work done}$$

The work done by a force depends on the size of the force and the distance moved. One joule of work is done when a force of one newton causes an object to move a distance of one metre in the direction of the force. To calculate the work done by a force when it causes displacement of an object, use this equation:

work done, W = force applied, F × distance moved along the line of action of the force, s

(joules, J)   (newtons, N)   (metres, m)

Superhuman force

Imagine pulling a lorry over 40 m. On level ground, a pull force of about 2000 N is needed. The work done by the pulling force is 80 kJ (= 2000 N × 40 m). Very few people can manage to pull with such force. Don't even try it, though. The people who can do it are very, very strong and have trained specially for it.

Doing work

Carry out a series of experiments to calculate the work done in performing the tasks below. Use a newton-meter to measure the force applied, and use a metre ruler to measure the distance moved.

1. Drag a small box a measured distance across a rough surface.
2. Repeat the test above with two rubber bands wrapped around the box (Figure 3).

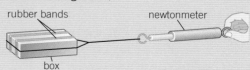

Figure 3 *At work*

- What is the resolution of your measuring instruments? Repeat your tests and comment on the precision of your repeat measurements. Can you be confident about the accuracy of your results?

Friction at work

Work done to overcome friction is mainly transferred to thermal energy stores by heating.

1 If you rub your hands together vigorously, they become warm. Your muscles do work to overcome the friction between your hands. The work you do is transferred as energy that warms your hands.

2 Brake pads on a vehicle become hot if the brakes are applied for too long. Friction between the brake pads and the wheel discs opposes the motion of the wheel. The force of friction does work on the brake pads and the wheel discs. As a result, energy is transferred from the kinetic energy store of the vehicle to the thermal energy store of the brake pads and the wheel discs. This makes them become hot and transfer energy by heating to the thermal energy store of the surrounding air.

3 Meteorites are small objects from space that enter the Earth's atmosphere and fall to the ground. As they pass through the atmosphere, friction caused by air resistance acts upon them. This results in energy being transferred from the meteorite's gravitational potential energy and kinetic energy stores to the meteorite's thermal energy store, causing the meteorite to heat up. If a meteorite becomes hot enough, it glows and becomes visible as a 'shooting star'. Very small objects can burn up completely. The surface of a space vehicle is designed to withstand the very high temperatures caused by this friction when it re-enters the Earth's atmosphere.

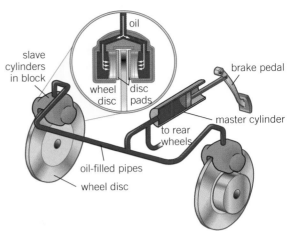

Figure 4 *Disc brakes*

1 **a** State what happens to the energy transferred:
   i by a rower rowing a boat [1 mark]
   ii by an electric motor used to raise a car park barrier. [1 mark]
 b Calculate how much work is done when a force of 3000 N pulls a truck through a distance of 50 m in the direction of the force. [1 mark]

2 A car is brought to a standstill when the driver applies the brakes.
 a Explain why the brake pads become warm. [2 marks]
 b The car travelled a distance of 20 metres after the brakes were applied. The braking force on the car during this time was 7000 N. Calculate the work done by the braking force. [1 mark]

3 **a** Calculate the work done when:
   i a force of 20 N makes an object move 4.8 m in the direction of the force [1 mark]
   ii an object of weight 80 N is raised through a height of 1.2 m. [1 mark]
 b When a cyclist brakes, his kinetic energy store is reduced from 1400 J to zero in a distance of 7.0 m. Calculate the braking force. [2 marks]

4 A student pushes a box at a steady speed a distance of 12 m across a level floor.
 a The student applied a horizontal force of 25 N to the box. Calculate the work done by the student. [1 mark]
 b Describe the energy transfers and changes to energy stores as the box moves. [3 marks]

Key points

- Work is done on an object when a force makes the object move.
- Energy transferred = work done.
- Work done is $W = F\,s$ where F is the force and s is the distance moved (along the line of action of the force).
- Work done to overcome friction is transferred as energy to the thermal energy stores of the objects that rub together and to the surroundings.

P1.4 Gravitational potential energy stores

Learning objectives

After this topic, you should know:

- what happens to the gravitational potential energy stores of an object when it moves up or down
- why an object moving up has an increase in its gravitational potential energy store
- why it is easier to lift an object on the Moon than on the Earth
- how to calculate the change in gravitational potential energy of an object when it moves up or down.

Changes in gravitational potential energy stores

Every time you lift an object up, you do some work. Some of your muscles transfer energy from the chemical energy store in the muscle to the gravitational energy store of the object. In calculations we refer to the energy in this store as gravitational potential energy E_p.

The force you need to lift an object at constant velocity is equal and opposite to the gravitational force on the object. So the upward force you need to apply to it is equal to the object's weight. For example, you need a force of 80 N to lift a box of weight 80 N.

- When an object is moved upwards, the energy in its gravitational potential energy store increases. This increase is equal to the work done on it by the lifting force to overcome the gravitational force on the object.

- When an object moves down, the energy in its gravitational potential energy store decreases. This decrease is equal to the work done by the gravitational force acting on it.

The work done when an object moves up or down depends on:

1 how far it is moved vertically (its change of height)
2 its weight.

Using work done = force applied × distance moved in the direction of the force:

$$\text{change in object's gravitational potential energy store (joules, J)} = \text{weight (newtons, N)} \times \text{change of height (metres, m)}$$

Worked example

A student of weight 300 N climbs on a platform that is 1.2 m higher than the floor. Calculate the increase in her gravitational potential energy store.

Solution

Increase of E_p = 300 N × 1.2 m = **360 J**

Worked example

A 2.0 kg object is raised through a height of 0.4 m. Calculate the increase in the gravitational potential energy store of the object. The gravitational field strength of the Earth at its surface is 9.8 N/kg.

Solution

Gain of E_p = mass × gravitational field strength × height gain
= 2.0 kg × 9.8 N/kg × 0.4 m
= **7.8 J**

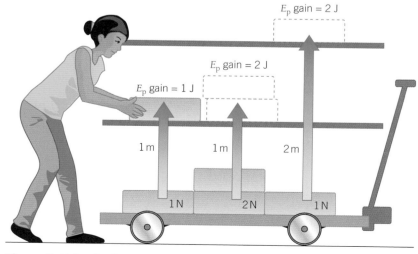

Figure 1 *Using joules*

Gravitational potential energy stores and mass

Astronauts on the Moon can lift objects much more easily than they can on the Earth. This is because the gravitational field strength on the Moon's surface is only about a sixth of the gravitational field strength on the Earth's surface.

You have previously learnt that the weight of an object in newtons is equal to its mass × the gravitational field strength. So, when an object is lifted or lowered, the change to its gravitational potential energy store is equal to its weight × its change of height. Therefore:

$$\underset{\substack{\text{change of}\\ \text{gravitational}\\ \text{potential}\\ \text{energy store, } \Delta E_p\\ \text{(joules, J)}}}{} = \underset{\substack{\text{mass, } m\\ \text{(kilograms, kg)}}}{} \times \underset{\substack{\text{gravitational}\\ \text{field strength, } g\\ \text{(newtons per}\\ \text{kilogram, N/kg)}}}{} \times \underset{\substack{\text{change of}\\ \text{height, } \Delta h\\ \text{(metres, m)}}}{}$$

1   a Describe the changes to the energy stores of a ball when it falls and rebounds without regaining its initial height. [4 marks]
    b When a ball of weight 1.4 N is dropped from rest from a height of 2.5 m above a flat surface, it rebounds to a height of 1.7 m above the surface.
       i Calculate the total energy lost from the ball's energy stores when it reaches this maximum rebound height. [2 marks]
       ii Name two causes of the energy transfer. [2 marks]

2   A student of weight 450 N steps on a box of height 0.2 m.
    a Calculate the increase in her gravitational potential energy store. [1 mark]
    b Calculate the work done by the student if she steps on and off the box 50 times. [1 mark]

3   a A weightlifter raises a steel bar of mass 25 kg through a height of 1.8 m. Calculate the change to the gravitational potential energy store of the bar. The gravitational field strength at the surface of the Earth is 9.8 N/kg. [2 marks]
    b The weightlifter then lowers the bar by 0.3 m and drops it so it falls to the ground. Assume that air resistance is unimportant. Calculate the change to its gravitational potential energy store in this fall. [2 marks]

4   You use energy when you hold an object stationary in your outstretched hand. Suggest what happens to the energy that must be supplied to your muscles to keep them contracted. 🖉 [3 marks]

Study tip

Watch out for objects going up a slope. To calculate the increase in their gravitational potential energy stores, you need the vertical height gained, *not* the distance along the slope.

Study tip

In physics, the greek letter delta (Δ) is used to represent the phrase 'change in'. For example, Δh can be used in place of 'change in height'.

Stepping up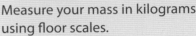

Measure your mass in kilograms using floor scales.

Step on and off a sturdy box or low platform.

Measure the height of the box.

Use the equation $\Delta E_p = m g \Delta h$, where $g = 9.8$ N/kg, to calculate the energy transferred to your gravitational potential energy store when you stepped on the box.

Safety: Make sure the box is secure and that you feel comfortable to do this.

Key points

● The gravitational potential energy store of an object increases when it moves up and decreases when it moves down.
● The gravitational potential energy store of an object increases when it is lifted up because work is done on it to overcome the gravitational force.
● The gravitational field strength at the surface of the Moon is less than on the Earth.
● The change in the gravitational potential energy store of an object is $\Delta E_p = m g \Delta h$

P1.5 Kinetic energy and elastic energy stores

Learning objectives

After this topic, you should know:

- what the amount of energy in a kinetic energy store depends on
- how to calculate the amount of energy in a kinetic energy store
- what an elastic potential energy store is
- how to calculate the amount of energy in an elastic potential energy store.

Investigating a catapult

Use rubber bands to catapult a trolley along a horizontal runway. Find out how the speed of the trolley depends on how much the catapult is pulled back before the trolley is released. For example, see if the distance needs to be doubled to double the speed. Figure 1 shows how the speed of the trolley can be measured.

Safety: Take care to ensure you do this safely. Protect your hands and feet, and the bench, from falling trolleys.

The energy an object has because of its motion depends on its mass and speed. This energy is called kinetic energy.

Investigating kinetic energy stores

Figure 1 shows how you can investigate how the kinetic energy store of an object depends on its speed.

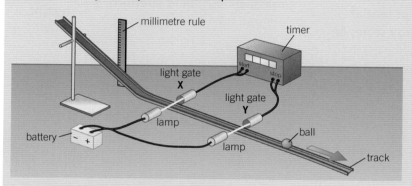

Figure 1 *Investigating changes in kinetic energy stores*

1 The ball is released on a slope from a measured height above the foot of the slope. You can calculate the decrease in its gravitational potential energy store by using the following equation:

change in gravitational potential energy store
= mass × gravitational field strength × change in height.

Due to conservation of energy, this decrease in the gravitational potential energy store is matched by an equal increase in its kinetic energy store.

2 The ball is timed, using light gates, over a measured distance between X and Y after the slope.

- Explain why light gates improve the quality of the data you can collect in this investigation.

Table 1 shows some sample results.

Table 1 *Sample measurements for a ball of mass 0.5 kg*

Height drop to foot of slope in m	0.05	0.10	0.16	0.20
Initial kinetic energy of ball in J	0.25	0.50	0.80	1.00
Time to travel 1.0 m from X to Y in s	0.98	0.72	0.57	0.50
Average speed of ball between X and Y in m/s	1.02			2.00

Work out the speed of the ball between X and Y in each case. The first and last values have been worked out for you. Can you see a link between speed and height drop? The results show that the greater the height drop, the faster is the speed. So it can be said that the kinetic energy store of the ball increases if the speed increases.

The kinetic energy equation

Table 1 shows that when the height drop is increased by four times from 0.05 m to 0.20 m, the speed doubles. The height drop is directly proportional to the speed squared, or (speed)². Because the height drop is a measure of the ball's kinetic energy store, it can be said that the ball's kinetic energy store is directly proportional to the square of its speed.

The amount of energy in the kinetic energy store of an object can be calculated using the kinetic energy equation below:

$$\text{kinetic energy, } E_k = \frac{1}{2} \times \text{mass, } m \times \text{speed}^2, v^2$$

(joules, J)   (kilograms, kg)   (metres per second, m/s)²

Using elastic potential energy

When you stretch a rubber band or a spring, the work you do is stored in it as **elastic potential energy**.

Figure 2 shows how the force F needed to stretch a spring varies with its extension e. The graph obeys the equation for Hooke's Law $F = k\,e$, where k is the spring constant.

For a spring stretched to an extension e, we can calculate the energy in its elastic potential energy store using the equation below:

$$\text{elastic potential energy, } E_e = \frac{1}{2} \times \text{spring constant, } k \times \text{extension}^2, e^2$$

(joules, J)   (newtons per metre, N/m)   (metres, m)²

1 **a** Calculate the kinetic energy store of:
     i a vehicle of mass 500 kg moving at a speed of 12 m/s   [2 marks]
     ii a football of mass 0.44 kg moving at a speed of 20 m/s.   [2 marks]
   b Calculate the velocity of a 500 kg vehicle with twice the kinetic energy store as calculated in **a i**.   [3 marks]

2 **a** A catapult is used to fire an object into the air. Describe the energy transfers when the catapult is:
     i stretched   [2 marks]   **ii** released.   [2 marks]
   b An object of weight 2.0 N fired vertically upwards from a catapult reaches a maximum height of 5.0 m. Calculate:
     i the increase in the gravitational potential energy store of the object   [1 mark]
     ii the speed of the object when it left the catapult.   [4 marks]

3 A car moving at a constant speed has 360 000 J in its kinetic energy store. When the driver applies the brakes, the car stops in a distance of 100 m.
   a Calculate the force that stops the vehicle.   [3 marks]
   b The speed of the car was 30 m/s when its kinetic energy store was 360 000 J. Calculate its mass.   [3 marks]

4 A mobility aid to assist walking uses a steel spring to store energy when the walker's foot goes down, and it returns energy as the foot is lifted. The spring has a spring constant of 250 N/m. Calculate the elastic potential energy stored in the spring when its extension is 0.21 m.   [2 marks]

Worked example

Calculate the kinetic energy stored in a vehicle of mass 500 kg moving at a speed of 12 m/s.

Solution

$$\text{Kinetic energy} = \frac{1}{2}mv^2$$
$$= 0.5 \times 500\,\text{kg} \times (12\,\text{m/s})^2$$
$$= \mathbf{36\,000\,J}$$

Go further!

In Figure 2, the force F increases as the extension e is increased. The average force when the spring is extended to extension e is $\frac{1}{2}F$, where $F = ke$. Therefore, the energy stored in the spring = work done = average force extension = $\frac{1}{2}Fe = \frac{1}{2}ke^2$.

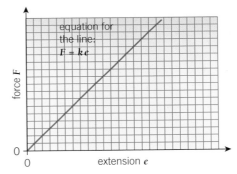

Figure 2 *Force versus extension for a spring. The spring constant k is the force per unit extension of the spring*

Key points

- The energy in the kinetic energy store of a moving object depends on its mass and its speed.
- The kinetic energy store of an object is $E_k = \frac{1}{2}mv^2$
- Elastic potential energy is the energy stored in an elastic object when work is done on the object.
- The elastic potential energy stored in a stretched spring is $E_e = \frac{1}{2}ke^2$, where e is the extension of the spring.

P1.6 Energy dissipation

Figure 1 *Using energy on a running machine*

Energy for a purpose

Where would people be without machines? Washing machines are used to clean your clothes. Machines in factories are used to make the things you buy. You might use exercise machines in a gym to keep yourself fit, and machines are used to get you from place to place.

A machine transfers energy for a purpose. Friction between the moving parts of a machine causes the parts to warm up. So, not all of the energy supplied to a machine is usefully transferred. Some of the energy is wasted.

- **Useful energy** is energy transferred to where it is wanted in the way that is wanted.
- **Wasted energy** is the energy that is not usefully transferred.

Whenever energy is transferred for a purpose in any system, some of the energy is transferred usefully. The rest is **dissipated** (spreads out) and may be stored in less useful ways. This energy is described as wasted energy because it is not transferred as useful energy. For example, when a jet plane takes off, its engines transfer energy from the chemical energy store in the fuel. Some of this energy increases the kinetic energy and the gravitational potential energy stores of the plane, which is the useful energy transfer pathway. The rest is wasted energy because some of it heats the plane and the surroundings and some is transferred to the surroundings by sound waves created by the engine vibrations.

Investigating friction

Friction in machines always causes energy to be wasted. Figure 2 shows two examples of friction in action.

In Figure 2a, friction acts between the drill bit and the wood. The bit becomes hot as it bores into the wood. Energy is transferred by an electric current to the thermal energy stores, heating up the drill bit and the wood.

When a bike rider brakes, friction acts between the brake blocks and the wheels (Figure 2b). Energy is wasted as it is transferred from the kinetic energy stores of the bike and the cyclist, to the thermal energy stores of the brake blocks and the wheels, which are heated by friction.

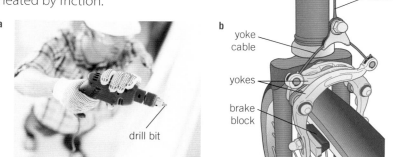

Figure 2 *Friction in action:* **a** *Using a drill,* **b** *Braking on a bicycle*

Friction at work

Next time you use a running machine, think about what happens to the energy transferred by your muscles. As you are exercising, energy is transferred to the thermal energy store of your muscles (so you get hot) and to the thermal energy store of the machine by the force you exert to overcome the friction in the machine.

Spreading out

● Wasted energy is dissipated (spreads out) to the surroundings, for example, the gears of a car get hot because of friction when the car is running. Here, energy is transferred from the kinetic energy store of the gear box to the thermal energy stores of the gear box and the surrounding air. The thermal energy stores of the gear box and the surrounding air therefore increase, as do their respective temperatures.

● Useful energy is eventually transferred to the surroundings too. For example, the useful energy supplied to turn the wheels of a car is eventually transferred from the kinetic energy stores of the wheels to the thermal energy stores of the tyres by heating – increasing the thermal energy stores of the tyres. This energy is then transferred to the thermal energy store of the road and surrounding air.

● Energy becomes less useful the more it spreads out. For example, the hot water from a central heating boiler in a building is pumped through pipes and radiators. The thermal energy store of the hot water decreases as it transfers energy by heating to the thermal energy stores of the radiators – heating the rooms in the building. But the energy supplied to heat these rooms will eventually be transferred to the surrounding air.

1 Copy and complete the table below.

Energy transfer by	Useful energy output	Wasted energy output
a An electric fan heater	warms the air and surrounding objects	
b A television		
c An electric kettle		
d Headphones		

[4 marks]

2 Describe what would happen, in terms of energy transfer and changes to energy stores, to:
  a a gear box that was insulated so it could not transfer energy by heating to the surroundings [3 marks]
  b the running shoes of a jogger if the shoes are well insulated [2 marks]
  c a blunt electric drill bit if you use it to drill into hard wood [2 marks]
  d the metal wheel discs of the brakes of a car when the brakes are applied. [2 marks]

3 **a** Describe the energy transfers and changes to energy stores of a pendulum as it swings from one side to the middle, and then to the opposite side. 🖊 [4 marks]
  b Explain why a swinging pendulum eventually stops. [3 marks]

4 A freewheeling cyclist on a level road gradually stops moving. Describe the energy transfers and changes to the energy stores of the cyclist and the bicycle. 🖊 [4 marks]

Key points

● Useful energy is energy in the place we want it and in the form we need it.
● Wasted energy is the energy that is not useful energy and is transferred by an undesired pathway.
● Wasted energy is eventually transferred to the surroundings, which become warmer.
● As energy dissipates (spreads out), it gets less and less useful.

P1.7 Energy and efficiency

Learning objectives

After this topic, you should know:

- what is meant by efficiency
- what is the maximum efficiency of any energy transfer
- how machines waste energy
- **(H)** how energy transfers can be made more efficient.

When you lift an object, energy stored in your muscles is transferred to the gravitational potential energy store of the object. The amount transferred depends on the object's weight and how high you lift it.

- Weight is measured in newtons (N). The weight of a 1 kilogram object on the Earth's surface is about 10 N.
- Energy is measured in joules (J). The energy needed to lift a weight of 1 newton by a height of 1 metre is equal to 1 joule.

The energy supplied to the device is called the input energy. The useful energy transferred by the device is called the useful output energy.

Because energy cannot be created or destroyed:

Input energy (energy supplied, J) = useful output energy (useful energy transferred, J) + energy wasted (J)

For any device that transfers energy:

$$\text{efficiency} = \frac{\text{useful output energy transferred by the device (J)}}{\text{total input energy supplied to the device (J)}}$$

Efficiency

Efficiency can be written as a decimal number (that is always less than 1) or as a percentage.

For example, a light bulb with an efficiency of 0.15 would radiate 15 J of energy as light for every 100 J of energy you supply to it.

- Its efficiency (as a number) $= \frac{15}{100}$
  $= 0.15$
- Its percentage efficiency
  $= 0.15 \times 100\% = 15\%$

Worked example

An electric motor is used to raise an object. The object's gravitational potential energy store increases by 60 J when the motor is supplied with 200 J of energy by an electric current. Calculate the percentage efficiency of the motor.

Solution

Total energy supplied to the device = 200 J

Useful energy transferred by the device = 60 J

Percentage efficiency of the motor

$= \dfrac{\text{useful output energy transferred by the device}}{\text{total input energy supplied to the device}} \times 100\%$

$= \dfrac{60\,J}{200\,J} \times 100\% = 0.30 \times 100\% = 30\%$

Efficiency limits

No device can be more than 100% efficient, because you can never get more energy from a machine than you put into it.

Investigating efficiency

Figure 1 shows how you can use an electric winch to raise a weight. You can use the joulemeter to measure the energy supplied.

- If you double the weight for the same increase in height, do you need to supply twice as much energy to do this task?

To calculate the change in the gravitational potential energy store of the weight use the following equation:

gravitational potential energy = weight in newtons × height increase in metres.

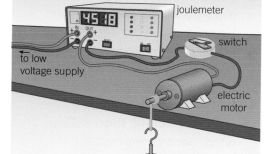

Figure 1 An electric winch

Use this equation and the joulemeter measurements to work out the percentage efficiency of the winch.

● Determine how the efficiency depends on the weight of the object being raised.

Plot a graph of your results, and use it to discuss how the efficiency of the winch changes with weight.

● Make some more measurements to find out whether lubricating the axle of the electric motor with a few drops of suitable oil makes the winch more efficient. Switch the motor off when you lubricate it.

Safety: Protect the floor and your feet. Stop the winch before the masses wrap round the pulley.

Higher

Improving efficiency

Table 1 *Increasing the efficiency of devices*

	Why devices waste energy	How to reduce the problem
1	Friction between the moving parts causes heating.	Lubricate the moving parts to reduce friction.
2	The resistance of a wire causes the wire to get hot when a current passes through it.	In circuits, use wires with as little electrical resistance as possible.
3	Air resistance causes a force on a moving object that opposes its motion. Energy transferred from the object to the surroundings by this force is wasted.	Streamline the shapes of moving objects to reduce air resistance.
4	Sound created by machinery causes energy transfer to the surroundings.	Cut out noise (e.g., tighten loose parts to reduce vibration).

1 a Compare the useful output energy by a machine and the input energy supplied to the machine. [1 mark]
  b Explain why the percentage efficiency of a machine can never be:
    i more than 100% [2 marks] ii equal to 100%. [4 marks]

2 An electric motor is used to raise a weight. When you supply 60 J of energy to the motor, the weight gains 24 J of gravitational potential energy. Calculate:
  a the energy wasted by the motor [1 mark]
  b the efficiency of the motor. [2 marks]

3 A machine is 25% efficient. If the total energy supplied to the machine is 3200 J, calculate how much useful energy can be transferred. [2 marks]

4 An electric fan heater contains a motor that blows hot air into the room. Describe the changes in the energy stores of the heater. 🖊 [4 marks]

P1.8 Electrical appliances

Learning objectives

After this topic, you should know:

- how energy is supplied to your home
- why electrical appliances are so useful
- what most everyday electrical appliances are used for
- how to choose an electrical appliance for a particular job.

Synoptic link

You should know that electricity is not an energy store, but rather it is a flow of charge that transfers energy from one energy store to another. You will learn more about electricty in Chapter P4.

Everyday electrical appliances

The energy you use in your home is mostly supplied by electricity, gas, and oil. Although all three of these energy supplies can be used for cooking and heating, your electricity supply is vital because you use electrical appliances for so many purposes every day. The charge that flows through these appliances transfers energy to them, which they then transfer usefully. But some of the energy you supply to them is wasted.

Figure 1 *Electrical appliances – how many can you see in this photo?*

Table 1 *Comparing energy use in electrical appliances*

Appliance	Useful energy	Energy wasted
Light bulb	Light emitted from the glowing filament.	Energy transfer from the filament heating the surroundings.
Electric heater	Energy heating the surroundings.	Light emitted from the glowing element.
Electric toaster	Energy heating bread.	Energy heating the toaster case and the air around it.
Electric kettle	Energy heating water.	Energy heating the kettle itself.
Hairdryer	Kinetic energy of the air driven by the fan. Energy heating air flowing past the heater filament.	Sound of fan motor (energy heating the motor heats the air going past it, so is not wasted). Energy heating the hairdryer itself.
Electric motor	Kinetic energy of objects driven by the motor. Gravitational potential energy of objects lifted by the motor.	Energy heating the motor and energy transferred by the sound waves generated by the motor.

Clockwork radio

People without electricity supplies can now listen to radio programmes – thanks to the British inventor Trevor Baylis. In the early 1990s, he invented the clockwork radio. When you turn a handle on a clockwork radio, you wind up a clockwork spring in it, and increase the elastic potential energy of the spring. When the spring unwinds, energy from its elastic potential energy store is transferred to its kinetic energy store and it turns a small electric generator in the radio. It doesn't need batteries or mains electricity. So people in remote areas where there is no mains electricity can listen to their radios without having to walk miles for a replacement battery. But they do have to wind up the spring every time its store of energy has been used.

Figure 2 *Clockwork radios are now mass produced and sold all over the world*

Choosing an electrical appliance

You use electrical appliances for many purposes. Each appliance is designed for a specific purpose, and it should waste as little energy as possible. Suppose you were a rock musician at a concert. You would need appliances that transfer the variations in sound waves to electricity and then back to sound waves. But you wouldn't want the appliances to transfer lots of energy to the thermal energy store of the surroundings or themselves. See if you can spot some of these appliances in Figure 3.

Figure 3 *On stage*

1 Match each electrical appliance in the list below with the energy transfer A or B it is designed to bring about.
Energy transfer   **A**   Energy transferred by an electric current →
energy transferred by sound waves
           B   Energy transferred by an electric current →
kinetic energy store
   a Electric drill   [1 mark]
   b Food mixer   [1 mark]
   c Electric bell   [1 mark]

2   **a** Explain why a clockwork radio needs to be wound up before it can be used.   [2 marks]
   b Describe the changes in energy stores that take place in a clockwork radio when it is wound up and then switched on.   [2 marks]
   c Give one advantage and one disadvantage of a clockwork radio compared with a battery-operated radio.   [2 marks]

3 An electric dishwasher heats water and sprays the hot water at the dishes. It then pumps the water out.
   a Describe how energy is usefully used in the machine.   [1 mark]
   b Describe how energy is wasted by the machine.   [2 marks]

4 A laptop battery stores energy with an efficiency of 80% when it is recharged from a low-voltage power supply. When it is connected to the laptop, it transfers energy to the laptop with an efficiency of 60%.
   a Calculate the overall percentage efficiency of the laptop battery.   [2 marks]
   b Calculate the overall percentage of energy wasted.   [1 mark]

Key points

- Electricity and gas and/or oil supply most of the energy you use in your home.
- Electrical appliances can transfer energy in the form of useful energy at the flick of a switch.
- Uses of everyday electrical appliances include heating, lighting, making objects move (using an electric motor), and producing sound and visual images.
- An electrical appliance is designed for a particular purpose and should waste as little energy as possible.

P1.9 Energy and power

Learning objectives

After this topic, you should know:

- what is meant by power
- how to calculate the power of an appliance
- how to calculate the efficiency of an appliance in terms of power
- how to calculate the power wasted by an appliance.

Figure 1 *A lift motor*

When you use a lift to go up, a powerful electric motor pulls you and the lift upwards. The electric current through the lift motor transfers energy to the gravitational potential energy store of the lift when the lift goes up at a steady speed. You also get energy (from the electric current) transferred to the thermal energy store of the motor and the surroundings due to friction between the moving parts of the motor. In addition, energy is transferred to the thermal energy store of the surroundings by sound waves created by the lift machinery.

- The energy you supply to the motor per second is the **power** supplied to it.
- The more powerful the lift motor is, the faster it moves a particular load.

The more powerful an appliance is, the faster the rate at which it transfers energy.

The power of an appliance is measured in watts (W) or kilowatts (kW).

1 watt is equal to the rate of transferring 1 joule of energy in 1 second (i.e., 1 W = 1 J/s)

1 kilowatt is equal to 1000 watts (i.e., 1000 joules per second or 1 kJ/s).

You can calculate power using the equation:

$$\underset{(\text{watts, W})}{\textbf{power, } P} = \frac{\textbf{energy transferred to appliance, } E \text{ (joules, J)}}{\textbf{time take for energy to be transferred, } t \text{ (seconds, s)}}$$

Worked example

A motor transfers 10 000 J of energy in 25 s. Work out its power.

Solution

$$P = \frac{E}{t}$$

$$P = \frac{10\,000\,\text{J}}{25\,\text{s}} = 400\,\text{W}$$

Power ratings

Table 1 shows some typical values of power ratings for different energy transfers.

Table 1

Appliance	Power rating
A torch	1 W
An electric light bulb	100 W
An electric cooker	10 000 W = 10 kW (where 1 kW = 1000 watts)
A railway engine	1 000 000 W = 1 megawatt (MW) = 1 million watts
A Saturn V space rocket	100 MW
A very large power station	10 000 MW
World demand for power	10 000 000 MW
A star like the Sun	100 000 000 000 000 000 000 MW

Orders of magnitude

The power ratings in Table 1 are called 'order of magnitude' values. This means that the values are estimates to the nearest power of ten.

The symbol ~ is used for an order of magnitude estimate. For example, if ten million people switch their electric kettles on at the same time (e.g., at half-time in a big TV football match), the jump in the demand for electric power would be ~ 10 000 MW (10^4 MW).

Muscle power

How powerful is a weightlifter?

A 30 kg dumbbell has a weight of 300 N. Raising it by 1 m would increase its gravitational potential energy store by 300 J. A weightlifter could lift it in about 0.5 seconds. The rate of energy transfer would be 600 J/s (= 300 J ÷ 0.5 s). So the weightlifter's power output would be about 600 W in total! The power output of two weightlifters could be compared by measuring how long they each take to lift the same weight through the same vertical height. The energy transferred is the same. So the one who takes the less time has a bigger power output.

Efficiency and power

For any appliance:

- its useful power out (or output power) is the useful energy per second transferred by it
- its total power in (or input power) is the energy per second supplied to it.

In Topic P1.7, you learnt that, for an appliance:

$$\text{efficiency} = \frac{\text{useful energy transferred by the device}}{\text{total energy supplied to it}} \, (\times 100\%)$$

Because power = energy per second transferred or supplied, this efficiency equation can be rewritten as

$$\textbf{efficiency} = \frac{\textbf{useful power out}}{\textbf{total power in}} \, \textbf{(× 100\%)}$$

Wasted power

In any energy transfer, the energy wasted = the input energy supplied − the useful output energy. Because power is energy transferred per second:

$$\textbf{power wasted = total power in − useful power out}$$

1 **a** State which of the following is more powerful.
   i A torch bulb or a mains filament lamp.   [1 mark]
   ii A 3 kW electric kettle or a 10 000 W electric cooker.   [1 mark]
  b There are about 20 million occupied homes in England. If a 3 kW electric kettle was switched on in 1 in 10 homes at the same time, work out how much power would need to be supplied.   [1 mark]

2 The total power supplied to a lift motor is 5000 W. In a test, the motor transfers 12 000 J of energy in 20 s to the gravitational potential energy store of the lift.
  a Calculate how much energy is supplied to the motor by the current through it in 20 s.   [2 marks]
  b Calculate its efficiency in the test.   [2 marks]

3 A machine has an input power rating of 100 kW. If the useful energy transferred by the machine in 50 seconds is 1500 kJ, calculate:
  a its output power in kilowatts   [2 marks]
  b its percentage efficiency.   [2 marks]

4 **a** Describe the energy transfers and changes to the energy stores of an electric hot-water shower when it is in operation.   [5 marks]
  b A 12 kW electric shower is used 4 times in a day for 20 minutes each time. Calculate the energy supplied to it by electricity in one day.   [2 marks]

Figure 2 *Muscle power*

Worked example

The useful power out of an electric motor is 20 W, and the total power into it is 80 W. Calculate:

a the percentage efficiency of the motor

b the power wasted by the motor.

Solution

a percentage efficiency =
$$\frac{\text{useful power out}}{\text{total power in}} \times 100\%$$
$$= \frac{20\,W}{80\,W} \times 100\% = 25\%$$

b wasted power = 80 W − 20 W = 60 W

Key points

- Power is rate of transfer of energy.
- The power of an appliance is $P = \dfrac{E}{t}$.
- efficiency of an appliance
$$= \frac{\text{useful power out}}{\text{total power in}} \, (\times 100\%)$$
- power wasted by an appliance = total power input − useful power output

Summary questions

1 An electric fan heater contains an electric heater and a fan driven by an electric motor.

  a Describe the energy transfers that take place in the fan heater when it is operating normally. [5 marks]

  b With reference to the fan heater, explain what is meant by:

    i useful energy [3 marks]

    ii wasted energy. [3 marks]

2 A bungee jumper jumps from a platform and transfers 12 000 J from his gravitational potential energy store before the rope attached to him becomes taut and starts to stretch. He then transfers a further 24 000 J from his gravitational potential energy store before he stops falling and begins to rise.

  a Describe the energy transfers and changes to the jumper's energy stores:

    i after he jumps before the rope starts to stretch [2 marks]

    ii after the rope starts to stretch until he stops falling. [3 marks]

  b Calculate the maximum value of his kinetic energy store during his descent. [2 marks]

3 A train on a straight level track is pulled at a constant speed of 23 m/s by an engine with an output power of 700 kW.

  a **i** Calculate how much energy is transferred from the train to the surroundings in 300 s. [2 marks]

    ii Calculate how far the train travels in 300 s. [1 mark]

    iii Show that the resistive force on the engine is approximately equal to 30 000 N. [2 marks]

    iv Explain why the driving force of the engine is equal and opposite to the resistive force on the train. [2 marks]

  b The train then moves on to an inclined section of the railway line where the track rises by 1 m for every kilometre of track. Explain why the output power of the engine needs to be increased to maintain the same speed of 23 m/s. [3 marks]

4 A student pushes a trolley of weight 150 N up a slope of length 20 m. The slope is 1.2 m high.

  a Calculate the increase in the gravitational potential energy store of the trolley. [1 mark]

  b The student pushed the trolley up the slope with a force of 11 N. Show that the work done by the student was 220 J. [1 mark]

  c If the student pushed the trolley up the slope at constant speed, explain why all the work done by the student was not transferred to the gravitational potential energy store of the trolley. [3 marks]

5 **a** A stone is fired into the air from a catapult and falls to the ground some distance away. Describe the energy transfers that take place after the catapult is released. [4 marks]

  b A stone of mass 0.015 kg is catapulted into the air at a speed of 25 m/s. It reaches a height of 20 m before it descends and hits the ground some distance away.

    i Calculate its initial kinetic energy. [2 marks]

    ii Calculate the increase in the gravitational potential energy store of the stone when it reached its maximum height ($g = 9.8$ N/kg). [2 marks]

    iii Estimate its speed when it was at its maximum height. [4 marks]

6 A parachutist of total mass 75 kg jumps from an aeroplane moving at a speed of 60 m/s at a height of 900 m above the ground.

  a Calculate her kinetic energy when she left the aeroplane. [2 marks]

  b Her parachute reduced her speed of descent to 5 m/s.

    i Calculate her kinetic energy at this speed. [2 marks]

    ii Calculate the decrease in her gravitational potential energy store as a result of her descent. [2 marks]

  c Calculate the work done by air resistance during her descent. [2 marks]

7 On a building site, a 600 W electric winch and a pulley were used to lift bricks from the ground. The winch raised a load of 500 N through a height of 3.0 m in 25 s. During this time, the average power supplied to the winch was 600 W.

  a **i** Calculate how much useful energy was transferred by the motor. [2 marks]

    ii Calculate the percentage efficiency of the system. [4 marks]

  b Explain why the efficiency of a winch can never be as much as 100%. [3 marks]

Practice questions

01.1 Calculate the work done by a long distance runner during a 10 km race. The mass of the runner is 52 kg. Gravitational field strength is 9.8 N/kg   [3 marks]

01.2 Suggest a reason why the actual amount of work done is likely to be higher than the calculated value.   [1 mark]

02 An independent researcher investigates the batteries in a set of electric shavers. She starts with the batteries fully discharged and then transfers 25 000 joules of energy to the chemical energy store of each battery. The shavers are then allowed to run until they stop.

Shaver	A	B	C
Time to stop – test 1 (in mins)	20.0	23.0	18.0
Time to stop – test 2 (in mins)	20.5	22.5	19.0
Mean time			

02.1 Calculate the mean time to stop for each shaver.   [3 marks]

02.2 Give one reason why it was good scientific practice to repeat the tests.   [1 mark]

02.3 Give one reason why the same amount of energy was supplied each time.   [1 mark]

02.4 Suggest why it is important that the investigation is carried out by an independent researcher rather than the manufacturers of the shavers.   [1 mark]

02.5 Name the shaver with the greatest power.   [1 mark]

02.6 Calculate the power of shaver B.   [3 marks]

03 An electric motor is used to power a car-lifting machine.

03.1 Calculate the increase in the gravitational potential energy store of a car lifted 2.5 m off the floor. The mass of the car is 950 kg. Gravitational field strength is 9.8 N/kg   [2 marks]

03.2 Calculate the power of the lifting machine if it takes 10 seconds to raise the car.   [2 marks]

03.3 Calculate the efficiency of the lifting machine if 3.6 kW of power were supplied. Give your answer to 2 significant figures.   [3 marks]

03.4 Grease is used to reduce friction in the lifting machine. Suggest one effect friction has when two surfaces are rubbed together.   [1 mark]

04 **Figure 1** shows how the gravitational potential energy store of a 22 kg rock changes as it is lifted up.

Figure 1

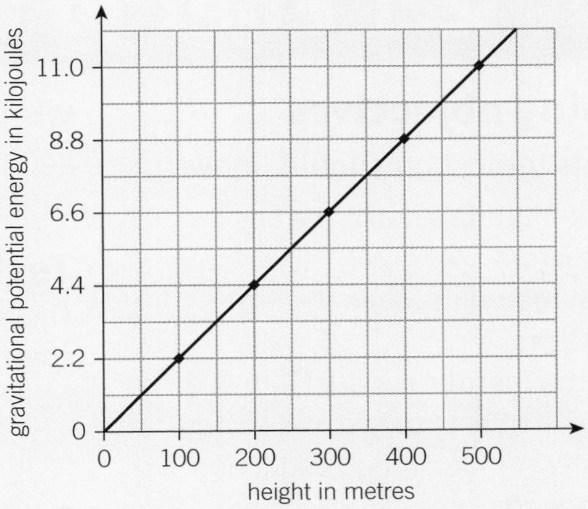

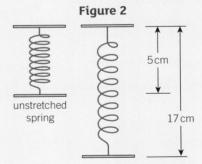

04.1 Describe the relationship between the height of the rock and the transfer of energy to the gravitational potential energy store of the rock.   [2 marks]

04.2 Use the information in the graph to determine the energy stored in the gravitational potential energy store of a 22 kg rock lifted 400 m up a cliff face.   [1 mark]

04.3 Calculate the velocity of the rock just before it hits the ground after falling 400 m.   [2 marks]

05 A trampoline has 32 equal springs attached to the sides. The springs are stretched when someone bounces on the trampoline. **Figure 2** shows one spring before and after it is stretched.

Figure 2

unstretched spring   5 cm   17 cm

05.1 Calculate the total energy in the elastic potential energy stores of the 32 stretched springs.

The spring constant is 5000 N/m. Use the equation $E_e = \frac{1}{2}ke^2$ [3 marks]

05.2 Give one reason why the springs must not exceed their elastic limit when being used.   [1 mark]

05.3 Suggest a reason why the springs are tested beyond the normal working load.   [1 mark]

Learning objectives

After this topic, you should know:

- which materials make the best conductors
- which materials make the best insulators
- how the thermal conductivity of a material affects the rate of energy transfer through it by conduction
- how the thickness of a layer of material affects the rate of energy transfer by conduction through it.

When you have a barbecue, you need to know which materials are good conductors and which ones are good insulators. If you can't remember, you're likely to burn your fingers!

Testing rods of different materials as conductors

The rods need to be the same width and length for a fair test. Each rod is coated with a thin layer of wax near one end. The uncoated ends are then heated together.

Look at Figure 1. The wax melts fastest on the rod that best conducts energy.

- Metals conduct energy better than non-metals.
- Copper is a better conductor than steel.
- Glass conducts better than wood.

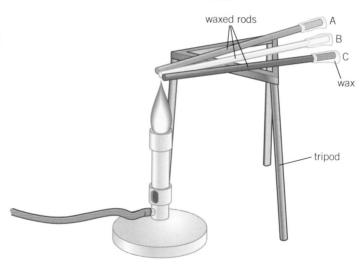

Figure 1 *Comparing conductors*

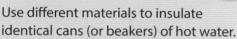

Testing sheets of materials as insulators

Use different materials to insulate identical cans (or beakers) of hot water.

When choosing your materials, consider which properties will make the materials good thermal insulators, for example, the thickness of the material or the colour of the material.

The volume of water and its temperature at the start should be the same.

Use a thermometer to measure the water temperature after a fixed time.

- Use your results to work out which is the best insulator.

The table below gives the results of comparing two different materials using the method above.

Material	Starting temperature in °C	Temperature after 300 s in °C
paper	40	32
felt	40	36

Safety: Take care if you are using very hot water.

Thermal conductivity

In Figure 1, each rod has the same temperature difference between its ends. Each rod is the same length and diameter. The energy transfer by conduction through a material depends on its **thermal conductivity**. The greater the thermal conductivity of a material, the more energy per second it transfers by conduction. So, in Figure 1, if

- A conducts better than C, and
- C conducts better than B, then
- the thermal conductivity of A is higher than the thermal conductivity of C, and
- the thermal conductivity of C is higher than the thermal conductivity of B.

Insulation matters

Materials that are good insulators are necessary to keep you warm in winter, whether you are at home or outdoors. Good insulators need to be materials that have low thermal conductivity, so energy transfer through them is as low as possible.

The energy transfer per second through a layer of insulating material depends on:

- the temperature difference across the material
- the thickness of the material
- the thermal conductivity of the material.

To reduce the energy transfer as much as possible in any given situation:

1 the thermal conductivity of the insulating material should be as low as possible

2 the thickness of the insulating layer should be as thick as is practically possible.

Figure 2 shows a layer of insulating material being fitted in the loft of a house. The insulating material chosen has a much lower thermal conductivity than the roof material. Several layers of this material fitted on the loft floor will reduce the energy transfer through the roof significantly. Insulating buildings is covered further in Topic P2.5.

Figure 2 *Insulating a loft. The air trapped between fibres makes fibreglass a good insulator*

1 **a** Explain why steel pans have handles made of plastic or wood.
    [2 marks]

   b Suggest which material, felt or paper, is the better insulator. Give a reason for your answer.
    [2 marks]

2 **a** Choose a material you would use to line a pair of winter boots. Give a reason for your choice of material.
    [2 marks]

   b Describe how you could carry out a test on three different lining materials. Assume you have a thermometer, a stopwatch, and you can wrap the lining round a container of hot water.
    [3 marks]

3 Describe an investigation you would carry out to find out how the thickness of a layer of insulating material affects the energy transfer through it. 🖊
    [5 marks]

4 In Figure 1, A is a copper rod, B is a glass rod, and C is a steel rod. Determine which rod keeps its wax in the solid state for the longest time. Explain your answer.
    [2 marks]

Key points

- Metals are the best conductors of energy.
- Non-metal materials such as wool and fibreglass are the best insulators.
- The higher the thermal conductivity of a material, the higher the rate of energy transfer through it.
- The thicker a layer of insulating material, the lower the rate of energy transfer through it.

P2.2 Infrared radiation

Learning objectives

After this topic, you should know:

- what infrared radiation is
- how infrared radiation depends on the temperature of an object
- what is meant by black body radiation
- **H** what happens to the temperature of an object if it absorbs more radiation than it emits.

Synoptic link

You'll learn more about electromagnetic waves in Chapter 13.

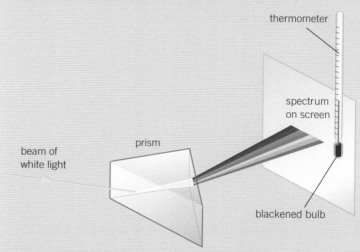

Figure 2 *Keeping watch in darkness*

Energy from the Sun

When you are in sunlight, you are absorbing infrared radiation from the Sun. Infrared radiation and visible light are parts of the electromagnetic spectrum. So too are radio waves, microwaves, ultraviolet rays and X-rays. Electromagnetic waves are electric and magnetic waves that travel through space. The wavelength of light increases across the visible spectrum from blue to red light. Infrared waves are longer in wavelength than visible light waves.

The Sun emits all types of electromagnetic radiation. Fortunately, the Earth's atmosphere blocks most of the types of radiation that are harmful to people. But it doesn't block infrared radiation or light from the Sun.

Detecting infrared radiation

You can use a thermometer with a blackened bulb to detect infrared radiation. Figure 1 shows how to do this.

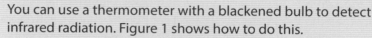

Figure 1 *Detecting infrared radiation*

The glass prism splits a narrow beam of white light into the colours of the spectrum.

The thermometer reading rises when it is placed just beyond the red part of the spectrum. Some of the infrared radiation in the beam goes there. Your eyes can't detect it, but the thermometer can.

Infrared radiation is beyond the red part of the visible spectrum.

- Suggest what would happen to the thermometer reading if you moved it away from the screen.

Radiation and surface temperature

If you want to see animals and people in the dark, you need to use special cameras. These cameras detect **infrared radiation**. All objects give out (emit) infrared radiation.

The higher the temperature of an object, the more infrared radiation it emits in a given time.

All bodies (objects), no matter what their temperature is, emit and absorb infrared radiation. A body at constant temperature emits infrared radiation at the same rate as it absorbs it.

A perfect black body is an object that absorbs all the radiation that hits it. It doesn't reflect any radiation, and it doesn't transmit any radiation (i.e., no radiation passes through it). A good absorber is also a good emitter, so a perfect black body is also the best possible emitter. The radiation emitted by a perfect black body is called **black body radiation**. No other object emits or absorbs radiation as effectively as a black body.

An object that has a constant temperature emits radiation across a continuous range of wavelengths.

Figure 4 shows how the intensity of black body radiation varies with wavelength. The intensity of the radiation is highest at a certain wavelength, which depends on the temperature.

If the temperature of the object is increased, the intensity of the radiation it emits is greater at every wavelength – as shown in Figure 4. Figure 4 also shows that the peak of the higher radiation curve is at a shorter wavelength than the peak of the lower curve. This is because the shorter the wavelength of the radiation, the greater the increase in intensity at that wavelength. Therefore, the peak intensity is at a shorter wavelength than it was at the lower temperature.

Observing a filament lamp

Look at a low-voltage filament lamp when its potential difference is slowly increased to its operating potential difference. At first, the filament only emits infrared radiation. Then it glows dull red because it emits light as well. Its dull red glow changes to orange-red and then yellow-white as the filament becomes hotter. This is because it emits more radiation at shorter wavelengths as it gets hotter.

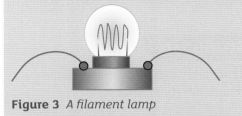

Figure 3 *A filament lamp*

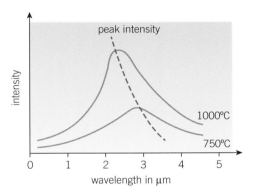

Figure 4 *Black body radiation*

1 Explain what infrared radiation is. [1 mark]

2 **a** An infrared camera on a satellite shows that more infrared radiation is emitted at night from a city than from the surrounding rural area. Explain what this tells you about the city compared with the rural area. Give a reason for your answer. [2 marks]

   b Explain how you could tell if an electric iron is hot without touching it. [1 mark]

3 **a** When an iron nail was heated in the flame of a Bunsen burner, the colour of the nail became dull-red then orange-red. Explain why the colour of the nail changed in this way. [2 marks]

   b The colour of a star depends on its surface temperature. A student notices that three stars, **X**, **Y**, and **Z**, near to each other have different colours. **X** is yellow, **Y** is red, and **Z** is white.
   i Write down and explain which star is hottest. [2 marks]
   ii Write down and explain which star is coolest. [2 marks]

Key points

- All objects emit and absorb infrared radiation.
- The hotter an object is, the more infrared radiation it emits in a given time.
- Blackbody radiation is radiation emitted by a body that absorbs all the radiation incident on it.

P2.3 More about infrared radiation

Learning objectives

After this topic, you should know:

- what happens to the the temperature of an object if it absorbs more radiation than it emits
- how the temperature of the Earth is affected by the balance of absorbed and emitted radiation.

Absorption and emission of infrared radiation

Every object absorbs and emits infrared radiation, whatever its temperature is. If an object has a constant temperature, the object emits infrared radiation at the same rate as it absorbs it. When an object absorbs radiation faster than it emits radiation, its temperature increases.

Rescue teams use light-coloured, shiny blankets to keep accident survivors warm (Figure 1). A light, shiny outer surface emits a lot less radiation than a dark, matt (non-glossy) surface. This keeps the patient warm, as less infrared radiation is emitted than if an ordinary blanket had been used.

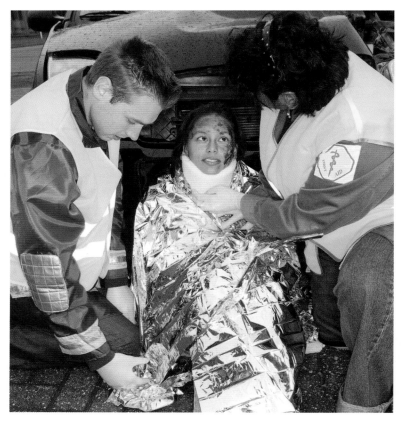

Figure 1 *An emergency blanket in use*

Radiation and the Earth's temperature

The temperature of the Earth depends on lots of factors, such as the rate that light and infrared radiation from the Sun are:

- reflected back into space or absorbed by the Earth's atmosphere or by the Earth's surface
- emitted from the Earth's surface and from the Earth's atmosphere into space.

If the Earth had no atmosphere, the temperature on the surface would plunge to about −180 °C at night, the same as the Moon's surface at night. This would happen because the surface would not be receiving any radiation from the Sun – it would be emitting radiation into space.

Some gases in the atmosphere, such as water vapour, methane, and carbon dioxide (greenhouse gases) absorb longer wavelength infrared radiation from the Earth and prevent it escaping into space. These gases absorb the radiation and then emit it back to the surface (Figure 2). This process makes the Earth warmer than it would be if these gases were not in its atmosphere.

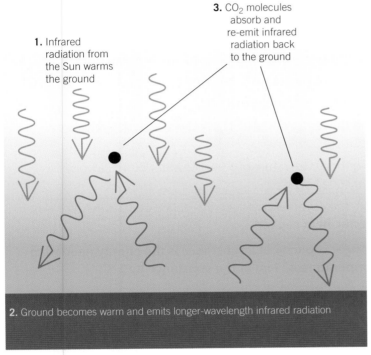

3. CO_2 molecules absorb and re-emit infrared radiation back to the ground

1. Infrared radiation from the Sun warms the ground

2. Ground becomes warm and emits longer-wavelength infrared radiation

Figure 2 *The absorption and emission of infrared radiation*

1 The surface of the Earth absorbs radiation from the Sun and it emits radiation.
   a State one similarity and one difference in the radiation the Earth emits and the radiation it absorbs from the Sun. [2 marks]
   b Describe what happens to the radiation the surface of the Earth emits on a very clear night. [2 marks]

2 On a hot day in summer, the interior of a parked car in sunlight with its windows closed becomes unbearably hot.
   a Explain why the temperature inside the car becomes much higher than the outside temperature. [2 marks]
   b Explain why the inside of the car would not become as hot if it had been parked in a shaded area. [2 marks]

3 Explain why the presence of greenhouse gases in the atmosphere makes the Earth warmer than it would be if there were no greenhouse gases in the atmosphere. [3 marks]

Key points

- The temperature of an object increases if it absorbs more radiation than it emits.
- The Earth's temperature depends on a lot of factors, including the absorption of infrared radiation from the Sun, and the emission of radiation from the Earth's surface and atmosphere.

P2.4 Specific heat capacity

Learning objectives

After this topic, you should know:

- what is meant by the specific heat capacity of a substance
- how to calculate the energy changes that occur when an object changes temperature
- how the mass of a substance affects how quickly its temperature changes when it is heated
- how to measure the specific heat capacity of a substance.

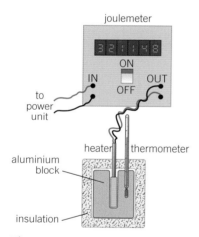

joulemeter

ON
IN   OUT
OFF
to
power
unit

heater   thermometer
aluminium
block

insulation

Figure 1 *Heating an aluminium block*

A car in strong sunlight can become very hot. A concrete block of equal mass would not become as hot. Metal heats up more easily than concrete. Investigations show that when a substance is heated, its temperature rise depends on:

- the amount of energy supplied to it
- the mass of the substance
- what the substance is.

The following results were obtained using two different amounts of water. They show that:

- heating 0.1 kg of water by 4 °C required an energy transfer of 1600 J
- heating 0.2 kg of water by 4 °C required an energy transfer of 3200 J

Using these results, you can say that:

- Increasing the temperature of 1.0 kg of water by 4 °C requires a transfer of 16 000 J of energy
- Increasing the temperature of 1.0 kg of water by 1 °C involves a transfer of 4000 J of energy.

More accurate measurements would give 4200 J per kg per °C for water. This is its **specific heat capacity**.

The specific heat capacity of a substance is the energy needed to raise the temperature of 1 kg of the substance by 1 °C.

The unit of specific heat capacity is the joule per kilogram degree Celsius (J/kg °C).

For a known change of temperature of a known mass of a substance:

| **energy transferred, ΔE** (joules, J) | = | **mass, m** (kilograms, kg) | × | **specific heat capacity, c** (joule per kilogram per degree Celsius, J/kg °C) | × | **temperature change, $\Delta \theta$** (degree Celsius, °C) |

The energy transferred to the substance increases the thermal energy store of the substance by an equal amount.

To find the specific heat capacity of a substance, rearrange the above equation into the form:

$$c = \frac{\Delta E}{m \, \Delta \theta}$$

Measuring specific heat capacity

Use the arrangement shown in Figure 1 to heat a metal block of known mass. You will need to use a thermometer and a top-pan balance.

Use the energy meter (or joulemeter) to measure the energy supplied to the block. Use the thermometer to measure its temperature rise.

What changes to energy stores occur as result of the transfer of energy to the block?

To find the specific heat capacity of aluminium, insert your measurements into the equation:

$$c = \frac{\Delta E}{m \, \Delta \theta}$$

Replace the block with an equal mass of water in a suitable container. Measure the temperature rise of the water when the same amount of energy is supplied to it by the heater.

Your results should show that aluminium heats up more quickly than water.

Safety: Wear eye protection and take care with a hot immersion heater.

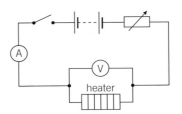

Figure 2 *Circuit diagram*

Synoptic links

You will see in Topic P5.5 that the energy supplied by electricity $E =$ heater potential difference $V \times$ heater current $I \times$ heating time t.

Storage heaters

A storage heater uses electricity at night (off-peak) to heat special bricks or concrete blocks in the heater. Energy transfer from the bricks keeps the room warm. The bricks have a high specific heat capacity, so they store lots of energy. They warm up slowly when the heater element is on, and cool down slowly when it is off.

Table 1 *The specific heat capacity for some other substances*

Substance	water	oil	aluminium	iron	copper	lead	concrete
Specific heat capacity in J/kg °C	4200	2100	900	390	385	130	850

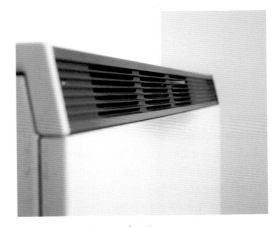

Figure 3 *A storage heater*

1 A small bucket of water and a large bucket of water are left in strong sunlight. Which one warms up faster? Give a reason for your answer. [2 marks]

2 Use the information in Table 1 above to answer this question.
   a Explain why a mass of lead heats up more quickly than an equal mass of aluminium. [2 marks]
   b Calculate the energy needed:
      i to raise the temperature of 0.20 kg of aluminium from 15 °C to 40 °C [2 marks]
      ii to raise the temperature of 0.40 kg of water from 15 °C to 40 °C [2 marks]
      iii to raise the temperature of 0.40 kg of water in an aluminium container of mass 0.20 kg from 15 °C to 40 °C. [3 marks]
   c A copper water tank of mass 20 kg contains 150 kg of water at 15 °C. Calculate the energy needed to heat the water and the tank to 55 °C. [5 marks]

3 State *two* ways in which a storage heater differs from a radiant heater. [2 marks]

4 Design an experiment to measure the specific heat capacity of oil using the arrangement in Figure 1. ✏ [6 marks]

Key points

- The specific heat capacity of a substance is the amount of energy needed to change the temperature of 1 kg of the substance by 1 °C.
- Use the equation $\Delta E = m \, c \, \Delta \theta$ to calculate the energy needed to change the temperature of mass m by $\Delta \theta$.
- The greater the mass of an object, the more slowly its temperature increases when it is heated.
- To find the specific heat capacity c of a substance, use a joulemeter and a thermometer to measure ΔE and $\Delta \theta$ for a measured mass m, then use $c = \frac{\Delta E}{m \, \Delta \theta}$.

P2.5 Heating and insulating buildings

Learning objectives

After this topic, you should know:

- how homes are heated
- how you can reduce the rate of energy transfer from your home
- what cavity wall insulation is.

Reducing the rate of energy transfers at home

Houses are heated by electric or gas heaters, oil or gas central heating systems, or by solid fuel in stoves or in fireplaces. Whichever form of heating you have in your home, the heating bills can be expensive. When your home heating system is transferring energy into your home to keep you warm, energy is also transferring to the surroundings outside your home. Figure 1 shows some of the measures that can be taken to reduce the rate of energy transfer from a home, and so reduce home heating bills.

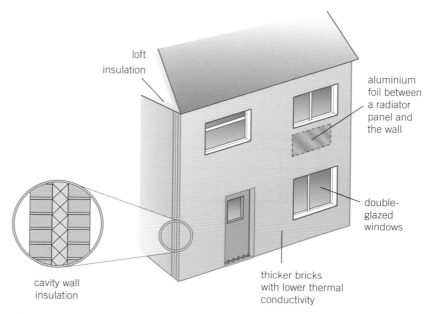

loft insulation

aluminium foil between a radiator panel and the wall

double-glazed windows

thicker bricks with lower thermal conductivity

cavity wall insulation

Figure 1 *Saving money*

Go further!

A duvet is a bed cover filled with 'down' or soft feathers, or some other suitable insulator such as wool. Because the filling material traps air, a duvet on a bed reduces the rate at which energy is transferred from you as you sleep. The tog rating of a duvet depends on the thickness of the material and on its thermal conductivity. It tells you how effective it is as an insulator. The thicker the material, or the lower its thermal conductivity, the better it is as an insulator, and so the higher its tog rating.

- Loft insulation such as fibreglass reduces the rate of energy transfer through the roof. Fibreglass is a good insulator. The air between the fibres also helps to reduce the rate of energy transfer by conduction. The greater the number of layers of insulation, the thicker the insulation will be. So the rate of energy transfer through the roof will be less.

- Cavity wall insulation reduces the rate of energy transfer through the outer walls of the house. The cavity of an outer wall is the space between the two layers of brick that make up the wall. The insulation is pumped into the cavity. It is a better insulator than the air it replaces. It traps the air in small pockets, reducing the rate of energy transfer by conduction.

- Aluminium foil between a radiator panel and the wall reflects radiation away from the wall and so reduces the rate of energy transfer by radiation.

- Double-glazed windows have two glass panes with dry air or a vacuum between the panes. The thicker the glass and the lower its thermal conductivity is, the slower the rate of transfer of energy through it by conduction will be. Dry air is a good insulator, so it reduces the rate of energy transfer by conduction. A vacuum also prevents energy transfer by convection.

- If the external walls of a warm building have thicker bricks and lower thermal conductivity, the rate of transfer of energy from the inside of the building to the outside will be lower and the cost of heating will be less.

Solar panels

Heating a home using electricity or gas can be expensive. Solar panels absorb infrared radiation from the Sun and are used to generate electricity directly (solar cell panels) or to heat water directly (solar heating panels). In the northern hemisphere, a solar panel is usually fitted on a roof that faces south so that it absorbs as much infrared radiation from the Sun as possible.

Synoptic link

You will learn more about solar panels in Topic P3.3.

1 **a** Explain why cavity wall insulation is better than air in the cavity between the walls of a house. [2 marks]

   b Explain why fixing aluminium foil to the wall behind a radiator reduces energy transfer through the wall. [2 marks]

2 Some double-glazed windows have a plastic frame and a vacuum between the panes.

   a Explain why a plastic frame is better than a metal frame. [2 marks]

   b State why a vacuum between the panes is better than air. [1 mark]

3 Two manufacturers advertise double-glazed windows of the same size and with dry air between the panes at the same price, but with a different gap width between the glass panes. Explain which one you would choose. [2 marks]

4 A manufacturer of loft insulation claimed that each roll of loft insulation would save £10 per year on fuel bills. A householder bought six rolls of the loft insulation at £15 per roll and paid £90 to have the insulation fitted in her loft.

   a Calculate how much it cost to buy and install the loft insulation. [2 marks]

   b Calculate what the saving each year would be on fuel bills. [1 mark]

Key points

- Electric and/or gas heaters and gas or oil-fired central heating or solid-fuel stoves are used to heat houses.
- The rate of energy transfer from houses can be reduced by using:
  - loft insulation
  - cavity wall insulation
  - double-glazed windows
  - aluminium foil behind radiators
  - external walls with thicker bricks and lower thermal conductivity.
- Cavity wall insulation is insulation material that is used to fill the cavity between the two brick layers of an external house wall.

Summary questions

1 a i Explain why a white hat is better to wear outdoors in summer than a black hat. [2 marks]

 ii Describe the type of surface that is better for a flat roof: matt or smooth, dark or shiny. Explain your answer. [2 marks]

b A solar heating panel is used to heat water. Some panels have a transparent cover and a matt black base. Others have a matt black cover and a shiny base.

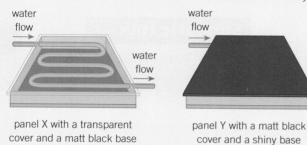

panel X with a transparent cover and a matt black base

panel Y with a matt black cover and a shiny base

Suggest which of these two designs, Panel X or Panel Y, is better. Give reasons for your answer. [4 marks]

2 A heat sink is a metal plate or clip fixed to an electronic component to stop it overheating.

a When the component becomes hot, how does energy transfer from:

 i where the component is in contact with the plate to the rest of the plate? [1 mark]

 ii the plate to the surroundings? [1 mark]

b Describe the purpose of the metal fins on the plate. [2 marks]

c Heat sinks are made from metals such as copper or aluminium. Copper is approximately three times more dense than aluminium, and its specific heat capacity is about twice as large. Discuss how these physical properties are relevant to the choice of whether or not to use copper or aluminium for a heat sink in a computer. [4 marks]

3 a Explain why woolly clothing is very effective at keeping people warm in winter. [3 marks]

b Wearing a hat in winter is a very effective way of keeping your head warm. Describe how a hat helps to reduce the rate of energy transfer from your head. [3 marks]

c Keeping your ears warm is important too. Explain why energy transfer from your ears to the surroundings can be significant in winter. [3 marks]

4 Marathon runners at the end of a race are often supplied with a shiny emergency blanket to stop them becoming cold. These silvery blankets are very light in weight because they are made from plastic film with a reflective metallic coating inside.

a Name the form of energy transfer that is reduced by the reflective coating inside the emergency blanket. [1 mark]

b Explain why an emergency blanket helps to stop the runners becoming cold. [2 marks]

5 A meteorite loses about 60 MJ/kg of energy from its kinetic and gravitational potential energy stores when it falls to the ground from space. The specific heat capacity of a meteorite is about 400 J/kg °C.

a Estimate its maximum temperature rise if just 1% of the kinetic and gravitational potential energy is transferred as thermal energy of the meteorite. [4 marks]

b The melting point is about 2500 °C. Discuss whether or not the 1% assumption in **a** is realistic. [2 marks]

6 A 5 kW electric shower heats the water flowing through it from 15 °C to 40 °C when the water flow rate is 1.5 kg per minute.

a Calculate the energy per second used to heat the water. The specific heat capacity of water is 4200 J/kg °C. [2 marks]

b Calculate the percentage efficiency of the shower heater. [3 marks]

7 Penguins huddle together to keep warm. Design an investigation to model the effect of penguins huddling together. Use beakers of hot water to represent the penguins. [5 marks]

Practice questions

01 A student had read about an outdoor ice container. The article described ways of slowing down the rate at which the ice melts on hot summer days. She decided to investigate using the apparatus in **Figure 1**.

Figure 1

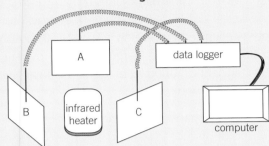

A is copper
B is glass
C is plastic

01.1 Why did the student use an infrared heater in the investigation? [1 mark]

01.2 Name two control variables in the investigation. [2 marks]

01.3 The student used a temperature probe attached to a data logger instead of a thermometer. Suggest how this improved the investigation. [2 marks]

01.4 A politician has suggested glaciers should be covered in insulation to slow down their rate of melting. Do you agree with this suggestion? Explain your answer [2 marks]

02 **Table 1** shows the thermal conductivity of three metals used in the manufacture of saucepans.

Table 1

Metal	Thermal conductivity in W/m °C
copper	380
stainless Steel	54
aluminium	250

02.1 Choose one metal for the base of a saucepan that would give the best thermal efficiency. Give a reason for your choice. [2 marks]

02.2 Describe how to check a saucepan is hot without touching it or using a thermometer. [2 marks]

03.1 A hot water bottle made of rubber is filled with 0.65 kg of hot water. The temperature of the water is 90 °C. Calculate the temperature of the hot water bottle after 163 800 J of energy are transferred during the night. Specific heat capacity of water is 4200 J/kg °C [3 marks]

03.2 A new type of bed warmer is sealed, filled with polymer gel, and heated using an electric insert. The bed warmer can control the temperature of the polymer gel. Suggest two advantages of using the new bed warmer rather than a traditional hot water bottle. [2 marks]

04 A student investigated the insulation properties of two materials, **A** and **B**. The apparatus she used is shown in Figure 2.

Figure 2

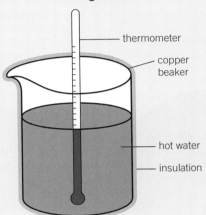

Her method was as follows:
1 Wrap a 2 cm layer of material **A** around the beaker.
2 Fill the beaker with 200 ml of hot water and record the temperature of the water.
3 Record the temperature of the water after 10 minutes.
4 Wrap a second 2 cm layer of material **A** around the beaker.
5 Repeat stages 2 and 3.
6 Replace material **A** with material **B** and repeat stages 1–5.

Table 2 shows the results of the investigation.

Table 2

Material	Number of layers	Water temperature at the start, in °C	Water temperature after 10 mins, in °C
A	1	82.5	66.0
B	1	83.0	71.5
A	2	81.5	72.0
B	2	75.0	67.5

04.1 Calculate the temperature change for each test. [2 marks]

04.2 Which material was the better insulator? Give a reason for your answer. [2 marks]

04.3 Suggest **two** ways the student could have improved the investigation. [2 mark]

P 3 Energy resources

3.1 Energy demands

Learning objectives

After this topic, you should know:

- how most of your energy demands are met today
- what other energy resources are used
- how nuclear fuels are used in power stations
- what other fuels are used to generate electricity.

Most of the energy you use comes from burning fossil fuels, mostly gas or oil or coal. The energy in homes, offices, and factories is mostly supplied by gas or by electricity generated in coal or gas-fired power stations. Oil is needed to keep road vehicles, ships, and aeroplanes moving. Burning one kilogram of fossil fuel releases about 30 million joules of energy. You use about 5000 joules of energy each second, which is about 150 thousand million joules each year. But because of the inefficiencies in how energy is distributed and used, a staggering 10 000 kg of fuel is used each year to supply the energy needed just for you!

Figure 1 shows how the global demand for energy is met. Fossil fuels are extracted from underground or under the sea bed and then transported to oil refineries and power stations. Much of the electricity you use is generated in fossil-fuel power stations. Instead of fossil fuels, some power stations use biofuels or nuclear fuel. Fossil fuels and nuclear fuel are non-renewable because they can not be replaced. As you will learn in Topic P3.4, their use is causing major environmental problems and increasing the levels of greenhouse gases, such as carbon dioxide, in the atmosphere. Some of the electricity you use is from renewable energy resources such as wind energy, hydroelectricity and solar energy, which you'll learn more about in Topics P3.2 and P3.3.

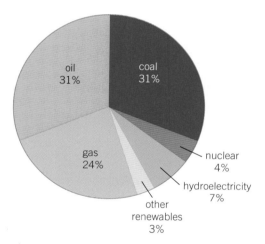

Figure 1 *World energy demand and sources of energy in 2013*

Inside a power station

In coal- or oil-fired power stations, and in most gas-fired power stations, the burning fuel heats water in a boiler. This produces steam. The steam drives a turbine that turns an electricity generator. Coal, oil, and gas are fossil fuels, which are fuels that come from long-dead animals and plants.

Biofuels

Methane gas can be collected from cows or animal manure, from sewage works, decaying rubbish, and other sources. It can be used in small gas-fired power stations. Methane from these sources is an example of a biofuel.

A **biofuel** is any fuel taken from living or recently living organisms. Animal waste is an example of a biofuel. Biofuels can be used instead of fossil fuel in modified engines for transport and in generators at power stations. Biodiesel uses waste vegetable oil and plants such as rapeseed. Other examples of biofuels are ethanol (from fermented sugar cane), straw, nutshells, and woodchip.

A biofuel is:

- **renewable** because its biological source either regrows (vegetation) or is continually produced (sewage and rubbish). This means it is used at the same rate that it is replaced.

- **carbon-neutral** because, in theory, the carbon that the living organism takes in from the atmosphere as carbon dioxide can balance the amount that is released when the biofuel is burnt.

Figure 2 *Using biofuel to generate electricity*

Pie chart labels: oil 31%, coal 31%, gas 24%, nuclear 4%, hydroelectricity 7%, other renewables 3%

Nuclear power

Nuclear fuel takes energy from atoms. Figure 3 shows that every atom contains a positively charged **nucleus** surrounded by electrons.

The fuel in a nuclear power station is uranium (or plutonium). The uranium fuel is in sealed cans in the core of the reactor. The nucleus of a uranium atom is unstable and can split in two. Energy is transferred from the nucleus when this happens. Because there are lots of uranium atoms in the core, it becomes very hot.

The energy of the core is transferred by a fluid (called the coolant) that is pumped through the core.

● The coolant is very hot when it leaves the core. It flows through a pipe to a heat exchanger, then back to the **reactor core**.

● The energy transferred by the coolant is used to turn water into steam in the heat exchanger. The steam drives turbines that turn electricity generators.

Table 1 *Comparing nuclear power and fossil fuel power*

	Nuclear power station	**Fossil fuel power station**
Fuel	Uranium or plutonium	Coal, oil, or gas
Energy released per kg of fuel	≈ 300 000 MJ (= about 10 000 × energy released per kg of fossil fuel)	≈ 30 MJ
Waste	Radioactive waste that needs to be stored for many years	Non-radioactive waste
Greenhouse gases, e.g., carbon dioxide	No – because uranium releases energy without burning	Yes – because fossil fuels produce gases such as carbon dioxide when they burn

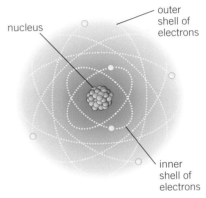

Figure 3 *The structure of the atom*

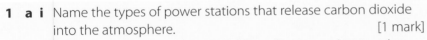

1 **a i** Name the types of power stations that release carbon dioxide into the atmosphere. [1 mark]

   ii Name the type of power station that does not release carbon dioxide into the atmosphere. [1 mark]

   b Nuclear fuel releases about 10 000 times as much energy as the same mass of fossil fuel. Give one disadvantage of nuclear fuel compared with other types of fuel. [2 marks]

2 **a** Give one advantage and one disadvantage of:

   i an oil-fired power station compared with a nuclear power station [1 mark]

   ii a gas-fired power station compared with a coal-fired power station. [1 mark]

   b Look at Table 1. Calculate how many kilograms of fossil fuel would give the same amount of energy as 1 kilogram of uranium fuel. [1 mark]

3 **a** Explain why ethanol is described as a biofuel. [2 marks]

   b Ethanol is also described as carbon-neutral. Explain what a carbon-neutral fuel is. [2 marks]

4 Global energy usage is currently about 5.0×10^{20} joules per year. The global population is about 6×10^9 people. Estimate how much energy per second each person uses on average. [4 marks]

Key points

● Your energy demands are met mostly by burning oil, coal, and gas.

● Nuclear power, biofuels, and renewable resources provide energy to generate some of the energy you use.

● Uranium or plutonium is used as the fuel in a nuclear power station. Much more energy is released per kilogram from uranium or plutonium than from fossil fuels.

● Biofuels are renewable sources of energy. Biofuels such as methane and ethanol can be used to generate electricity.

P3.2 Energy from wind and water

Learning objectives

After this topic, you should know:

- what a wind turbine is made up of
- how waves can be used to generate electricity
- the type of power station that uses water running downhill to generate electricity
- how the tides can be used to generate electricity.

Figure 1 *A wind farm is a group of wind turbines*

The mass m of wind (air) passing through a wind turbine each second is proportional to the wind speed v. If the wind speed doubles, the mass of wind passing through the wind turbine each second also doubles. As kinetic energy $= \frac{1}{2}mv^2$, the kinetic energy of the wind passing through each second is therefore 2^3 times greater, because m increases by ×2 and v^2 increases by ×4. In other words, the power of the wind is proportional to v^3.

Strong winds can cause lots of damage on a very stormy day. Even when the wind is much weaker, it can still turn a wind turbine. Energy from the wind and other sources such as waves and tides is called renewable energy. That's because such natural sources of energy can never be used up because they are always being replenished (i.e., replaced) by natural processes.

As well as this, no fuel is needed to produce electricity from these natural sources, so they are carbon-free to run.

Wind power

A wind turbine is an electricity generator at the top of a narrow tower. The force of the wind drives the turbine's blades around. This turns a generator. The power generated increases as the wind speed increases. Wind turbines are unreliable because when there is little or no wind they do not generate any electricity.

Wave power

A wave generator uses the waves to make a floating generator move up and down. This motion turns the generator so it generates electricity. A cable between the generator and the shoreline delivers electricity to the grid system.

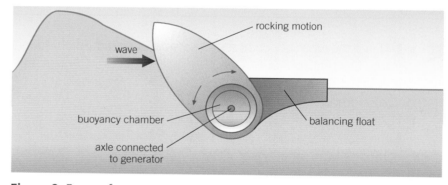

Figure 2 *Energy from waves*

Wave generators need to withstand storms, and they don't produce a constant supply of electricity. Also, lots of cables (and buildings) are needed along the coast to connect the wave generators to the electricity grid. This can spoil areas of coastline. Tidal flow patterns might also change, affecting the habitats of marine life and birds.

Hydroelectric power

Hydroelectricity can be generated when rainwater that's collected in a reservoir (or water in a pumped storage scheme) flows downhill. The flowing water drives turbines that turn electricity generators at the bottom of the hill.

Tidal power

A tidal power station traps water from each high tide behind a barrage. The high tide can then be released into the sea through turbines. The turbines drive generators in the barrage.

One of the most promising sites for a tidal power station in Britain is the Severn estuary. This is because the estuary rapidly becomes narrower as you move up-river away from the open sea. So it funnels the incoming tide and makes it higher.

In some coastal areas, electricity is generated by the tidal flow passing through undersea turbines on the sea bed. Underwater cables are used to connect these turbines to the national grid.

1 Hydroelectricity, tidal power, wave power, and wind power are all renewable energy resources.
   a Explain what a renewable energy resource is. [2 marks]
   b Name the renewable energy resource listed above that:
      i does not need energy from the Sun [1 mark]
      ii does not need water and is unreliable. [1 mark]

2 a Use the table below for this question. The output of each source is given in millions of watts (MW).
      i Calculate how many wind turbines would give the same total power output as a tidal power station. [1 mark]
      ii Calculate how many kilometres of wave generators would give the same total output as a hydroelectric power station. [1 mark]
   b Use the words below to complete the location column in the table. [2 marks]

   coastline   estuaries   hilly or coastal areas   mountain areas

	Output	Location	Total cost in £ per MW
Hydroelectric power station	500 MW per station		50
Tidal power station	2000 MW per station		300
Wave power generators	20 MW per kilometre of coastline		100
Wind turbines	2 MW per wind turbine		90

3 The last column of the table above shows an estimate of the total cost per MW of generating electricity using different renewable energy resources. The total cost for each resource includes its running costs and the capital costs to set it up.
   a The capital cost per MW of a tidal power station is much higher than that of a hydroelectric power station. Give one reason for this difference. [2 marks]
   b i Name the energy resource that has the lowest total cost per MW. [1 mark]
      ii Give two reasons why this resource might be unsuitable in many locations. [2 marks]

4 a Explain what a pumped storage scheme is. [2 marks]
   b Describe the main benefit to electricity users of a pumped storage scheme. [2 marks]

Figure 3 *A hydroelectric power station. Some hydroelectric power stations are designed as pumped storage schemes. When electricity demand is low, electricity can be supplied from other power stations and electricity generators to pumped storage schemes to pump water uphill into a reservoir. When demand is high, the water can be allowed to run downhill to generate electricity*

Figure 4 *A tidal power station*

Key points

- A wind turbine is an electricity generator on top of a tall tower.
- Waves generate electricity by turning a floating generator.
- Hydroelectricity generators are turned by water running downhill.
- A tidal power station traps each high tide and uses it to turn generators.

P3.3 Power from the Sun and the Earth

Learning objectives

After this topic, you should know:

- what solar cells are and how they are used
- the difference between a panel of solar cells and a solar heating panel
- what geothermal energy is
- how geothermal energy can be used to generate electricity.

Solar radiation transfers energy to you from the Sun. That can sometimes be more energy than is healthy if you get sunburnt. But the Sun's energy can be used to generate electricity using solar cells. The Sun's energy can also be used to heat water directly in solar heating panels.

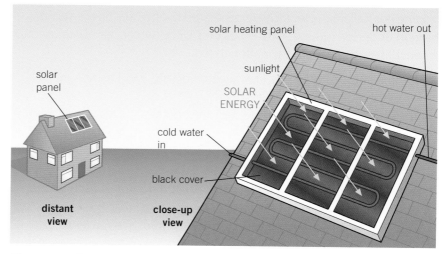

Figure 1 *Solar water heating*

Today's solar cells convert less than 10% of the solar energy they absorb into the energy transferred by electricity. They can be connected together to make solar cell panels.

- They are useful where only small amounts of electricity are needed (e.g., in watches and calculators) or in remote places (e.g. on small islands in the middle of an ocean).
- They are very expensive to buy but they cost nothing to run.
- Lots of them are needed – and plenty of sunshine – to generate enough power to be useful. Solar panels can be unreliable in areas where the Sun is often covered by clouds.

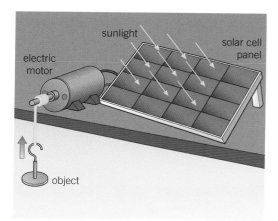

Figure 2 *Solar cells at work*

A solar heating panel heats water that flows through it. On a sunny day in Britain, a solar heating panel on a house roof can supply plenty of hot water for domestic use (Figure 1).

A solar power tower uses thousands of flat mirrors to reflect sunlight on to a big water tank at the top of a tower (Figure 3). The mirrors on the ground surround the base of the tower.

- The water in the tank is turned to steam by the heating effect of the solar radiation directed at the water tank.
- The steam is piped down to ground level, where it turns electricity generators.
- The mirrors are controlled by a computer so that they track the Sun.

Figure 3 *A solar power tower*

A solar power tower in a hot dry climate can generate more than 20 MW of electrical power, which is enough to power a few thousand homes.

Geothermal energy

Geothermal energy comes from energy released by radioactive substances deep within the Earth.

- The energy transferred from these radioactive substances heats the surrounding rock.
- So energy is transferred by heating towards the Earth's surface.

Geothermal power stations can be built in volcanic areas or where there are hot rocks deep below the surface. Water gets pumped down to these rocks to produce steam. Then the steam that is produced drives electricity turbines at ground level (Figure 4).

In some areas, buildings can be heated using geothermal energy directly. Heat flow from underground is sometimes called ground source heat. It can be used to heat water in long underground pipes. The hot water is then pumped around the buildings. In some big eco-buildings, this geothermal heat flow is used as under-floor heating.

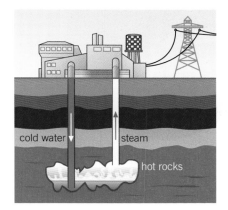

Figure 4 *A geothermal power station*

1 **a** What is the source of geothermal energy? [1 mark]
   b State why geothermal energy is more reliable than solar energy for heating water. [1 mark]

2 A satellite in space uses a solar cell panel for electricity. The panel generates 300 W of electrical power and has an area of $10\,m^2$.
   a Each cell generates 0.2 W. Calculate how many cells are in the panel. [2 marks]
   b The satellite carries batteries that are charged by electricity from the solar cell panels. State why batteries are carried as well as solar cell panels. [1 mark]

3 A certain geothermal power station has a power output of 200 000 W.
   a Calculate how many kilowatt-hours of energy the power station generates in 24 hours. [2 marks]
   b Give one advantage and one disadvantage of a geothermal power station compared with a wind turbine. [2 marks]

4 A solar water panel heats the water flowing through from 14 °C to 35 °C when water flows through it at a rate of 0.010 kilograms per second.
   a Calculate the energy per second transferred by the hot water from the solar panel. [2 marks]
   b Estimate the output temperature of the hot water if the flow rate is reduced to 0.007 kg/s. The specific heat capacity of water is 4200 J/kg °C [3 marks]

P3.4 Energy and the environment

Learning objectives

After this topic, you should know:

- what fossil fuels do to your environment
- why people are concerned about nuclear power
- the advantages and disadvantages of renewable energy resources
- how to evaluate the use of different energy resources.

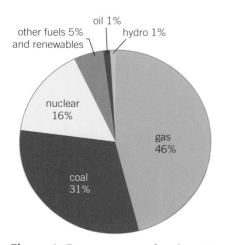

Figure 1 *Energy sources for electricity*

Figure 2 *Greenhouse gases from fossil fuels*

Synoptic links

Look back at Topic P2.3 for more about how greenhouse gases cause global warming.

Can you get energy without creating any problems? Figure 1 shows the energy sources people use today to generate electricity. What effect does each one have on your environment?

Fossil fuel problems

When coal, oil, or gas is burnt, greenhouse gases such as carbon dioxide are released. The amount of these gases in the atmosphere is increasing, and most scientists believe that this is causing more global warming and climate change. Some electricity comes from oil-fired power stations. People use much more oil to produce fuels for transport.

Burning fossil fuels can also produce sulfur dioxide. This gas causes acid rain. The sulfur can be removed from a fuel before burning it, to stop acid rain. For example, natural gas has its sulfur impurities removed before it is used.

Fossil fuels are non-renewable. Sooner or later, people will have used up the Earth's reserves of fossil fuels. Alternative sources of energy will then have to be found. But how soon? Oil and gas reserves could be used up within the next 50 years. Coal reserves will last much longer.

Carbon capture and storage (CCS) technology could be used to stop carbon dioxide emissions into the atmosphere from fossil fuel power stations. Old oil and gas fields could be used for carbon dioxide storage.

Nuclear versus renewable

People need to use fewer fossil fuels in order to stop global warming. Should people rely on nuclear power or on renewable energy in the future?

Nuclear power

Advantages

- No greenhouse gases (unlike fossil fuel).
- Much more energy is transferred from each kilogram of uranium (or plutonium) fuel than from fossil fuel.

Disadvantages

- Used fuel rods contain radioactive waste, which has to be stored safely for centuries.
- Nuclear reactors are safe in normal operation. However, an explosion in a reactor could release radioactive material over a wide area. This would affect this area, and the people living there, for many years.

Renewable energy sources and the environment

Advantages

- They will never run out because they are always being replenished by natural processes.
- They do not produce greenhouse gases or acid rain.
- They do not create radioactive waste products.

- They can be used where connection to the National Grid is uneconomical. For example, solar cells can be used for road signs and to provide people with electricity in remote areas.

Disadvantages

- Renewable energy resources are not currently able to meet the world demand. So fossil fuels are still needed to provide some of the energy demand.
- Wind turbines create a whining noise that can upset people nearby, and some people consider them unsightly.
- Tidal barrages affect river estuaries and the habitats of creatures and plants there.
- Hydroelectric schemes need large reservoirs of water, which can affect nearby plant and animal life. Habitats are often flooded to create dams.
- Solar cells need to cover large areas to generate large amounts of power.
- Some renewable energy resources are not available all the time or can be unreliable. For example, solar power is not produced at night and is affected by cloudy weather. Wind power is reduced when there is little or no wind, and hydroelectricity is affected by droughts if reservoirs dry up.

Figure 3 *The effects of acid rain*

1 **a** Name the type of fuel that is used to generate most of Britain's electricity.   [1 mark]
   b Give two problems caused by burning fossil fuel.   [2 marks]
   c State two advantages and two disadvantages of using renewable energy sources instead of fossil fuels.   [4 marks]

2 Match each energy source with a problem it causes:

Energy source	Problem	
a Coal	**A** Noise	
b Hydroelectricity	**B** Acid rain	
c Uranium	**C** Radioactive waste	
d Wind power	**D** Takes up land	[2 marks]

3 **a** Name three possible renewable energy resources that could be used to generate electricity for people on a remote flat island in a hot climate.   [3 marks]
   b Name three types of power stations that do not release greenhouse gases into the atmosphere.   [3 marks]

4 A tidal power station, a nuclear power station, or 1000 wind turbines can each supply enough power to meet the electricity needs of a large city on an estuary. Describe the advantages and disadvantages of each type of power station for this purpose. 🖊 [5 marks]

Key points

- Fossil fuels produce increased levels of greenhouse gases, which could cause global warming.
- Nuclear fuels produce radioactive waste.
- Renewable energy resources will never run out, they do not produce harmful waste products (e.g., greenhouse gases or radioactive waste), and they can be used in remote places. But they cover large areas, and they can disturb natural habitats.
- Different energy resources can be evaluated in terms of reliability, environmental effects, pollution, and waste.

P3.5 Big energy issues

Learning objectives

After this topic, you should know:

- how best to use electricity supplies to meet variations in demand
- how the economic costs of different energy resources compare
- which energy resources need to be developed to meet people's energy needs in future.

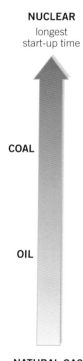

NUCLEAR
longest
start-up time

COAL

OIL

NATURAL GAS
shortest
start-up time

Figure 2 *Start-up times of different types of power station*

Supply and demand

The demand for electricity varies during each day. It is also higher in winter than in summer. Electricity generators need to match these changes in demand.

Power stations can't just start up instantly. The start-up time depends on the type of power station.

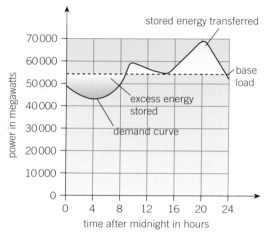

Figure 1 *Example of electricity demand*

Renewable energy resources are unreliable. The amount of electricity they generate depends on the conditions.

Table 1 *Reliability problems with renewable energy resources*

Hydroelectric	Upland reservoir could run dry
Wind, waves	Wind and waves too weak on very calm days
Tidal	Height of tide varies both on a monthly and yearly cycle
Solar	No solar energy at night, and variable during the day

The variable demand for electricity is met by:

- using nuclear and coal-fired power stations to provide a constant amount of electricity (the base load demand)
- using gas-fired power stations and pumped-storage schemes to meet daily variations in demand and extra demand in winter
- using renewable energy resources when demand is high and when the conditions for renewable energy generation are suitable (e.g., use of wind turbines in winter and when wind speeds are high enough)
- using renewable energy resources when demand is low to store energy in pumped storage schemes.

Cost comparisons

The overall cost of a new energy facility involves capital costs to build it, running costs for fuel and maintenance, and more capital costs to take it out of use at the end of its working lifetime.

You can use Table 2 to compare the capital and overall costs of each resource with the associated costs of gas. In practice, capital costs are estimated in terms of cost per kilowatt of power, but running costs are estimated in terms of cost per kilowatt hour of energy.

Table 2 *Cost comparisons (2012). ᵃ coal/gas with carbon capture storage(CCS); ᵇ nuclear (excluding decommissioning costs); ᶜ offshore wind; ᵈ solar cell panels.*

Resource	gasᵃ	coalᵃ	uraniumᵇ	hydroelectric	windᶜ	solarᵈ
Capital costs	1.0–2.0	5.5	5.0	3.0	6.0	4.0
Overall costs	1.0–1.2	1.5	1.1	0.7	2.2	1.9

Key points from Table 2 include:

● Capital costs are lowest for gas-fired power stations and greatest for wind power and nuclear power, including decommissioning costs (i.e., taking stations out of use).

● Overall costs including fuel costs are the lowest for hydroelectricity, and greatest for offshore wind farms.

The costs of new energy facilities are usually passed on to consumers through increased fuel bills. Energy-saving schemes such as low-energy light bulbs in your home would reduce the need for more power stations. Schemes such as improved home insulation reduce the demand for non-renewable energy resources (e.g., gas). Home owners pay upfront and eventually get their money back through reduced fuel bills.

1 a Name the type of power station that can be started fastest. [1 mark]
b Name the type of power station that does not produce greenhouse gases or radioactive waste. [1 mark]
c Name the types of renewable resource that are unreliable. [1 mark]
d Name the type of renewable resource that can be used to store energy at times of low electricity demand. [1 mark]

2 People need to cut back on fossil fuels to reduce the production of greenhouse gases. What could happen if the only energy people used was:
a renewable energy [1 mark] **b** nuclear power? [1 mark]

3 State why nuclear power stations are unsuitable for meeting daily variations in the demand for electricity. [1 mark]

4 State what pumped storage schemes are and why they are useful. [3 marks]

5 Using the data in Table 2, compare the use of fossil fuel resources with renewable resources.
a Name the resource that has the lowest capital costs. [1 mark]
b Hydroelectricity has the lowest overall costs. Give a reason why its overall cost is less than:
i wind or solar power [1 mark]
ii the non-renewable resources listed in Table 2. [2 marks]
c The overall cost of a nuclear power station is about 9 pence per kW h. Nuclear reactor decommissioning costs have been estimated by the UK government at £1000 million per reactor. Estimate the extra overall cost per kW h of decommissioning a 4000 MW reactor that has a lifetime of 30 years. [3 marks]

Key points

● Gas-fired power stations and pumped-storage stations can meet variations in demand.
● Nuclear power stations are expensive to build, run, and decommission. Carbon capture of fossil fuel emissions is likely to be very expensive. Renewable resources are cheap to run but expensive to install.
● Nuclear power stations, fossil-fuel power stations that use carbon capture technology, and renewable energy resources are all likely to be needed for future energy supplies.

P3 Energy resources

Summary questions

1 a i Explain what is meant by a renewable energy source. [2 marks]

  ii State with the aid of a suitable example what is meant by renewable fuel. [2 marks]

  b Discuss whether or not renewable fuels contribute to greenhouse gases in the atmosphere. [2 marks]

2 a Compare a tidal power station and a hydroelectric power station. Give two similarities and two differences. [4 marks]

  b Name the renewable energy resource that transfers:

  i the kinetic energy of moving air to energy transferred by an electric current [1 mark]

  ii the gravitational potential energy of water running downhill into energy transferred by an electric current [1 mark]

  iii the kinetic energy of water moving up and down to energy transferred by an electric current. [1 mark]

3 a i Name the energy resource that does not produce greenhouse gases and uses energy which is from inside the Earth. [1 mark]

  ii Name the energy resource that uses running water and does not produce greenhouse gases. [1 mark]

  iii Name the energy resource that releases greenhouse gases and causes acid rain. [1 mark]

  iv Name the energy resource that does not release greenhouse gases but does produce waste products that need to be stored for many years. [1 mark]

  b Wood can be used as a fuel. State whether it is

  i renewable or non-renewable [1 mark]

  ii a fossil fuel or a non-fossil fuel. [1 mark]

4 a Figure 1 shows a landscape showing three different renewable energy resources, numbered 1 to 3.

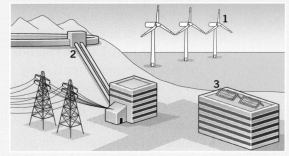

Figure 1 *Renewable energy*

Match each type of energy resource with one of the labels below.

hydroelectricity   solar energy   wind energy   [2 marks]

b Determine which of the three resources shown is not likely to produce as much energy as the others if the area is:

  i hot, dry, and windy [1 mark]

  ii wet and windy. [1 mark]

5 a Use the data in Table 2 in Topic P3.5 to discuss whether or not wind turbines are less expensive to build than nuclear power stations, and if they are also cheaper to run. [4 marks]

  b Discuss the reliability and environmental effects of the non-fossil fuel resources listed in Table 2 in Topic P3.5. [6 marks]

6 A hydroelectric power station has an upland reservoir that is 400 m above the power station. The power station is designed to produce 96 MW of electrical power with an efficiency of 60%.

  a Estimate the loss of gravitational potential energy per second when the hydroelectric power station generates 96 kW of power. [2 marks]

  b Use your estimate to calculate the volume of water per second that flows from the reservoir through the power station generators when 96 kW of power is generated. The density of water is 1000 kg/m³. [3 marks]

7 The national demand for electricity varies during the day and also from winter to summer. Different types of power stations are used to meet these variations in demand. The lowest level of demand for power (the base load demand) is during the daytime in summer.

  a State one reason why the demand for electricity is lower in summer than in winter. [1 mark]

  b Different types of power stations are listed below.

coal	oil	gas	nuclear
geothermal		hydroelectric	

  i State and explain which two types of power station from the options above should be connected to the grid system to supply extra power if the demand suddenly increases. [4 marks]

  ii Which type of power station is the least suitable for connecting to the grid to meet a sudden increase in demand? Give a reason for your answer. [2 marks]

  c Discuss whether wind turbines or solar panels could be used to meet a sudden increase in demand for energy. [4 marks]

Practice questions

01.1 Which of the following statements are good reasons for using renewable energy resources?

A supplies of renewable energy are unlimited

B renewable energy cannot generate electricity all the time

C renewable energy can be replenished

D renewable energy cannot supply electricity to millions of homes

E renewable energy does not produce carbon dioxide [2 marks]

02.1 **Figure 1** shows a lagoon tidal barrage system used to generate electricity.

Figure 1

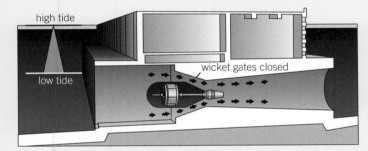

The sentences are not in the correct order. Write sentences A to E in the correct order, with A first.

A When the tide is high the gates open and water rushes through the turbines into the lagoon.

B The water flowing out again turns the turbines and generates electricity.

C The water coming in turns the turbines and generates electricity.

D When the tide is low the gates are opened and the water rushes out of the lagoon.

E The lagoon fills up with water and the gates are closed. [4 marks]

02.2 The lagoon tidal barrage system is claimed to be able to generate electricity for 14 hours a day. Give two reasons why it is necessary to connect houses to the National Grid rather than to the lagoon tidal barrage system alone. [2 marks]

02.3 Suggest a reason, apart from cost, why people may object to the planned system. [1 mark]

03 A group of students investigated whether changing the direction of a solar panel affected the amount of electricity generated. The tests were performed on a sunny day at midday. The students method is given below.

1 Fix the solar panel to a board and incline the panel 45° to the horizontal.

2 Attach a voltmeter to the solar panel.

3 Point the solar panel to the North and record the voltage reading.

4 Repeat the test pointing the solar panel in different directions.

The students recorded their results in **Table 1**.

Table 1

Direction	N	NE	E	SE	S	SW	W	NW
Volts	1.2	1.9	2.5	3.9	4.3	3.7	2.4	1.8

03.1 Name the independent and dependent variable. [2 marks]

03.2 Suggest what advice you would give to a school about to install solar panels. [1 mark]

03.3 One student stated that the voltage readings would be higher if the investigation was carried out on the roof. Is the student correct? Give a reason for your decision. [2 marks]

03.4 The students want to check if their results are repeatable. They intend to carry out the investigation using the same apparatus at the same location. State one other factor they must keep the same. [1 mark]

04.1 The UK Government has set a target to use renewable energy for 20% of the country's energy use by 2020. Explain why it is important to increase the use of renewable energy sources to generate electricity. [3 marks]

04.2 The UK Government has agreed to build a nuclear power plant by 2025. It is estimated that the power plant will provide at least 7% of the electricity needed in the UK.
Give two advantages and one disadvantage of using nuclear energy to produce electricity. [3 marks]

05 Electricity can be generated using ethanol as an energy source. Ethanol can be produced from sugar cane. Brazil is the biggest grower of sugar cane, and the biggest exporter of ethanol made from sugar cane in the world.

05.1 Evaluate the benefits of growing and using ethanol made from sugar cane. Your answer should include environmental as well as economic reasons. [3 marks]

2 Particles at work

All substances are made of atoms. Most atoms are stable and remain stable. Without this, the world as we know it wouldn't exist, and neither would we.

Every atom contains a nucleus surrounded by tiny particles called electrons. Atoms can lose or gain electrons, with different results. For example:

- Materials have different properties when the electrons in their atoms are shared in different ways.

- Metals conduct electricity because they contain electrons that have broken away from atoms inside the metal.

- Radioactive substances are made of atoms with unstable nuclei that emit harmful radiation when they become stable.

In this section you will learn about atoms as you learn about materials, electricity, and radioactivity.

Key questions

- What is an electric current?
- How do series and parallel circuits differ?
- What do we mean by density and elasticity?
- What is the half-life of a radioactive isotope?

Making connections

- The electricity you use at home is produced by generators in power stations and is used by electric motors in appliances such as washing machines. You will learn how generators and motors work in **P15 Electromagnetism**.

- Strong lightweight materials are used for many purposes, including safety equipment such as cycle helmets. Density and elasticity are important properties of such materials. You will meet the use of these materials in **P10 Force and motion**.

- Nuclear fusion inside a star releases energy which heats the outside of the star and makes it give out light. You will learn more about stars in **P16 Space**.

I already know...

There are two types of electric charge.

Potential difference is measured in volts and current is measured in amperes.

A cell or battery pushes electrons round a circuit.

Power is how much energy is transferred per second.

Mass is the amount of matter in a substance and is measured in kilograms.

Gas particles move about very quickly and collide with the surface of the gas container.

The nucleus of an atom is composed of protons and neutrons.

Energy is released when hydrogen nuclei fuse together in the Sun.

I will learn...

How to calculate the charge flow in an electric circuit.

How to work out the resistance and potential difference in an electric circuit.

How mains electricity differs from electricity supplied by batteries.

How to calculate the power of an electrical appliance.

What we mean by density and how we can measure it.

How to explain why the pressure of a gas increases when it is heated in a sealed container.

How an unstable nucleus changes when it becomes stable and why the radiation it gives out is harmful.

What nuclear fission and fusion are.

Required Practicals

Practical		Topic
3	Investigating resistance	P4.3 P4.6
4	Investigating different components	P4.4
5	Measuring the density of a solid object and of a liquid	P6.1

Learning objectives

After this topic, you should know:

- what happens when insulating materials are rubbed together
- what is transferred when objects become charged
- what happens when charges are brought together
- recognise that the force between charged objects is a non-contact force.

> ### Study tip
>
> Objects become electrically charged by gaining electrons (so becoming more negative) or losing electrons (so becoming more positive).

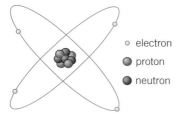

- ○ electron
- ● proton
- ● neutron

Figure 1 *Inside an atom*

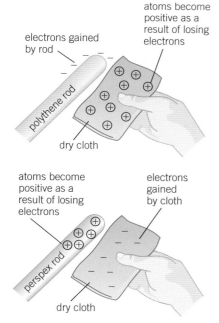

Figure 2 *Charging by friction*

Have you ever stuck a balloon on a ceiling? All you need to do is rub the balloon on your clothing before you touch it on the ceiling. The rubbing action charges the balloon with static electricity. In other words, the balloon becomes electrically charged. The charge on the balloon attracts it to the ceiling.

Inside the atom

Protons and **neutrons** make up the nucleus of an atom. Electrons move about in the space around the nucleus.

- A proton has a positive charge.
- An electron has an equal negative charge.
- A neutron is uncharged.

An uncharged atom has equal numbers of electrons and protons. Only electrons can be transferred to or from an atom. A charged atom is called an **ion**.

1. Adding electrons to an uncharged atom makes it negative (because the atom then has more electrons than protons).

2. Removing electrons from an uncharged atom makes it positive (because the atom has fewer electrons than protons).

Charging by friction

Some insulators become charged when you rub them with another insulator such as a dry cloth. When you rub the insulator, electrons are transferred from one of the materials to the other material.

- Rubbing a polythene rod with a dry cloth transfers electrons to the surface atoms of the rod from the cloth. So the polythene rod becomes negatively charged.

- Rubbing a perspex rod with a dry cloth transfers electrons from the surface atoms of the rod on to the cloth. So the perspex rod becomes positively charged. Its positive charge is equal to the negative charge of the dry cloth.

The force between two charged objects

Two charged objects exert a non-contact force on each other because of their charge. This is because a charged object (X) creates an **electric field** around itself. A second charged object (Y) in the field experiences a force because of the field. The field, and therefore the force between the two charged objects, becomes stronger as the distance between the objects decreases.

You can draw lines of force to represent an electric field. Figure 3 shows the electric field near an isolated positively charged sphere. Each line of force is the path that a small positive charge Q would follow because of the electric field.

The lines point away from the centre of the charged sphere because the force on Q is directed away from the sphere. If the sphere was negatively charged, the lines would point towards the centre of the sphere.

If the two objects are oppositely charged, electrons in the air molecules between the two objects experience a force towards the positive object. If the field is too strong, sparking happens, because some electrons are pulled out of air molecules by the force of the field. These electrons hit other air molecules and knock electrons out of them, creating a sudden flow of electrons between the two charged objects.

Figure 4 shows how you can investigate the force between two charged objects A and B. The experiment shows that two objects that are not in contact with each other have either:

● the same type of charge (i.e., like charges) and repel each other
● different types of charge (i.e., unlike charges) and attract each other.

Like charges repel. Unlike charges attract.

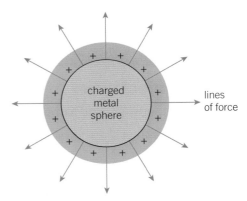

Figure 3 *The electric field near a positively charged sphere*

1 **a** In terms of electrons, explain why:
  i a polythene rod becomes negatively charged when rubbed with a dry cloth [1 mark]
  ii a perspex rod becomes positively charged when rubbed with a dry cloth. [1 mark]
  b Glass is charged positively when it is rubbed with a dry cloth. Explain whether glass gains or loses electrons when it is charged. [2 marks]

2 Look at the setup in Figure 4. Describe what will happen to the suspended Rod A when Rod B is near the positive end of A and when:
  a B is negative [1 mark]
  b B is positive. [1 mark]

3 When rubbed with a dry cloth, perspex becomes positively charged. Polythene and ebonite become negatively charged. Determine whether or not attraction or repulsion takes place when:
  a a perspex rod is held near a polythene rod [1 mark]
  b a perspex rod is held near an ebonite rod [1 mark]
  c a polythene rod is held near an ebonite rod. [1 mark]

4 **a** After two rods X and Y made of different insulating materials are charged, they are found to repel each other. Describe what this tells you about the charge on the two rods. [1 mark]
  b You are given a rod R that is known to charge negatively. Describe how you would determine the type of charge on X and on Y. 🖉 [4 marks]

5 **a** Explain why you may get an electric shock if you walk on a nylon carpet and then touch a metal radiator. [3 marks]
  b Explain why you sometimes get an electric shock when you get out of a car seat. [3 marks]

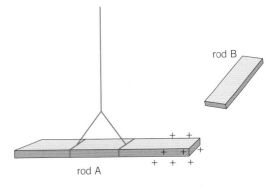

Figure 4 *Investigating the force between charged objects*

Key points

● Some insulating materials become charged when rubbed together.
● Electrons are transferred when objects become charged:
  ▪ Insulating materials that become positively charged when rubbed lose electrons.
  ▪ Insulating materials that become negatively charged when rubbed gain electrons.
● Like charges repel. Unlike charges attract.
● The force between two charged objects is a non-contact force.

P4.2 Current and charge

Learning objectives

After this topic, you should know:

- how electric circuits are shown as diagrams
- the difference between a battery and a cell
- what determines the size of an electric current
- how to calculate the size of an electric current from the charge flow and the time taken.

If you experience a power cut at night, an electric torch can be very useful. But it needs to be checked regularly to make sure it works. Figure 1 shows what is inside a torch. The circuit shows how the torch bulb is connected to the switch and the two cells.

A circuit diagram shows you how the components in a circuit are connected together. Each component has its own symbol. Figure 2 shows the symbols for some of the components you will meet in this course. The function of each component is also described. You need to recognise these symbols and remember what each component is used for – otherwise you will get mixed up. More importantly, if you mix them up when building a circuit you could get a big shock.

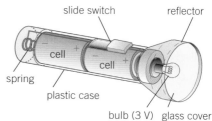

Figure 1 *An electric torch*

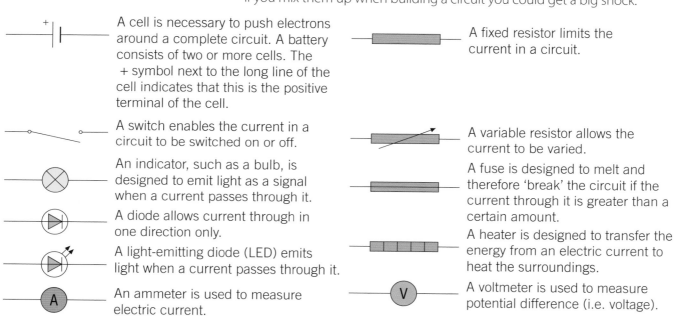

A cell is necessary to push electrons around a complete circuit. A battery consists of two or more cells. The + symbol next to the long line of the cell indicates that this is the positive terminal of the cell.

A switch enables the current in a circuit to be switched on or off.

An indicator, such as a bulb, is designed to emit light as a signal when a current passes through it.

A diode allows current through in one direction only.

A light-emitting diode (LED) emits light when a current passes through it.

An ammeter is used to measure electric current.

A fixed resistor limits the current in a circuit.

A variable resistor allows the current to be varied.

A fuse is designed to melt and therefore 'break' the circuit if the current through it is greater than a certain amount.

A heater is designed to transfer the energy from an electric current to heat the surroundings.

A voltmeter is used to measure potential difference (i.e. voltage).

Figure 2 *Components and symbols*

Electric current

An electric current is a flow of charge. When an electric torch is on, millions of **electrons** pass through the torch bulb and through the cell every second (Figure 3). Each electron carries a negative charge. Metals contain lots of electrons that move about freely between the positively charged metal ions. These electrons stop the ions moving away from each other. The electrons pass through the bulb because it is in a circuit and the bulb filament is metal. The current in the circuit transfers energy from the cell to the torch bulb.

The size of an electric current is the rate of flow of electric charge. This is the flow of charge per second. The bigger the number of electrons that pass through a component each second, the bigger is the current passing through it.

Figure 3 *Electrons on the move*

In a circuit that is a single closed loop like in Figure 3, the current at any point in the circuit is the same as the current at any other point in the circuit. This is because the number of electrons per second that pass through any part of the circuit is the same as at any other part.

Electric charge is measured in coulombs (C). Electric current is measured in amperes (A), sometimes abbreviated as 'amps'.

An electric current of 1 ampere is a rate of flow of charge of 1 coulomb per second. When there is a constant current in a wire (or component) in a circuit for a given time,

<table>
<tr><td>charge flow, Q</td><td>=</td><td>current, I</td><td>×</td><td>time taken, t</td></tr>
<tr><td>(couloumbs, C)</td><td></td><td>(amperes, A)</td><td></td><td>(seconds, s)</td></tr>
</table>

Circuit tests

Connect a variable resistor in series with the torch bulb and a cell (Figure 4). Adjust the slider of the variable resistor. This alters the amount of current flowing through the bulb and so affects its brightness.

- In Figure 4, the torch bulb goes dim when the slider is moved one way. Describe what happens if the slider is moved back again.
- Describe what happens if you include a diode in the circuit.

More about diodes

You would damage a portable radio if you put the batteries in the wrong way around, unless a diode were in series with the battery. The diode allows current through only when it is connected as shown in Figure 5.

1 Name the numbered components in the circuit diagram in Figure 6.

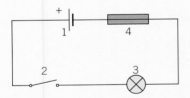

Figure 6   [2 marks]

2 a Redraw the circuit diagram in Figure 6 with a diode in place of the switch so it allows current through. [1 mark]
b Name the further component you would need in this circuit to change the current in it. [1 mark]
c When the switch is closed in Figure 6, a current of 0.25 A passes through the lamp. Calculate the charge that passes through the lamp in 60 seconds. [2 marks]

3 a State what an ammeter is used for. [1 mark]
b State what a variable resistor is used for. [1 mark]
4 a Draw a circuit diagram for the electric torch in Figure 1. [2 marks]
b Describe the energy transfers of an electron around the circuit when the switch is closed. 🖊 [3 marks]

Worked example

A charge of 16.0 C passes through a bulb in 5.0 seconds. Calculate the current through the bulb.

Solution

$$I = \frac{Q}{t} = \frac{16.0\,C}{5.0\,s} = 3.2\ \mathbf{A}$$

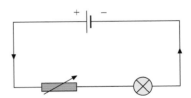

Figure 4 *Using a variable resistor*

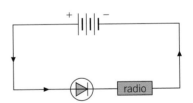

Figure 5 *Using a diode*

Key points

- Every component has its own agreed symbol. A circuit diagram shows how components are connected together.
- A battery consists of two or more cells connected together.
- The size of an electric current is the rate of flow of charge.
- The equation for the electric current of a circuit is $I = \frac{Q}{t}$
- The above equation can be rearranged to find charge flow or time: $Q = I\,t$ or $t = \frac{Q}{I}$

P4.3 Potential difference and resistance

Learning objectives

After this topic, you should know:

- what is meant by potential difference
- what resistance is and what its unit is
- what Ohm's law is
- what happens when you reverse the potential difference across a resistor.

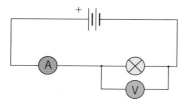

Figure 1 *Using an ammeter and a voltmeter*

Worked example

The energy transferred to a bulb is 320 J when 64 C of charge passes through it. Calculate the potential difference across the bulb.

Solution

$V = \dfrac{E}{Q} = \dfrac{320\,\text{J}}{64\,\text{C}} = 50\,\text{V}$

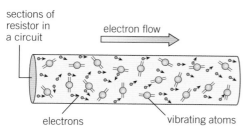

sections of resistor in a circuit

electron flow

electrons

vibrating atoms

Figure 2 *Electrons moving through a resistor*

Rearranging equations

Rearranging the equation

$R = \dfrac{V}{I}$ gives:

$V = I \times R$ or $I = \dfrac{V}{R}$

Potential difference

Look at the circuit in Figure 1. The battery forces electrons to pass through the ammeter and the bulb.

- The ammeter measures the current through the torch bulb. It is connected in **series** with the bulb, so the current through them is the same. The ammeter reading tells you the current in amperes (or milliamperes (mA) for small currents, where 1 mA = 0.001 A).

- The voltmeter measures the **potential difference** (p.d. or voltage) across the torch bulb. This is the energy transferred to the bulb or the work done on it by each coulomb of charge that passes through it. The unit of potential difference is the volt (V).

- The voltmeter is connected in **parallel** with the torch bulb, so it measures the potential difference across it. The voltmeter reading tells you the potential difference in volts (V).

When charge flows steadily through an electronic component:

$$\text{potential difference across a component, } V \text{ (volts, V)} = \frac{\text{energy transferred, } E \text{ (joules, J)}}{\text{charge, } Q \text{ (coulombs, C)}}$$

Resistance

Electrons passing through a torch bulb have to push their way through lots of vibrating atoms in the metal filament. The atoms resist the passage of electrons through the torch bulb.

The **resistance** of an electrical component is defined as:

$$\text{resistance, } R \text{ (ohms, } \Omega) = \frac{\text{potential difference, } V \text{ (volts, V)}}{\text{current, } I \text{ (amperes, A)}}$$

The unit of resistance is the ohm (Ω). Note that a resistor in a circuit limits the current. For a given potential difference, the larger the resistance of a resistor, the smaller the current.

Current–potential difference graphs

Look at Figure 3. Part **a** shows a circuit that can be used to investigate how the current in a wire depends on the potential difference across the wire. The graph in part **b** shows the measurements and is a straight line through the origin. This means that the current is directly proportional to the potential difference. In other words, the resistance (= potential difference ÷ current of the wire) is constant. This was first discovered for a wire at constant temperature by Georg Ohm and is called Ohm's law:

The current through a resistor at constant temperature is directly proportional to the potential difference across the resistor.

A wire is called an ohmic conductor because its resistance stays constant as the current changes, provided its temperature is constant. Figure 4 shows that, reversing the potential difference (and therefore the current) makes no difference to the shape of the line. The resistance is the same whichever direction the current is in.

The gradient of the line depends on the resistance of the resistor. The greater the resistance of the resistor, the less steep the line.

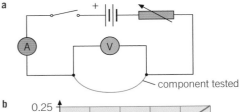

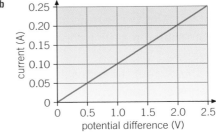

How does the resistance of a wire depend on its length?
Use the circuit in Figure 3 to measure the p.d. for the same current across different lengths of wire.

● Plot a graph of resistance on the y-axis against length on the x-axis.

● Use your graph to draw conclusions about how resistance varies with length.

Safety: Make sure the wire does not get hot. If it does get hot, reduce the current or switch the circuit off, and ask your teacher to check it.

Figure 3 *Investigating the resistance of a wire* **a** *Circuit diagram* **b** *A current–potential difference graph for a wire*

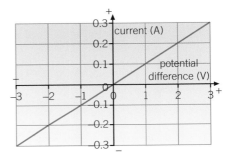

1 **a** The current through a wire is 0.5 A when the potential difference across it is 4.0 V. Calculate the resistance of the wire. [1 mark]
   b Calculate the resistance of the wire that gave the results in the graph in Figure 3. [2 marks]

2 Calculate the missing value in each line of this table.

Resistor	Current in amperes	Potential difference in volts	Resistance in ohms
W	2.0	12.0	
X	4.0		20
Y		6.0	3.0

[3 marks]

Figure 4 *A current–potential difference graph for a resistor*

3 A torch bulb lights normally when the current through it is 0.015 A and the potential difference across it is 12.0 V.
   a Calculate the resistance of the torch bulb when it lights normally. [1 mark]
   b When the torch bulb lights normally, calculate:
   i the charge passing through the torch bulb in 1200 s [1 mark]
   ii the energy delivered to the torch bulb in this time. [1 mark]

4 **a** Calculate the resistance of the wire that gave the results shown in Figure 4. [2 marks]
   b i Calculate the current through the wire when the p.d. across it is 1.6 V. [2 marks]
   ii Calculate the p.d. across the wire when the current through it is 0.42 A. [2 marks]

Key points

● potential difference across a component, $V = \dfrac{\text{energy transferred}, E}{\text{charge}, Q}$

● resistance, $R = \dfrac{\text{potential difference}, V}{\text{current}, I}$

● Ohm's law states that the current through a resistor at constant temperature is directly proportional to the potential difference across the resistor.

● Reversing the potential difference across a resistor reverses the current through it.

P4.4 Component characteristics

Learning objectives

After this topic, you should know:

- what happens to the resistance of a filament lamp as its temperature increases
- how the current through a diode depends on the potential difference across it
- what happens to the resistance of:
  - a temperature-dependent resistor as its temperature increases
  - a light-dependent resistor as the light level increases.

Have you ever switched a light bulb on only to hear it pop and fail? Electrical appliances can fail at very inconvenient times. Most electrical failures happen because too much current passes through a component in the appliance.

Investigating different components

The resistance of a wire is independent of the current passing through it. You can use the circuit in Figure 1 to find out if the resistance of other components in the circuit depends on the current. You can also see if reversing the component in the circuit has any effect.

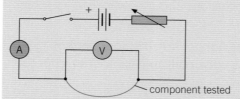

component tested

Figure 1

Make your own measurements using a resistor, a filament lamp, and a diode.

Plot your measurements on a current–potential difference graph. Plot the reverse measurements on the negative section of each axis.

Safety: Make sure the wire does not get hot. If it does get hot, reduce the current or switch the circuit off, and ask your teacher to check it.

Using current–potential difference graphs
A filament lamp

Figure 2 shows the graph for a torch bulb (i.e., a low-voltage filament lamp).

- The line curves away from the y-axis. So, the current is *not* directly proportional to the potential difference. The filament lamp is a non-ohmic conductor.

- The equation for the resistance of an appliance is:

$$\text{resistance, } R \text{ (ohms, } \Omega) = \frac{\text{potential difference, } V \text{ (volts, V)}}{\text{current, } I \text{ (amperes, A)}}$$

- The resistance increases as the current increases. So, the resistance of a filament lamp increases as the filament temperature increases. The atoms in the metal filament vibrate more as the temperature increases. So they resist the passage of the electrons through the filament more. The resistance of any metal increases as its temperature increases.

- Reversing the potential difference reverses the current and makes no difference to the shape of the curve. The resistance is the same for the same current, regardless of its direction.

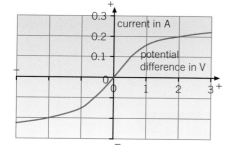

Figure 2 *A current–potential difference graph for a filament lamp*

When a filament light bulb fails, it usually happens when you switch it on. Because resistance is low when the bulb is off, a big current passes through it when you switch it on. If the current is too big, it burns the filament out.

The diode

Figure 3 shows a current–potential difference graph for a diode. The current through a diode flows in one direction only, called the forward direction.

- In the forward direction, the line curves towards the *y*-axis. So the current is not directly proportional to the potential difference. The resistance changes as the current changes. A **diode** is a non-ohmic conductor.

- In the reverse direction, the current is virtually zero. So the diode's resistance in the reverse direction is a lot higher than its resistance is in the forward direction.

A **light-emitting diode (LED)** is a diode that emits light when a current passes through it in the forward direction. LEDs are used as indicators in many electronic devices such as battery chargers and alarm circuits.

Thermistors and light-dependent resistors

Thermistors and light-dependent resistors are used in sensor circuits.

A **thermistor** is a temperature-dependent resistor, and its resistance *decreases* if its temperature increases (and increases if the temperature decreases).

The resistance of a **light-dependent resistor (LDR)** *decreases* if the light intensity increases (and increases if the light intensity decreases).

1 **a** Identify the type of component that has a resistance that:
   i decreases as its temperature increases [1 mark]
   ii depends on which way around it is connected in a circuit [1 mark]
   iii increases as the current through it increases. [1 mark]
  b Calculate the resistance of the filament lamp that gave the graph in Figure 2 at:
   i 0.1 A **ii** 0.2 A. [2 marks]

2 A thermistor is connected in series with an ammeter and a 9.0 V battery (Figure 5).

  a At 15 °C, the current through the thermistor is 0.6 A, and the potential difference across it is 9.0 V. Calculate its resistance at this temperature. [1 mark]
  b Name and explain what happens to the ammeter reading if the thermistor's temperature is increased. [3 marks]

3 The thermistor in the Figure 5 is replaced by a light-dependent resistor (LDR). Name and explain what happens to the ammeter reading when the LDR is covered. [3 marks]

4 For the diode that gave the graph in Figure 3, when the potential difference is increased steadily from zero, describe how:
  a the current changes [2 marks]
  b the resistance changes. [2 marks]

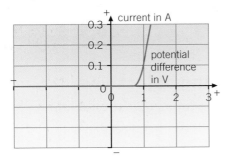

Figure 3 *A current–potential difference graph for a diode*

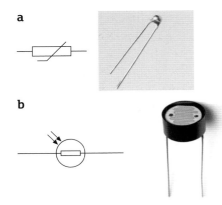

Figure 4 **a** *A thermistor and its circuit symbol*
b *An LDR and its circuit symbol*

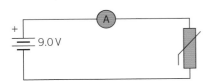

Figure 5 *See Summary question 3*

Key points

- The resistance of an appliance is $R = \dfrac{V}{I}$
- A filament lamp's resistance increases if the filament's temperature increases.
- Diode: forward resistance low; reverse resistance high.
- A thermistor's resistance decreases if its temperature increases.
- An LDR's resistance decreases if the light intensity on it increases.

P4.5 Series circuits

Learning objectives

After this topic, you should know:

- about the current, potential difference, and resistance for each component in a series circuit
- about the potential difference of several cells in series
- how to calculate the total resistance of two resistors in series
- why adding resistors in series increases the total resistance.

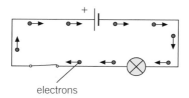

Figure 1 *A torch bulb circuit*

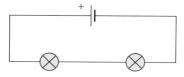

Figure 2 *Bulbs in series*

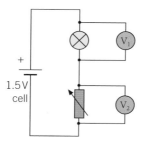

Figure 3 *Voltage tests*

Table 1 *The voltmeter readings for each setting add up to 1.5 V. This is the potential difference of the cell*

Filament lamp	Voltmeter V_1 in volts	Voltmeter V_2 in volts
normal	1.5	0.0
dim	0.9	0.6
very dim	0.5	1.0

Circuit rules

In the torch circuit in Figure 1, the bulb, the cell, and the switch are connected in series with each other. The same number of electrons passes through each component every second. So the same current passes through every component.

In a series circuit, the same current passes through each component.

In Figure 2, each electron from the cell passes through two bulbs. The electrons are pushed through each bulb by the cell. The potential difference (p.d.) of the cell is a measure of the energy transferred from the cell by each electron that passes through it. Because each electron in the circuit in Figure 2 passes through both bulbs, the potential difference of the cell is shared between the bulbs. This rule applies to any series circuit.

In a series circuit, the total potential difference of the power supply is shared between the components.

Cells in series

What happens if you use two or more cells in series in a circuit? As long as you connect the cells so that they act in the same direction, each electron gets a push from each cell. So an electron would get the same push from a battery of three 1.5 V cells in series as it would from a single 4.5 V cell. In other words, as long as the cells act in the same direction:

The total potential difference of cells in series is the sum of the potential difference of each cell.

Investigating potential differences in a series circuit

Figure 3 shows how to test the potential difference rule for a series circuit. The circuit consists of a filament lamp in series with a variable resistor and a cell. You can use the variable resistor to see how the voltmeter readings change when you change the current. Make your own measurements.

- Compare your measurements with the data in Table 1.

Safety: Make sure the wire does not get hot. If it does get hot, reduce the current or switch the circuit off, and ask your teacher to check it.

The resistance rule for components in series

In Figure 3, suppose the current through the bulb is 0.1 A when the bulb is dim. Using data from Table 1:

- the resistance of the bulb would be 9 Ω (= 0.9 V ÷ 0.1 A),
- the resistance of the variable resistor at this setting would be 6 Ω (= 0.6 V ÷ 0.1 A).

If you replaced these two components with a single resistor, its resistance would need to be 15 Ω for the same current of 0.1 A. This is because 1.5 V ÷ 0.1 A is equal to 15 Ω. This resistance is the sum of the resistance of the two components. The same rule applies to any series circuit.

The total resistance of two (or more) components in series is equal to the sum of the resistance of each component.

So for two components of resistances R_1 and R_2 in series (Figure 4):

$$\text{total resistance, } R_{\text{total}} \, (\Omega) = R_1 + R_2$$

R_1 (Ω) is the resistance of the first component and R_2 (Ω) is the resistance of the second component.

Adding more resistors in series increases the total resistance of the circuit. This is because the total potential difference is shared between more resistors, and as a result the potential difference across each of them is less than before. The current through the resistors is therefore less than before, and as the total potential difference is unchanged, the total resistance is therefore greater.

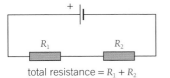

total resistance = $R_1 + R_2$

Figure 4 *Resistors in series*

Worked example

A 4.5 V battery is connected to a 1.0 Ω resistor and a 5.0 Ω in series with each other.

Calculate:

a the total resistance of the two resistors

b the current through the resistors.

Solution

a Total resistance = $1.0\,\Omega + 5.0\,\Omega = 6.0\,\Omega$

b Current = $\dfrac{\text{battery potential difference}}{\text{total resistance}}$

$= \dfrac{4.5\ V}{6.0\ \Omega} = 0.75\ A$

1 **a** In Figure 2, if the potential difference of the cell is 1.2V and the potential difference across one bulb is 0.8V, work out the potential difference across the other bulb. [2 marks]

  b In Figure 3, the bulb emits light when the resistance of the variable resistor is 5.0 Ω and the potential difference across the variable resistor is 1.0V. Calculate the current through the bulb and the potential difference across it. [2 marks]

2 A 1.5V cell is connected to a 3.0 Ω resistor and 2.0 Ω resistor in series with each other.

  a Draw the circuit diagram for this arrangement. [1 mark]

  b Calculate:

   i the total resistance of the two resistors [1 mark]

   ii the current through the resistors. [2 marks]

  c The 3.0 Ω resistor is replaced by a different resistor X and the current changes to 0.25 A. Calculate the resistance of X. [2 marks]

3 For the circuit in Figure 5, each cell has a potential difference of 1.5V.

  a Calculate:

   i the total resistance of the two resistors [1 mark]

   ii the total potential difference of the two cells. [1 mark]

  b Show that the current through the battery is 0.25 A. [2 marks]

  c Calculate the potential difference across each resistor. [2 marks]

  d If a 3 Ω resistor R is connected in series with the two resistors, calculate:

   i their total resistance [1 mark]

   ii the current through the resistors [2 marks]

   iii the potential difference across each resistor. [3 marks]

4 Explain why the resistance of several resistors in series is increased if an additional resistor is connected in series with them. ⊘ [4 marks]

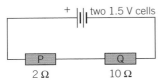

two 1.5 V cells

P 2 Ω Q 10 Ω

Figure 5 *See Summary question 3*

Key points

- For components in series:
  - the current is the same in each component
  - the total potential difference is shared between the components
  - adding their resistances gives the total resistance.
- For cells in series, acting in the same direction, the total potential difference is the sum of their individual potential differences.
- Total resistance $R_{\text{total}} = R_1 + R_2$.
- Adding more resistors in series increases the total resistance because the current through the resistors is reduced and the total potential difference across them is unchanged.

Learning objectives

After this topic, you should know:

- about the currents and potential differences for components in a parallel circuit
- how to calculate the current through a resistor in a parallel circuit
- why the total resistance of two resistors in parallel is less than the resistance of the smallest individual resistor
- Adding resistors in parallel decreases the total resistance.

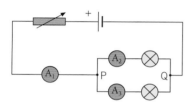

Figure 1 *At a junction*

Table 1

Ammeter readings in amperes		
A_1	A_2	A_3
0.50	0.30	0.20
0.30	0.20	0.10
0.18	0.12	0.06

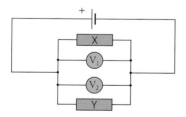

Figure 2 *Components in parallel*

Figure 1 shows how you can investigate the current through two bulbs in parallel with each other. You can use ammeters in series with the bulbs and the cell to measure the current through each component. The bulbs are in separate branches of the circuit.

In each case, the reading of ammeter A_1 is equal to the sum of the readings of ammeters A_2 and A_3. This shows that the current through the cell is equal to sum of the currents through the two bulbs. This rule applies wherever components are in parallel.

The total current through the whole circuit is the sum of the currents through the separate branches.

Potential difference in a parallel circuit

Figure 2 shows two resistors X and Y in parallel with each other. A voltmeter is connected across each resistor. The voltmeter across resistor X shows the same reading as the voltmeter across resistor Y. This is because each electron from the cell either passes through X or passes through Y. So it delivers the same amount of energy from the cell, whichever resistor it goes through. In other words:

For components in parallel, the potential difference across each component is the same.

Parallel routes

You can think of a parallel route as a kind of bypass. A heart bypass is another route for the flow of blood. A road bypass is a road that passes a town centre instead of going through it. In the same way, for components in parallel, charge flows separately through each component. The total flow of charge is the sum of the flow through each component. The flow of charge per second through each component is the current through it. So, for components that are in parallel, the total current is the sum of the currents through each component.

Calculations on parallel circuits

Components in parallel have the same potential difference across them. The current through each component depends on the resistance of the component.

- The bigger the resistance of the component, the smaller the current through it. The component that has the biggest resistance passes the smallest current.

- You can calculate the current passing through each component using the equation:

$$\textbf{current, } I \text{ (amperes, A)} = \frac{\textbf{potential difference, } V \text{ (volts, } V)}{\textbf{component resistance, } R \text{ (ohms, } \Omega)}$$

Rules for resistors in parallel

Adding more resistors in parallel decreases the total resistance. This is because the total potential difference is the same across each resistor. So adding an extra resistor in parallel increases the total current entering the combination. Because the total resistance is equal to the battery potential difference ÷ the total current entering the combination, the total resistance is less than it was before the extra resistor was added.

The total resistance of two (or more) components in parallel is less than the resistance of the resistor with the least resistance.

Testing resistors in series and parallel

Use the circuit in Figure 3 on P4.3 to measure the resistance of two resistors individually, and then in series with each other. The resistance of the resistors in series should equal the sum of their individual resistances.

Then measure the resistance of two resistors individually and then in parallel with each other. Discuss if your results agree with the statement above this practical box.

Safety: Make sure the wire does not get hot. If it does get hot, reduce the current or switch the circuit off, and ask your teacher to check it.

1 **a** In Table 1, if ammeter A_1 reads 0.40 A and A_2 reads 0.1 A, calculate what A_3 would read. [1 mark]
   b A 3 Ω resistor and a 6 Ω resistor are connected in parallel in a circuit. Identify the resistor that passes the most current. [1 mark]
   c In the circuit shown in Figure 3, calculate what the resistance of a single resistor would be that could replace the three parallel resistors across the 6 V battery and allow the same current to pass through the battery. [2 marks]

2 A 6.0 V battery is connected across a 12 Ω resistor in parallel with a 24 Ω resistor.
   a Draw the circuit diagram for this circuit. [1 mark]
   b i Show that the current through the 12 Ω resistor is 0.50 A [1 mark]
   ii Calculate the current through the 24 Ω resistor. [1 mark]
   c Calculate the current passing through the cell. [2 marks]

3 **a** In a circuit similar to Figure 3, three resistors $R_1 = 2\,\Omega$, $R_2 = 3\,\Omega$, and $R_3 = 6\,\Omega$ are connected to each other in parallel and to a 6 V battery.
   Draw the circuit diagram and calculate:
   i the current through each resistor [3 marks]
   ii the current through the battery. [1 mark]
   b The 6 Ω resistor in the figure is replaced by a 4 Ω resistor, calculate the new battery current. [2 marks]
4 Explain why the equivalent resistance of the three resistors in the circuit in Question **3** is less than 2 Ω. [4 marks]

Worked example

The circuit diagram in Figure 3 shows three resistors $R_1 = 1\,\Omega$, $R_2 = 2\,\Omega$, and $R_3 = 6\,\Omega$ connected in parallel to a 6 V battery.

Calculate:

a the current through each resistor
b the current through the battery.

Solution

a $I_1 = \dfrac{V_1}{R_1} = \dfrac{6}{1} = \mathbf{6\,A}$

$I_2 = \dfrac{V_2}{R_2} = \dfrac{6}{2} = \mathbf{3\,A}$

$I_3 = \dfrac{V_3}{R_3} = \dfrac{6}{6} = \mathbf{1\,A}$

b The total current from the battery
$= I_1 + I_2 + I_3 = 6\,A + 3\,A + 1\,A = \mathbf{10\,A}$

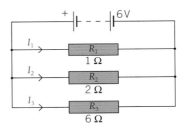

Figure 3

Key points

- For components in parallel:
  - the total current is the sum of the currents through the separate components
  - the potential difference across each component is the same.
- The bigger the resistance of a component, the smaller the current that will pass through that component.
- The current through a resistor in a parallel circuit is $I = \dfrac{V}{R}$
- Adding more resistors in parallel decreases the total resistance because the total current through the resistors is increased and the total potential difference across them is unchanged.

P4 Electric circuits

Summary questions

1 Write and explain how the resistance of a filament lamp changes when the current through the filament is increased. [3 marks]

2 Match each component in the list to each statement **a** to **d** that describes it.

> diode filament lamp resistor thermistor

   a Its resistance increases if the current through it increases. [1 mark]
   b The current through it is proportional to the potential difference across it. [1 mark]
   c Its resistance decreases if its temperature is increased. [1 mark]
   d Its resistance depends on which way around it is connected in a circuit. [1 mark]

3 **a** Sketch a circuit diagram to show two resistors **P** and **Q** connected in series to a battery of two cells in series with each other in the same direction. [1 mark]
   b In the circuit in part **a**, resistor **P** has a resistance of 4.0 Ω, resistor **Q** has a resistance of 6.0 Ω, and each cell has a potential difference of 1.5 V. Calculate:
   i the total potential difference of the two cells [1 mark]
   ii the total resistance of the two resistors [1 mark]
   iii the current in the circuit [2 marks]
   iv the potential difference across each resistor. [2 marks]

4 **a** Sketch a circuit diagram to show two resistors **R** and **S** in parallel with each other connected to a single cell. [1 mark]
   b In the circuit in part **a**, resistor **R** has a resistance of 8.0 Ω, resistor **S** has a resistance of 4.0 Ω, and the cell has a potential difference of 2.0 V. Calculate:
   i the current through resistor **R** [2 marks]
   ii the current through resistor **S** [2 marks]
   iii the current through the cell in the circuit. [1 mark]

5 The figure shows a light-dependent resistor (LDR) in series with a 200 Ω resistor, a 3.0 V battery, and an ammeter.

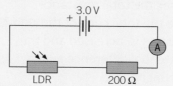

   a With the LDR in daylight, the ammeter reads 0.010 A.
   i Calculate the potential difference across the 200 Ω resistor when the current through it is 0.010 A. [1 mark]
   ii Show that the potential difference across the LDR is 1.0 V when the ammeter reads 0.010 A. [1 mark]
   iii Calculate the resistance of the LDR in daylight. [1 mark]
   b i If the LDR is then covered, explain whether the ammeter reading increases or decreases or stays the same. [2 marks]
   ii Explain how the resistance of the LDR can be calculated from the current I, the battery potential difference V, and the resistance R of the LDR. [2 marks]

6 In the figure to Question **5**, the LDR is replaced by a 100 Ω resistor and a voltmeter connected in parallel with this resistor.
   a Draw the circuit diagram for this circuit. [1 mark]
   b Calculate:
   i the total resistance of the two resistors in the circuit [1 mark]
   ii the current through the ammeter [2 marks]
   iii the voltmeter reading [2 marks]
   iv the potential difference across each resistor. [2 marks]

7 The figure shows a light-emitting diode (LED) in series with a resistor and a 3.0 V battery.

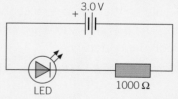

   a The LED in the circuit emits light. The potential difference across it when it emits light is 0.6 V.
   i Explain why the potential difference across the 1000 Ω resistor is 2.4 V. [2 marks]
   ii Calculate the current in the circuit. [2 marks]
   b The current through the LED must not exceed 15 mA or it will be damaged. If the resistor in the figure is replaced by a different resistor **R**, calculate what should be the minimum resistance of **R**. [2 marks]
   c If the LED in the circuit is reversed, what would be the current in the circuit? Give a reason for your answer. [3 marks]

8 **a** Design a temperature sensor that will switch a buzzer on if the temperature is too low. [2 marks]
   b Explain how your circuit works. [3 marks]

Practice questions

01.1 Complete the sentences using words from the box. You may use the words once, more than once, or not at all.

protons	atoms	negatively	equally
	electrons	positively	

When two different materials are rubbed together, _____ are rubbed off one material and on to the other.
The material that gains _____ becomes _____ charged.
The material that loses _____ becomes _____ charged. [5 marks]

01.2 A positively charged ball **a** is in a fixed position and a second charged ball **b** attached to string is brought close to the fixed ball.

Figure 1

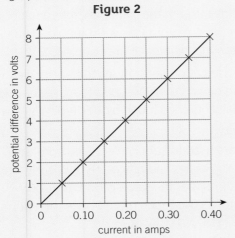

Describe the type of force and direction of ball **b**. [2 marks]

01.3 Describe the effect, if any, of moving the second ball to position **x**. [1 mark]

02.1 **Figure 2** shows a current–potential difference graph of a fixed resistor.

Figure 2

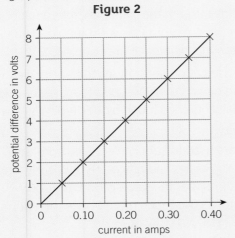

Calculate the resistance of the fixed resistor R_1 and give the unit. [2 marks]

02.2 The fixed resistor R_1 is placed in series with another fixed resistor R_2.

Figure 3

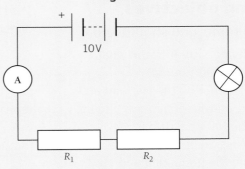

Calculate the total resistance of resistor R_1 and resistor R_2.
Resistance of resistor $R_2 = \frac{2}{3}R_1$ [2 marks]

02.3 A voltmeter is used to measure the potential difference across the bulb. Draw the circuit with the voltmeter in the correct position. [2 marks]

02.4 Calculate the charge and give the unit when a current of 1.5 A flows around the circuit for 20 seconds. [3 marks]

02.5 Calculate how much energy is transferred to the lamp in 20 seconds. Use your answer from **2.4**. [2 marks]

03 A student measures the resistance of a coil of wire using the equipment in **Figure 4**.

Figure 4

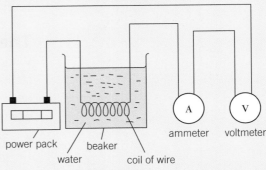

03.1 Give one mistake the student has made in setting up the equipment. [1 mark]

03.2 Give one reason why the coil of wire is in water. [1 mark]

03.3 Describe how the student measures the resistance of the wire using the equipment given. [3 marks]

03.4 Suggest one safety precaution the student should take. [1 mark]

Learning objectives

After this topic, you should know:

- what direct current is and what alternating current is
- what is meant by the live wire and the neutral wire of a mains circuit
- what the National Grid is
- how to use an oscilloscope to measure the frequency and peak potential difference of an alternating current.

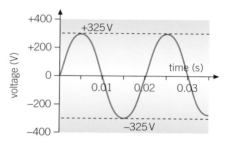

Figure 1 *Mains voltage against time*

The battery in a torch makes the current go around the circuit in one direction only. The current in the circuit is called a **direct current** (d.c.) because it is in one direction only.

When you switch a light on at home, you use **alternating current** (a.c.) because mains electricity is an a.c. supply. An alternating current repeatedly reverses its direction. It flows one way then the opposite way in successive cycles. Its frequency is the number of cycles it passes through each second.

In the UK, the mains frequency is 50 cycles per second (or 50 Hz). The time taken for 1 cycle = 0.02 s (= 1 ÷ frequency). A light bulb works just as well at this frequency as it would with a direct current.

Mains circuits

Every mains circuit has a **live wire** and a **neutral wire**. The current through a mains appliance alternates. That is because the mains supply provides an alternating potential difference between the two wires. In other words, the polarity of the potential difference repeatedly reverses its direction. In comparison, potential differences in direct current circuits do not change direction.

The neutral wire is earthed at the local electricity substation. The potential difference between the live wire and the earth is usually called the potential or voltage of the live wire. The live wire is dangerous because its potential repeatedly changes from + to – and back every cycle. In UK homes, it reaches about 325 V in each direction (Figure 1).

The National Grid

When you use mains appliances, the electricity is supplied from power stations to homes and buildings through the National Grid – a nationwide network of cables and transformers. A typical power station generates electricity at an alternating potential difference of about 25 000 V.

- Step-up transformers are used at power stations to transfer electricity to the National Grid. These transformers are used to make the size of the alternating potential difference much bigger, typically from 25 000 V to about 132 000 V).

- Step-down transformers are used to supply electricity from the National Grid to consumers. Homes and offices in the UK are supplied with mains electricity that provides the same power as a 230 V direct-current supply. Factories use much more power than homes, so they are supplied with a p.d. of 100 kV or 33 kV.

By making the grid potential difference very large, much less current is needed to transfer the same amount of power. So the power loss due to the resistance heating in the cables is much reduced. This means that the National Grid is an efficient way to transfer power.

Investigating an alternating potential difference

An oscilloscope is used to show how an alternating potential difference changes with time.

1   Connect a signal generator to an oscilloscope (Figure 2).

   –   The waves on the oscilloscope screen are caused by the potential difference increasing and decreasing continuously. Adjusting the 'Y-gain' control changes how tall the waves are. Adjusting the 'time base' control changes how many waves fit across the screen.

   –   The peak potential difference (peak voltage) is the difference in volts between the highest level and the middle level of the waves. Increasing the potential difference of the a.c. supply makes the waves on the screen taller.

   –   Increasing the frequency of the a.c. supply increases the number of cycles you see on the screen. So the waves on the screen get squashed together.

2   Connect a battery to the oscilloscope. You should see a flat line at a constant potential difference.

   What difference on the oscilloscope screen is made by reversing the battery?

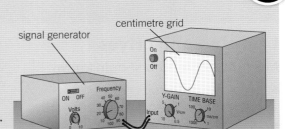

Figure 2 *Using an oscilloscope*

1   Choose the correct potential difference from the list for each appliance **a** to **d**.
   1.5 V   12 V   230 V   325 V   [4 marks]
   a   a car battery   **c**   a torch cell
   b   the mains voltage   **d**   the maximum potential of the live wire.

2   Describe how the trace on the screen in Figure 2 would change if the frequency of the a.c. supply was:
   a   increased without changing the peak potential difference [2 marks]
   b   reduced and the peak potential difference was doubled.   [2 marks]

3   Calculate the frequency in Figure 2 if one cycle measures 8 cm across the screen for a time base setting of 10 milliseconds per centimetre.
                                                                    [2 marks]

4   **a**   Describe how an alternating current differs from a direct current.
                                                                    [1 mark]

   b i   Figure 3 shows a diode and a resistor in series with each other and connected to an a.c. supply. State why the current in the circuit is a direct current, not an alternating current.   [1 mark]

Figure 3

ac supply

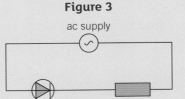

   c i   Sketch a graph to show how the current varies with time. [1 mark]
   ii   Suggest how your graph would differ if a resistor of greater resistance had been used.   [2 marks]

Go further!

You'll learn more about alternating current and oscilloscopes at A level, when you carry out experiments such as measuring the frequency of an alternating potential difference.

Key points

- Direct current (d.c.) flows in one direction only. Alternating current (a.c.) repeatedly reverses its direction of flow.
- A mains circuit has a live wire, which is alternately positive and negative every cycle, and a neutral wire at zero volts.
- The peak potential difference of an a.c. supply is the maximum voltage measured from zero volts.
- To measure the frequency of an a.c. supply, measure the time period of the waves, then use the equation

   $$\text{frequency} = \frac{1}{\text{time taken for 1 cycle}}$$

P5.2 Cables and plugs

Learning objectives

After this topic, you should know:

- what the casing of a mains plug or socket is made of and why
- what is in a mains cable
- the colours of the live, neutral, and earth wires
- why a three-pin plug includes an earth pin.

When you plug a heater with a metal case into a wall socket, the metal case is automatically connected to earth through a wire called the **earth wire**. This stops the metal case becoming live if the live wire breaks and touches the case. If the case did become live and you touched it, you would be electrocuted.

Plastic materials are good insulators. An appliance that has a plastic case is double-insulated, so it has no earth wire connection. All electrical appliances with a plastic case sold in the UK must be double-insulated.

Plugs, sockets, and cables

The outer casings of plugs, sockets, and cables of all mains circuits and appliances are made of hard-wearing electrical insulators. That's because plugs, sockets, and cables contain live wires. Most mains appliances are connected by a wall socket to the mains using a cable and a three-pin plug.

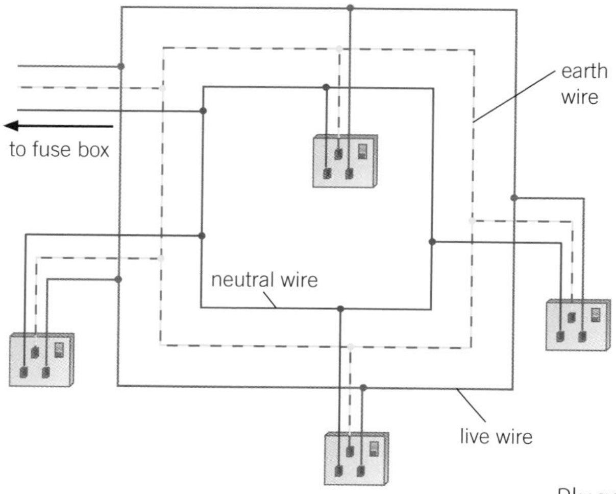

Sockets are made of stiff plastic materials with the wires inside them. Figure 1 shows part of a wall socket circuit. It has an earth wire as well as a live wire and a neutral wire.

- The earth wire of this circuit is connected to the ground at your home. It is at zero volts (0 V) and carries a current only if there is a fault.

- The longest pin of a three-pin plug is designed to make contact with the earth wire of a wall socket circuit. So, when you plug an appliance with a metal case into a wall socket, the case is automatically earthed.

Figure 1 *A wall socket circuit*

Plugs have cases made of stiff plastic materials. The live pin, the neutral pin, and the earth pin stick out through the plug case. Figure 2 shows inside a three-pin plug.

- The pins are made of brass because brass is a good conductor and doesn't rust or oxidise. Copper isn't as hard as brass even though it conducts better.

- The case material is an electrical insulator. The inside of the case is shaped so that the wires and the pins can't touch each other when the plug is sealed.

- The plug contains a **fuse** between the live pin and the live wire. If too much current passes through the wire in the fuse, it melts and cuts the live wire off.

- The brown wire is connected to the live pin.

- The blue wire is connected to the neutral pin.

- The green and yellow striped wire (of a three-core cable) is connected to the earth pin. A two-core cable doesn't have an earth wire.

Study tip

Make sure you know what's inside a three-pin plug and the colour of each wire.

Cables used for mains appliances (and for mains circuits) are made up of two or three insulated copper wires surrounded by an outer layer of rubber or flexible plastic material.

- Copper is used for the wires because it is a good electrical conductor and it bends easily.
- Plastic is a good electrical insulator, so if anyone touches the cable, it stops them from getting an electric shock.
- Two-core cables are used for appliances that have plastic cases (e.g. hairdryers, radios, mobile phone chargers).
- Cables of different thicknesses are used for different purposes. For example, the cables that join together the wall sockets in your house must be much thicker than the cables that join together the light fittings. This is because more current passes along wall socket cables than along lighting circuits, so the wires in them have to be much thicker so they have less resistance. This stops the heating effect of the current making the wires too hot.

Short circuits

If a live wire inside the appliance touches a neutral wire, a very big current passes between the two wires at the point of contact. This is called a short circuit. Provided the fuse blows, it cuts the current off.

Even if an appliance is switched off, never touch the wires inside the supply cable. People's bodies are at zero volts. If someone touches a live wire, a big potential difference will act across their body, causing a current to flow through them. They will suffer an electric shock, which could be lethal.

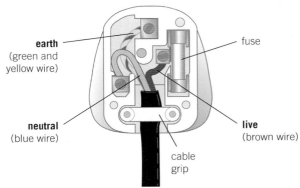
earth (green and yellow wire), fuse, neutral (blue wire), live (brown wire), cable grip

Figure 2 *Inside a three-pin plug*

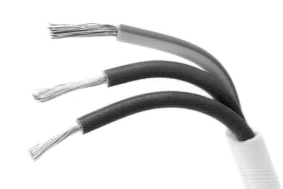

Figure 3 *Mains cables*

1 a Name the colour of each of the three wires in a mains plug. [1 mark]
  b i Explain why sockets are wired in parallel with each other.[2 marks]
    ii Explain why brass, an alloy of copper and zinc, is better than copper for the pins of a three-pin plug. [2 marks]
    iii Give a reason why cables that are worn away or damaged are dangerous. [1 mark]

2 a Match the list of parts 1 to 4 in a three-pin plug with the list of materials **A** to **D**.
  1 cable insulation   **A** brass
  2 case   **B** copper
  3 pin   **C** rubber
  4 wire   **D** stiff plastic [2 marks]
  b Give your choice of material for each part in **a**. [4 marks]

3 a Explain why each of the three wires in a three-core mains cable is insulated. [2 marks]
  b Describe how the metal case of an electrical appliance is connected to earth. [3 marks]

4 a Explain why the cables joining the wall sockets in a house need to be thicker than the cables joining the light fittings. [2 marks]
  b Describe the difference between a two-core cable and a three-core cable. [1 mark]
  c Explain what determines whether an appliance should have a two-core or a three-core cable. [2 marks]

Key points

- Sockets and plug cases are made of stiff plastic materials that enclose the electrical connections. Plastic is used because it is a good electrical insulator.
- A mains cable is made up of two or three insulated copper wires surrounded by an outer layer of flexible plastic material.
- In a three-pin plug or a three-core cable, the live wire is brown, the neutral wire is blue, and the earth wire is striped green and yellow.
- The earth wire is connected to the longest pin in a plug and is used to earth the metal case of a mains appliance.

P5.3 Electrical power and potential difference

Synoptic link

You first met this equation in Topic P1.9.

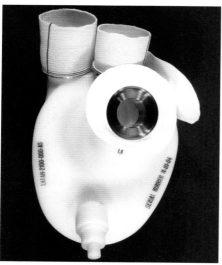

Figure 1 *An artificial heart. A surgeon fitting an artificial heart in a patient needs to make sure the battery will last a long time. Even so, the battery may have to be replaced every few years*

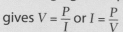

Rearranging equations
Rearranging the equation $P = I\,V$
gives $V = \dfrac{P}{I}$ or $I = \dfrac{P}{V}$

When you use an electrical appliance, the current through it transfers energy to it from the power source it is connected to. The power of the appliance, in watts, is the energy it transfers in joules per second. You can see this using the equation:

$$\textbf{power, } P \text{ (watts, W)} = \frac{\textbf{energy transferred, } E \text{ (joules, J)}}{\textbf{time, } t \text{ (seconds, s)}}$$

You can calculate the energy transferred E in a given time t if you know the power P, rearrange the above equation to give:

$$E = P\,t$$

> **Worked example**
> A 40 W light bulb is switched on for 30 minutes. Calculate the energy it transfers.
>
> **Solution**
> Time taken = $30 \times 60\,\text{s} = 1800\,\text{s}$
>
> $E = P \times t = 40\,\text{W} \times 1800\,\text{s} = 72\,000\,\text{J}$

Calculating power

Millions of electrons pass through the circuit of an artificial heart every second. Work is done by a battery in the artificial heart to force the electrons around the circuit. Each electron transfers a small amount of energy to the heart from the battery. So the total energy transferred to the artificial heart each second is big enough to allow the appliance to function.

For any electrical appliance:

- the current through it is the charge that flows through it each second
- the potential difference across it is the energy transferred to the appliance by each coulomb of charge that passes through it
- the power supplied to it is the energy transferred to it each second. This is the energy transferred by an electric current every second.

So the energy transfer to the appliance each second = the charge flow per second × the energy transfer per unit charge.

In other words:

$$\textbf{power supplied, } P = \textbf{current, } I \times \textbf{potential difference, } V$$
$$\text{(watts, W)} \qquad \text{(amperes, A)} \qquad \text{(volts, V)}$$

For example, the power supplied to:

- a 4 A, 12 V electric motor is 48 W (= 4 A × 12 V)
- a 0.1 A, 3 V torch lamp is 0.3 W (= 0.1 A × 3.0 V).

Choosing a fuse

Domestic appliances are often fitted with a 3 A, 5 A, or 13 A fuse. If you do not know which one to use for an appliance, you can calculate it by using the power rating of the appliance and its potential difference (voltage). The next time you change a fuse, do a quick calculation to make sure its rating is correct for the appliance.

Resistance heating

When an electric current passes through a resistor, the power supplied to the resistor heats it. The resistor heats the surroundings, so the energy supplied to it is dissipated to the surroundings.

For a current I in a resistor of resistance R:

● the potential difference V across the resistor $= I \times R$

● the power P supplied to the resistor $= I \times V = I \times I \times R = I^2 R$

power, P (W) = current^2, I^2 (A) × resistance, R (Ω)

This equation shows that the power supplied to a resistor is proportional to the square of the current. So, for example, if the current is doubled, the power becomes four times greater.

1 a The human heart transfers about 30 000 J of energy in about 8 hours. Estimate to an order of magnitude the power of the human heart. [2 marks]
  b Calculate the power supplied to a 5 A, 230 V electric heater. [2 marks]
  c Explain why a 13 A fuse would be unsuitable for a 230 V, 100 W table lamp. [2 marks]

2 a Calculate the power supplied to each of the following devices in normal use:
    i a 12 V, 5 A light bulb [2 marks]
    ii a 230 V, 12 A heater. [2 marks]
  b Calculate the type of fuse, 3 A, 5 A, or 13 A, that would you select for:
    i a 50 W, 12 V heater [2 marks]
    ii a 230 V, 750 W microwave oven. [2 marks]

3 a Explain why a 3 A fuse would be unsuitable for a 230 V, 800 W microwave oven. [2 marks]
  b The heating element of a 12 V heater has a resistance of 4.0 Ω. When the heating element is connected to a 12 V power supply, calculate:
    i the current through it [2 marks]
    ii the electrical power supplied to it [2 marks]
    iii the energy, in joules, transferred to the heating element by an electric current in 20 minutes. [2 marks]

4 A 6.0 kW electric oven is connected to a fuse box by a cable of resistance 0.25 Ω. When the cooker is switched on at full power, a current of 26 A passes through it.
  a i Calculate the potential difference between the two ends of the cable. [2 marks]
    ii Calculate the power wasted in the cable because of the heating effect of the current. [2 marks]
  b Calculate what percentage of the power supplied to the oven and the cable is wasted in the cable. [2 marks]

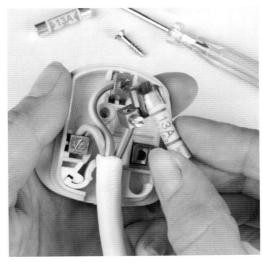

Figure 2 *Changing a fuse*

Worked example

a Calculate the normal current through a 500 W, 230 V heater.

b Determine which fuse, 1 A, 3 A, 5 A, or 13 A, you would use for the appliance.

Solution

a Current $= \dfrac{500\,W}{230\,V} = 2.2\,A$

b You would use a 3 A fuse because it would not melt when the current is 2.2 A, but it would melt if, due to a fault, the current exceeded 3 A. The 5 A and 13 A fuses would only melt if the current exceeded 5 A and 13 A, respectively.

Key points

● The power supplied to a device is the energy transferred to it each second.
● The energy transferred to a device is $E = P \times t$.
● The electrical power supplied to an appliance is equal to $P = I \times V$.
● The correct rating (A) for a fuse:
$$= \dfrac{\text{electrical power (watts)}}{\text{potential difference (volts)}}$$

P5.4 Electrical currents and energy transfer

Learning objectives

After this topic, you should know:

- how to calculate the flow of electric charge when you know the current and time
- what energy transfers happen when electrical charge flows through a resistor
- how the energy transferred by a flow of electrical charge is related to potential difference
- about the energy supplied by the battery in a circuit and the energy transferred to the electrical components.

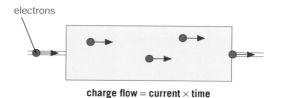

electrons

charge flow = current × time

Figure 1 *Charge and current*

Synoptic link

For more information on calculating the energy supplied to an electrical device, look at Topic P1.9.

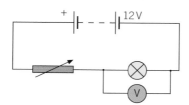

Figure 2 *Energy transfer in a circuit*

Calculating charge

When an electrical appliance is turned on, electrons are forced through it by the potential difference of its power supply unit. The potential difference causes charge to flow through the appliance, carried by the electrons.

You can calculate the charge flow using the equation:

charge flow, Q = current, I × time, t
(coulombs, C)   (amperes, A)   (seconds, s)

Energy and potential difference

When a resistor is connected to a battery, work is done by the battery to make electrons pass through the resistor. Each electron repeatedly collides with the vibrating metal ions of the resistor, transferring energy to the ions. So the ions of the resistor gain kinetic energy and vibrate even more. The resistor becomes hotter. The electrical work done by the battery is equal to the energy transferred to the resistor. In this way, energy is transferred from the chemical energy store in the battery to the resistor's store of thermal energy.

When charge flows through a resistor, energy is transferred to the resistor, so the resistor becomes hotter.

The energy transferred to a resistor E, in a given time t, can be calculated using either of the equations:

energy, E = charge flow, Q × potential difference, V

energy, E = power, P × time, t = potential difference, V × current, I × time, t

Energy transfer in a circuit

The circuit in Figure 2 shows a 12 V battery in series with a torch bulb and a variable resistor. When the voltmeter reads 10 V, the potential difference across the variable resistor is 2 V.

Each coulomb of charge:

- leaves the battery with 12 J of energy (because energy from the battery = charge × battery potential difference)
- transfers 10 J of energy to the torch bulb (because energy transfer to bulb = charge × potential difference across bulb)
- transfers 2 J of energy to the variable resistor.

The energy transferred to the bulb and the resistor increases their thermal energy stores. As a result, the bulb becomes hot and emits light and the variable resistor becomes warm so it heats the surroundings. So energy is transferred to the surroundings by both the bulb and the resistor.

Therefore, the energy from the battery = the energy transferred to the bulb + the energy transferred to the variable resistor.

Worked example

Calculate the energy transferred in a component when the charge passing through it is 30 C and the potential difference is 20 V.

Solution

Using the equation $E = V \times Q$ gives:

energy transferred = $20\,V \times 30\,C =$ **600 J**

1 **a** Calculate the charge flowing in 50 s when the current is 3 A.
[2 marks]

   b Calculate the energy transferred when the charge flow is 30 C and the potential difference is 4 V. [2 marks]

   c Calculate the energy transferred in 60 s when a current of 0.5 A passes through a 12 Ω resistor. [2 marks]

2 **a** Calculate the charge flow for:

    i a current of 4 A for 20 s [2 marks]

    ii a current of 0.2 A for 60 minutes. [2 marks]

   b Calculate the energy transfer:

    i for a charge flow of 20 C when the potential difference is 6.0 V
[2 marks]

    ii for a current of 3 A that passes through a resistor for 20 s, when the potential difference is 5 V. [2 marks]

3 In Figure 2, an ammeter is connected in the circuit in series with the battery. The variable resistor is then adjusted until the ammeter reading is 2.0 A. The voltmeter reading is then 9.0 V.

   a Calculate the charge that passes through the battery in 60 s.
[2 marks]

   b Calculate the energy transferred to or from each coulomb of charge when it passes through each component including the battery. [3 marks]

   c Show that the energy transferred from the battery in 60 s is equal to the sum of the energy transferred to the lamp and the variable resistor in this time. [2 marks]

4 In Figure 3 a 4.0 Ω resistor and an 8.0 Ω resistor in series with each other are connected to a 6.0 V battery.

Calculate:

   a the resistance of the two resistors in series [2 marks]

   b the current through the resistors [2 marks]

   c the potential difference across each resistor [2 marks]

   d the energy transferred to each resistor in 60 seconds [2 marks]

   e the energy supplied by the battery in 60 seconds. [2 marks]

Go further!

The charge of an electron was first measured by J J Thompson about 120 years ago. He showed that all atoms contain identical negative particles, which were later called electrons. Before that time, physicists didn't know about electrons. Thompson measured the charge of an electron and found that it is 0.000 000 000 000 000 00016 coulombs (= $1.6 \times 10^{-19}\,C$). Prove, for yourself, that when a current of 1 A is in a wire, 6.25 million million million electrons pass along the wire each second!

Study tip

Make sure you know and can use the relationship between charge, current, and time.

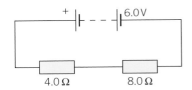

Figure 3 *See summary question 4*

Key points

- The charge flow is $Q = I \times t$.
- When charge flows through a resistor, energy transferred to the resistor makes it hot.
- The energy transferred to a component is $E = V \times Q$.
- When charge flows around a circuit for a given time, the energy supplied by the battery is equal to the energy transferred to all the components in the circuit.

P5.5 Appliances and efficiency

Learning objectives

After this topic, you should know:

- how to calculate the energy supplied to an electrical appliance from its current, its potential difference, and how long it is used for
- how to work out the useful energy output of an electrical appliance
- how to work out the output power of an electrical appliance
- how to compare different appliances that do the same job.

When you use an electric heater, how much energy is transferred from the mains? You can work this out if you know its power and how long you use it for.

For any appliance, the energy supplied to it depends on how long it has been switched on and the power supplied to it. For example:

- A 1 kilowatt heater uses the same amount of energy in 1 hour as a 2 kilowatt heater would use in half an hour.
- A 100 W lamp uses the same amount of energy in 30 hours as a 3000 W heater does in 1 hour.

You can use the equation below to work out the energy, in joules, transferred to a mains appliance in a given time:

energy transferred from the mains, E = **power, P** × **time, t**

(joules, J)   (watts, W)   (seconds, s)

To calculate the power supplied to an electrical appliance use the following equation:

power, P = **current, I** × **potential difference, V**

(watts, W)   (amperes, A)   (volts, V)

Electrical appliances and efficiency

The efficiency of any device can be calculated from the power supplied to it (its input power) and the useful energy per second it transfers (its output power) using the equation:

$$\text{efficiency} = \frac{\text{its output power}}{\text{its input power}} (\times\ 100\%)$$

Given the input power and efficiency of an appliance, the output power can be calculated by rearranging the above equation to give:

$$\text{output power} = \text{efficiency} \times \text{its input power}$$

Efficiency values can be expressed as a ratio or a percentage. For example, a percentage efficiency of 60% is equivalent to a ratio of 0.60.

The percentage efficiency of an electrical appliance is always less than 100%. The difference between the percentage efficiency of an appliance and 100% tells you the percentage of the energy supplied that is wasted.

Electrical appliances waste energy because the current in both the wires and the components of the appliance has a heating effect due to the resistance of the wires and the components. So they transfer energy by heating to the surroundings. Electrical devices with moving parts such as electric motors also waste energy due to friction between their moving parts. Friction between the moving parts heats them, so they also transfer energy by heating the surroundings.

Rearranging the equation

$$E = P\,t$$

To make power the subject, divide both sides by t. this gives:

$$P = \frac{E}{t}$$

Study tip

Remember that 1 kW = 1000 W

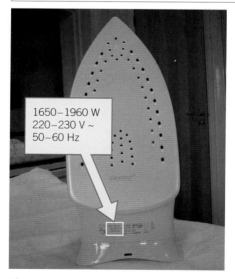

1650–1960 W
220–230 V ~
50–60 Hz

Figure 1 *Mains power*

Synoptic link

You met efficiency and power in Topic P1.9.

Worked example

A 230V, 12A electric motor in a machine has an efficiency of 60%. Calculate:

a the electrical power supplied to it

b the output power of the motor

c the energy per second wasted by the motor.

Solution

a power supplied = current × potential difference = 12 A × 230 V = 2760 W

b efficiency as a ratio = 60% ÷ 100% = 0.60
output power = efficiency × input power = 0.60 × 2760 W = 1660 W

c energy wasted per second = 2760 W − 1660 W = 1100 W

---

1 **a** Calculate how many joules of energy are used by:

   i a 5 W torch lamp in 50 minutes (3000 seconds)   [2 marks]

   ii a 100 W lamp in 24 hours.   [2 marks]

  b Calculate how much energy is transferred in each case below.

   i A 3 kilowatt electric kettle is used six times for 5 minutes each time.   [2 marks]

   ii A 1000 watt microwave oven is used for 30 minutes.   [2 marks]

2 Use the data in Table 1 to answer the following questions about different types of electric lamps.

Table 1

Type	Power in watts	Efficiency
Filament lamp	100	20%
Halogen bulb	100	25%
Low-energy compact fluorescent lamp bulb (CFL)	25	80%
Low-energy light-emitting diode (LED)	2	90%

  a How much energy is wasted in 1 second by:

   i the filament lamp?   [1 mark]   **ii** the CFL bulb?   [1 mark]

  b A student looks at Table 1 and claims that 50 LEDs would be needed to give the same output as the halogen bulb. Explain why this claim is incorrect and estimate how many LEDs would be needed for this purpose.   [3 marks]

3 An electric heater is left on for 4 hours. During this time it uses 36 MJ of energy.

  a Calculate the power of the heater.   [2 marks]

  b Calculate how long it would take for a 2000 W electric kettle to use 36 MJ of energy.   [2 marks]

4 The mains power supply of a computer provides a current of 1.5 A at 230 V.

  a Calculate the power supplied to the computer.   [2 marks]

  b In one month, the computer is used for 130 hours. Calculate how many joules of energy are supplied to the computer in this time.   [2 marks]

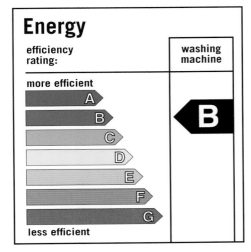

Figure 2 *Efficiency measures. All new appliances such as washing machines and freezers sold in the EU are labelled clearly with an efficiency rating. The rating is from A (very efficient) to G (lowest efficiency). Light bulbs are also labelled in this way on the packaging*

Key points

- A domestic electricity meter measures how much energy is supplied.
- The energy supplied to an appliance is $E = P\,t$.
- Useful energy used = efficiency × energy supplied.

Summary questions

1 a In a mains circuit, name the wire that:
  i is earthed at the local substation [1 mark]
  ii alternates in potential. [1 mark]
  b An oscilloscope is used to display the potential difference of an alternating potential difference supply unit. Write and explain how the trace would change if:
  i the potential difference is increased [2 marks]
  ii the frequency is increased. [2 marks]

2 a Explain why a mains appliance with a metal case is unsafe if the case is not earthed. [2 marks]
  b Write the colour of each wire in a mains circuit. [1 mark]

3 a Explain why the wall sockets in a room are connected in parallel with each other. [2 marks]
  b Explain why a fuse in the mains plug of an appliance is in series with the appliance. [2 marks]

4 a i Calculate the current in a 230V, 2.5 kW electric kettle. [2 marks]
  ii Write the fuse, 3 A, 5 A, or 13 A, that would you fit in the kettle plug. [1 mark]
  iii If the kettle is used on average six times a day for 5 minutes each time, calculate the energy in kWh it uses in 28 days. [2 marks]
  b A student uses a 4.0 A, 230V microwave oven for 10 minutes every day, and a 2500W electric kettle three times a day for 4 minutes each time. State and explain which appliance uses more energy in one day. [3 marks]
  c The electric kettle takes 300s to heat 1.5 kg of water from 15 °C to 100 °C. Calculate its efficiency. The specific heat capacity of water is 4200 J/kg°C. [4 marks]

5 A 5 Ω resistor is in series with a bulb, a switch and a 12V battery.
  a Draw the circuit diagram. [1 mark]
  b When the switch is closed for 60 seconds, a direct current of 0.6 A passes through the resistor. Calculate:
  i the energy supplied by the battery [2 marks]
  ii the energy transferred to the resistor [2 marks]
  iii the energy transferred to the bulb. [2 marks]
  c The bulb is replaced by a 25 Ω resistor.
  i Calculate the total resistance of the two resistors. [1 mark]
  ii Calculate the current in the battery. [2 marks]

iii Calculate the power supplied by the battery and the power delivered to each resistor. [3 marks]

6 A 12V, 36W bulb is connected to a 12V supply.
  a Calculate:
  i the current through the bulb [2 marks]
  ii the charge flow through the bulb in 200s. [2 marks]
  b i Show that 7200 J of energy is delivered to the bulb in 200 s. [2 marks]
  ii Calculate the energy delivered to the bulb by each coulomb of charge that passes through it. [2 marks]
  c A second 12V, 36W bulb is connected to the power supply in parallel with the first bulb.
  i Calculate the current through each bulb and through the battery. [2 marks]
  ii Show that the energy delivered per second to the two bulbs is equal to the energy supplied per second by the battery. [3 marks]

7 An electrician has the job of connecting a 6.6 kW electric oven to the 230V mains supply in a house.
  a Calculate the current needed to supply 6.6 kW of electrical power at 230V. [2 marks]
  b The table below shows the maximum current that can pass safely through five different mains cables. For each cable the cross-sectional area of each conductor is given in square millimetres (mm²).

Table 1

	Cross-sectional area of conductor in millimetres squared	Maximum safe current in amperes
A	1.0	14
B	1.5	18
C	2.5	28
D	4.0	36
E	6.0	46

  i To connect the oven to the mains supply, determine which cable the electrician should choose. Give a reason for your answer. [3 marks]
  ii Explain what would happen if she chose a cable with thinner conductors. [2 marks]

8 Design an experiment to investigate how the power supplied to a 12V filament lamp varies with current up to a maximum of 2 A. Assume you have a 12V, 2 A lamp, an ammeter, a voltmeter, and a variable resistor. In your answer, you should draw a circuit diagram for the lamp, describe the method you would follow, the measurements you would make, and how you would use your results to test your predictions in **a**.
[6 marks]

Practice questions

01.1 Complete the sentences using the words in the box. Each word can be used once, more than once, or not at all.

electrons	protons	current	the same
	a changing	volts	

When an electric _____ flows through a wire it is a flow of _____. [2 marks]
In a d.c. circuit the flow is always in _____ direction. [1 mark]
In an a.c. circuit the flow is always in _____ direction. [1 mark]

01.2 An oscilloscope is connected to a power supply.

Figure 1

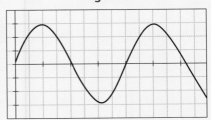

Determine the peak voltage and frequency of the supply. The Y-gain control is set at 6.5 volts/division and the time base is set at 10 milliseconds/division. [2 marks]

02.1 State the functions of the components of the three-pin plug shown in **Figure 2**. [5 marks]

Figure 2

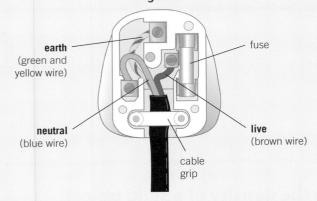

earth
(green and yellow wire)

fuse

neutral
(blue wire)

live
(brown wire)

cable grip

02.2 Explain how the fuse in three-pin plug protects the wires from overheating. [3 marks]

02.3 Explain why it is important to have an earth wire connected to a metal electric fan. [3 marks]

02.4 Give the advantages of using a circuit breaker compared to a fuse. [2 marks]

03 A student investigates the power rating of a resistor. The apparatus used is shown in **Figure 3**.

Figure 3

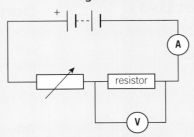

03.1 Give one reason why a variable resistor is used in the circuit. [1 mark]

03.2 State the resolution of the voltmeter. [1 mark]

Figure 4

Dagatron® 7203   V   Ω⊥2.0

03.3 Calculate the power of the fixed resistor. The potential difference across the resistor is 12V and the current through the resistor is 50mA. [3 marks]

03.4 Calculate the resistance of the fixed resistor. Use the information in **03.3**. [2 marks]

04 A rating plate for an electric toaster states: 230 volts, 50 Hz, 800 watts.

04.1 State the frequency of the mains supply. [1 mark]

04.2 Give the power rating of the toaster. [1 mark]

04.3 Calculate the current in amps flowing through the toaster. [2 marks]

04.4 The householder has a choice of fuses. Choose the correct fuse from the box to use in the plug attached to the toaster. Explain your choice.

3 A	5 A	10 A	13 A

[2 marks]

04.5 Calculate the amount of charge flowing through the toaster if it takes 2.5 minutes to toast two slices of bread. Give your answer to 3 significant figures. [3 marks]

05 A gardener uses a plastic electric lawn mower to cut the grass.

05.1 Describe what is meant in the handbook by "the lawn mower is double insulated". [2 marks]

05.2 Explain how the gardener may get an electric shock if the cable gets cut whilst using the lawnmower. [3 marks]

P 6 Molecules and matter

6.1 Density

Learning objectives

After this topic, you should know:

- how density is defined and its units of measurement
- how to measure the density of a solid object or a liquid
- how to use the density equation to calculate the mass or the volume of an object or a sample
- how to tell from its density if an object will float in water.

Figure 1 *Materials of different densities*

Synoptic link

The density of pure water is 1000 kg/m³. Objects that float in water have a density less than 1000 kg/m³. You will learn more about this in Topic P11.4.

Rearranging equations

Rearranging the density equation $\rho = \frac{m}{V}$ gives:

$m = \rho V$ or $V = \frac{m}{\rho}$

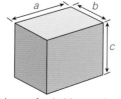

volume of cuboid = $a \times b \times c$

Figure 2 *The volume of a cuboid*

Density comparisons

Any builder knows that a concrete post is much heavier than a wooden post of the same size. This is because the **density** of concrete is much greater than the density of wood. A volume of one cubic metre of wood has a mass of about 800 kg. But a cubic metre of concrete has a mass of about 2400 kg. So the density of concrete is about three times the density of wood.

The density of a substance is defined as its mass per unit volume.

You can use the equation below to calculate the density ρ of a substance if you know the mass m and the volume V of a sample of it.

$$\underset{\text{(kilogram per cubic metre, kg/m}^3)}{\text{density, }\rho} = \frac{\text{mass, }m \text{ (kilograms, kg)}}{\text{volume, }V \text{ (metres}^3, \text{m}^3)}$$

Converting units and using standard form

$1\,\text{kg} = 1000\,\text{g} = 10^3\,\text{g}$

$1\,\text{m} = 100\,\text{cm} = 10^2\,\text{cm}$

$1\,\text{m}^3 = 1\,000\,000\,\text{cm}^3 = 10^6\,\text{cm}^3$

So $1000\,\text{kg/m}^3 = 1\,000\,000\,\text{g}/1\,000\,000\,\text{cm}^3 = 1\,\text{g/cm}^3$

Standard form is useful when you are working with very large numbers, particularly when you need to convert values to SI units (e.g., converting MJ to J) for a calculation. In standard form, a number is written as $A \times 10^n$, where n is the number of places you have had to move the decimal point to the left (or right for a negative power of ten) to get the decimal number A, which is greater than 1 and less than 10.

Worked example

A wooden post has a volume of 0.025 m³ and a mass of 20 kg. Calculate its density in kg/m³.

Solution

$$\text{density} = \frac{\text{mass}}{\text{volume}} = \frac{20\,\text{kg}}{0.025\,\text{m}^3} = 800\,\text{kg/m}^3$$

Measuring the density of a solid object

To measure the mass of the object, use an electronic balance. Make sure the balance reads zero before you place the object on it.

To find the volume of a regular solid, such as a cube or a cuboid, measure its dimensions using a millimetre ruler, vernier callipers, or a micrometer – whichever is the most appropriate. Use the measurements and the equation shown in Figure 2 to calculate its volume.

For a small irregular solid, lower it on a thread into a measuring cylinder partly filled with water. You can work out the volume of the object by the rise in the water level.

Measuring the density of a liquid

Use a measuring cylinder to measure the volume of a particular amount of the liquid.

Measure the mass of an empty beaker using a balance. Remove the beaker from the balance and pour the liquid from the measuring cylinder into the beaker. Use the balance again to measure the total mass of the beaker and the liquid. You can calculate the mass of the liquid by subtracting the mass of the empty beaker from the total mass of the beaker and the liquid.

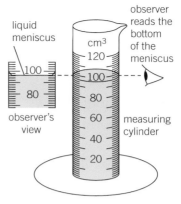

Figure 3 *Using a measuring cylinder*

Worked example

A measuring cylinder contained a volume of 120 cm^3 of a particular liquid. The liquid was then poured into an empty beaker of mass 51 g. The total mass of the beaker and the liquid was then found to be 145 g.

a Calculate the mass of the liquid in grams.

b Calculate the density of the liquid in kg/m^3.

Solution

a Mass of liquid = 145 − 51 = **94 g.**

b density = $\dfrac{\text{mass}}{\text{volume}}$ = $\dfrac{94\,\text{g}}{120\,\text{cm}^3}$ = $\dfrac{0.094\,\text{kg}}{0.000120\,\text{m}^3}$ = **780 kg/m^3**

1 A rectangular concrete slab is 0.80 m long, 0.60 m wide, and 0.05 m thick.
   a Calculate its volume in m^3. [1 mark]
   b The mass of the concrete slab is 60 kg. Calculate its density in kg/m^3. [2 marks]

2 A measuring cylinder contains 80 cm^3 of a particular liquid. The liquid is poured into an empty beaker of mass 48 g. The total mass of the beaker and the liquid was found to be 136 g.
   a Calculate the mass of the liquid in grams. [2 marks]
   b Calculate the density of the liquid in g/cm^3. [2 marks]

3 A rectangular block of gold is 0.10 m in length, 0.08 m in width, and 0.05 m in thickness.
   a i Calculate the volume of the block. [1 mark]
      ii The mass of the block is 0.76 kg. Calculate the density of gold. [2 marks]
   b A thin gold sheet has a length of 0.15 m and a width of 0.12 m. The mass of the sheet is 0.0015 kg. Use these measurements and the result of your density calculation in part a ii to calculate the thickness of the sheet. [3 marks]

4 Describe how you would measure the density of a metal bolt. You may assume the bolt will fit into a measuring cylinder of capacity 100 cm^3. [4 marks]

Density tests

For each of the tests, measure the mass and the volume of the object as explained. Then use the equation

density = $\dfrac{\text{mass}}{\text{volume}}$ to calculate the density of the object.

Safety: Take care not to spill any liquids and, if you do, let your teacher know.

Study tip

The instrument you choose to use to take a measurement is important – you should consider the resolution and range.

Instrument	resolution	range
metre rule mm scale	±0.5 mm	1 m
vernier callipers	±0.05 mm	about 100 mm
micrometer	±0.005 mm	about 30 mm

Key points

- density = $\dfrac{\text{mass}}{\text{volume}}$ (in kg/m^3)
- To measure the density of a solid object or a liquid, measure its mass and its volume, then use the density equation $\rho = \dfrac{m}{V}$.
- Rearranging the density equation gives $m = \rho V$ or $V = \dfrac{m}{\rho}$
- Objects that have a lower density than water (i.e., < 1000 kg/m^3) float in water.

P6.2 States of matter

Learning objectives

After this topic, you should know:

- the different properties of solids, liquids, and gases
- the arrangement of particles in a solid, a liquid, and a gas
- why gases are less dense than solids and liquids
- why the mass of a substance that changes state stays the same.

Everything around you is made up of matter and exists in one of three states – solid, liquid, or gas. The table below summarises the main differences between the three states of matter.

Table 1 *Differences between the three states of matter*

State	Flow	Shape	Volume	Density
solid	no	fixed	fixed	much higher than a gas
liquid	yes	fits container shape	fixed	much higher than a gas
gas	yes	fills container	can be changed	lower than a solid or a liquid

Change of state

A substance can change from one state to another, as shown in Figure 2. Changes of state are examples of **physical changes** because no new substances are produced. If a physical change is reversed, the substance recovers its original properties. You can change the state of a substance by heating or cooling the substance.

For example:

- when water in a kettle boils, the water turns into steam. Steam (also called water vapour) is water in its gaseous state
- when solid carbon dioxide (also called dry ice) warms up, the solid turns into gas directly
- when steam touches a cold surface, the steam condenses and turns into water.

Conservation of mass

When a substance changes state, the number of particles in the substance stays unchanged. So the mass of the substance after the change of state is the same as the mass of the substance before the change of state.

In other words, the mass of the substance is conserved when it changes its state.

For example:

- when a given mass of ice melts, the water it turns into has the same mass. So the mass of the substance stays unchanged.
- when water is boiled in a kettle and some of the water turns into steam, the mass of the steam produced is the same as the mass of water boiled away. So the mass of the substance is unchanged even though some of it (i.e., the steam) is no longer in the kettle.

Figure 1 *Spot the three states of matter*

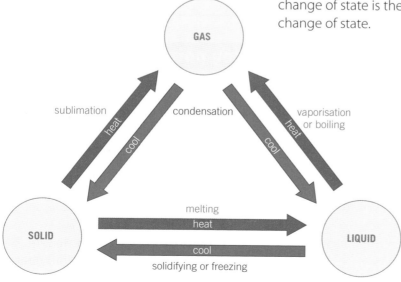

Figure 2 *Change of state*

The kinetic theory of matter

Solids, liquids, and gases are made of particles. Figure 4 shows the arrangement of the particles of a substance in its solid, liquid, and gas states. When the temperature of the substance is increased, the particles move faster.

- The particles of a substance in its solid state are held next to each other in fixed positions. They vibrate about their fixed positions, so the solid keeps its own shape.

- The particles of a substance in its liquid state are in contact with each other. They move about at random. So a liquid doesn't have its own shape, and it can flow.

- The particles of a substance in its gas state move about at random much faster than they do in a liquid. They are, on average, much further apart from each other than the particles of a liquid. So the density of a gas is much less than that of a solid or a liquid.

- The particles of a substance in its solid, liquid, and gas states have different amounts of energy. For a given amount of a substance, its particles have more energy in the gas state than they have in the liquid state, and they have more energy in the liquid state than they have in the solid state.

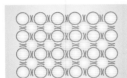   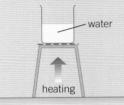

Figure 4 *The arrangement of particles of a substance in solid state, liquid state, and gas state*

1 Name the change of state that occurs when:
  a i wet clothing on a washing line dries out [1 mark]
    ii hailstones form [1 mark]
    iii snowflakes turn to liquid water. [1 mark]
  b When an ice cube in an empty beaker melts, the volume of water in the beaker just after the ice has melted is less than the volume of the ice cube. Explain what this tells you about the density of ice compared with the density of the water just after the ice has melted. [2 marks]

2 Give the scientific word for each of the following changes:
  a the windows in a bus full of people mist up [1 mark]
  b water vapour is produced from the surface of the water in a pan when the water is heated before it boils [1 mark]
  c ice cubes taken from a freezer thaw out [1 mark]
  d water put into a freezer gradually turns to ice. [1 mark]

3 Describe the changes that take place in the movement and arrangement of the particles in:
  a an ice cube when the ice melts [2 marks]
  b water vapour when it condenses on a cold surface. [3 marks]

4 Explain, using the kinetic theory of matter, why liquids and solids are much denser than gases. [4 marks]

Changing state

Heat some water in a beaker using a Bunsen burner (Figure 3).

- Water vapour leaves the surface of the water before the water boils.

- When the water boils, bubbles of vapour form inside the water and rise to the surface to release steam.

Switch the Bunsen burner off and hold a cold beaker or cold metal object above the boiling water. Observe the condensation of steam from the boiling water on the cold object.

water

heating

Figure 3 *Heating water to show a change of state*

Safety: Take care with boiling water, and wear eye protection.

Key points

- The particles of a *solid* are held next to each other in fixed positions. They are the least energetic of the states of matter.

- The particles of a *liquid* move about at random and are in contact with each other. They are more energetic than particles in a solid.

- The particles of a *gas* move about randomly and are far apart (so gases are much less dense than solids and liquids). They are the most energetic of the states of matter.

- When a substance changes state, its mass stays the same because the number of particles stays the same.

P6.3 Changes of state

Learning objectives

After this topic, you should know:

- what is meant by the melting point and the boiling point of a substance
- what is needed to melt a solid or to boil a liquid
- how to explain the difference between boiling and evaporation
- how to use a temperature–time graph to find the melting point or the boiling point of a substance.

Synoptic link

You will learn more about atmospheric pressure in Topic P11.3.

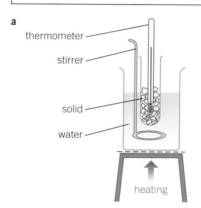

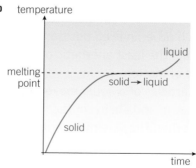

Figure 1a *Measuring the melting point of a substance* **b** *A temperature–time graph*

Study tip

Don't forget that during the time a pure substance is changing its state, its temperature does *not* change.

Melting points and boiling points

When pure ice is heated and melts, its temperature stays at 0 °C until all the ice has melted. When water is heated and boils at atmospheric pressure, its temperature stays at 100 °C.

For any pure substance undergoing a change of state, its temperature stays the same while the change of state is taking place. The temperature at which a solid changes to a liquid is called the **melting point**, and the temperature at which a liquid turns to a gas is called the **boiling point** of the substance. The temperature at which a liquid changes to a solid is called its **freezing point**. It is the same temperature as the melting point of the solid.

The melting point of a solid and the boiling point of a liquid are affected by impurities in the substance. For example, the melting point of water is lowered if you add salt to the water. This is why salt is added to the grit that's used for gritting roads in freezing weather – it means roads don't get icy until they are colder.

Table 1

Change of state	Initial and final state	Temperature
Melting	solid to liquid	melting point
Freezing (also called solidification)	liquid to solid	
Boiling	liquid to gas	boiling point
Condensation	gas to liquid	

Measuring the melting point of a substance

Place a substance in its solid state in a suitable test tube in a beaker of water (Figure 1a). Heat the water, and measure the temperature of the substance when it melts. If its temperature is measured every minute, you can plot the measurements on a graph (Figure 1b). The melting point is the temperature of the flat section of the graph because this is when the temperature stays the same during the time in which the substance is melting.

You can use the same arrangement without the beaker of water to find the boiling point of a liquid.

Safety: Wear eye protection.

Energy and change of state

Suppose a beaker of ice below 0 °C is heated steadily so that the ice melts and then the water boils. Figure 2 shows how the temperature changes with time.

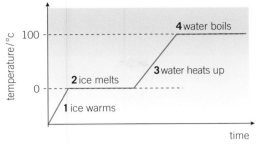

Figure 2 *Melting and boiling water*

The temperature of the water:

1 increases until it reaches 0 °C when the ice starts to melt at 0 °C, then

2 stays constant at 0 °C until all the ice has melted, then

3 increases from 0 °C to 100 °C until the water in the beaker starts to boil at 100 °C, then

4 stays constant at 100 °C as the water turns to steam (until all the water has boiled away, if the water continues to be heated).

The energy transferred to a substance when it changes its state is called **latent heat**. The energy transferred to the substance to melt or boil it is 'hidden' by the substance because its temperature does not change at the substance's melting point or at its boiling point.

Most pure substances produce a temperature–time graph with similar features to Figure 2. Note that:

● fusion is sometimes used to describe melting because different solids can be joined, or 'fused', together when they melt

● evaporation from a liquid happens at its surface when the liquid is below its boiling point. At its boiling point, a liquid boils because bubbles of vapour form inside the liquid and rise to the surface to release the gas.

1 Write three differences between evaporation and boiling. [3 marks]

2 A pure solid substance X was heated in a tube and its temperature was measured every 30 seconds.
The measurements are given in the table below.

Time in s	0	30	60	90	120	150	180	210	240	270	300
Temperature in °C	20	35	49	61	71	79	79	79	79	86	92

  a i Use the measurements in the table to plot a graph of temperature (y-axis) against time (x-axis). [3 marks]
    ii Use your graph to find the melting point of X. [1 mark]
  b Describe the physical state of the substance as it was heated from 60 °C to 90 °C. [3 marks]

3 Salt water has a lower freezing point than pure water. In icy conditions in winter, gritting lorries are used to scatter a mixture of salt and grit on roads. Explain the purpose of each of the two components of the mixture. [3 marks]

4 A substance has a melting point of 75 °C. Describe how the arrangement and motion of the particles changes as the substance cools from 80 °C to 70 °C. [4 marks]

P6.4 Internal energy

Learning objectives

After this topic, you should know:

- how increasing the temperature of a substance affects its internal energy
- how to explain the different properties of a solid, a liquid, and a gas
- how the energy of the particles of a substance changes when it is heated
- how to explain in terms of particles why a gas exerts pressure.

Synoptic link

For more about energy transfers, look back at Topic P1.1.

Synoptic link

For more about specific heat capacity, look back at Topic P2.4.

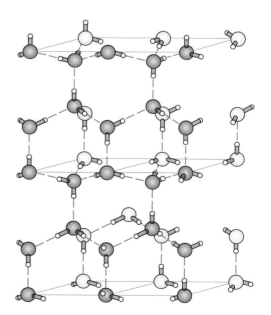

Figure 1 *Molecular model of ice*

When you switch a kettle on, the temperature of the water in the kettle increases until the water boils. The molecules in the water gain energy and move about faster as the temperature of the water increases. When the water boils, it means that the molecules have gained enough energy to move away from each other so that the water turns into vapour (steam).

The energy stored by the particles of a substance is called the substance's **internal energy**. This is the energy of the particles that is caused by their individual motion and positions. The internal energy of the particles is the sum of:

- the kinetic energy they have due to their individual motions relative to each other, and
- the potential energy they have due to their individual positions relative to each other.

So the internal energy of a substance is the total energy in the kinetic and potential energy stores of all the particles in the substance that is caused by their individual motions and positions.

Internal energy does *not* include gravitational potential energy or the kinetic energy that is caused by the motion of the whole substance.

Heating a substance changes the internal energy of the substance by increasing the energy of its particles. Because of this, the temperature of the substance increases or its physical state changes (i.e., it melts or boils).

- When the temperature of a substance increases (or decreases), the total kinetic energy of its particles increases (or decreases). For a given mass m of a substance of specific heat capacity c, the energy E needed to change its temperature by $\Delta\theta$ without a change of state is given by the specific heat equation $\Delta E = m\,c\,\Delta\theta$.
- When the physical state of a substance changes, the total potential energy of its particles changes. As you learnt in Topic P6.3, the term latent heat is used to describe the energy transferred to or from a substance when it changes state.

You'll learn more about latent heat in the next topic.

Comparing the particles in solids, liquids, and gases

In a solid, particles (i.e., atoms and molecules) are arranged in a three-dimensional structure.

- There are strong forces of attraction between these particles. These forces bond the particles in fixed positions.
- Each particle vibrates about an average position that is fixed.
- When a solid is heated, the particles' energy stores increase and they vibrate more. If the solid is heated up enough, the solid melts

(or sublimates) because its particles have gained enough energy to break away from the structure.

In a liquid, there are weaker forces of attraction between the particles than in a solid. These weak forces of attraction are not strong enough to hold the particles together in a rigid structure.

Figure 2 *Molecules in water*

- The forces of attraction are strong enough to stop the particles moving away from each other completely at the surface.

- When a liquid is heated, some of the particles gain enough energy to break away from the other particles. The molecules that escape from the liquid are in a gas state above the liquid.

In a gas, the forces of attraction between the particles are so weak, they are insignificant.

- The particles move about at high speed in random directions, colliding with each other and with the internal surface of their container. The pressure of a gas on a solid surface such as a container is caused by the force of impacts of the gas particles with the surface.

- When a gas is heated, its particles gain kinetic energy and on average move faster. This causes the pressure of the gas to increase because the particles collide with the container surface more often and with more force.

1 Explain the following statements in terms of particles.
   a A gas exerts a pressure on any surface it is in contact with. [3 marks]
   b Heating a solid makes it melt.   [3 marks]

2 Table 1 lists the properties of the molecules in four different substances. Write, with a reason, whether each substance is a solid, a liquid, or a gas, or doesn't exist.

   Table 1

	Distance between the molecules	Particle arrangement	Movement of the molecules	
a	close together	not fixed	move about	[1 mark]
b	far apart	not fixed	move about	[1 mark]
c	close together	fixed	vibrate	[1 mark]
d	far apart	fixed	vibrate	[1 mark]

3 Explain why the internal energy of a solid increases when it is heated at its melting point. [2 marks]

4 An ice cube at 0 °C is placed in a beaker of water to cool the water down. Describe the energy changes of the particles of the ice and the water that takes place. ⃰ [4 marks]

Synoptic link

For more about pressure, look back to Topic P11.1.

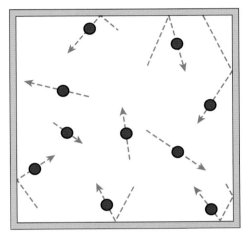

Figure 3 *Gas molecules in a box*

Study tip

Random means unpredictable or haphazard. The speed and direction of motion of a gas particle is unpredictable due to its random collisions with other particles.

Key points

- Increasing the temperature of a substance increases its internal energy.
- The strength of the forces of attraction between the particles of a substance explains why it is a solid, a liquid, or a gas.
- When a substance is heated:
  - if its temperature rises, the kinetic energy of its particles increases
  - if it melts or it boils, the potential energy of its particles increases.
- The pressure of a gas on a surface is caused by the particles of the gas repeatedly hitting the surface.

P6.5 Specific latent heat

Learning objectives

After this topic, you should know:

- what is meant by latent heat as a substance changes its state
- what is meant by specific latent heat of fusion and of vaporisation
- how to use specific latent heat in calculations
- how to measure the specific latent heat of ice and of water.

Study tip

Latent heat is the energy transferred when a substance changes its state.

Specific latent heat is the energy transferred per kilogram when a substance changes its state.

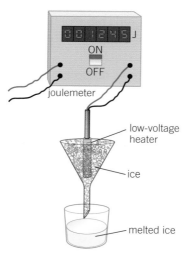

Figure 1 *Measuring the specific latent heat of fusion of ice*

Synoptic links

Instead of using a joulemeter, the energy supplied to the heater can be measured using the circuit and information in Topic P2.4.

Latent heat of fusion

When a solid substance is heated at its melting point, the substance melts and turns into liquid. Its temperature stays constant until all of the substance has melted. The energy supplied is called latent heat of fusion. It is the energy needed by the particles to break free from each other.

If the substance in its liquid state is cooled, it will solidify at the same temperature as its melting point. When this happens, the particles bond together into a rigid structure. Latent heat is transferred to the surroundings as the substance solidifies and the particles form stronger bonds.

The **specific latent heat of fusion** L_F of a substance is the energy needed to change the state of 1 kg of the substance from solid to liquid, at its melting point (i.e. without changing its temperature).

The unit of specific latent heat of fusion is the joule per kilogram (J/kg).

If energy E is transferred to a solid at its melting point, and mass m of the substance melts without change in temperature:

$$\text{specific latent heat of fusion, } L_F \text{ (J/kg)} = \frac{\text{energy, } E \text{ (joules, J)}}{\text{mass, } m \text{ (kilograms, kg)}}$$

You can rearrange this equation to $E = m L_F$

Specific latent heat of fusion of ice

In this experiment, a low-voltage heater is used to melt crushed ice in a funnel. The melted ice is collected using a beaker under the funnel (Figure 1). A joulemeter is used to measure the energy supplied to the heater.

1. With the heater off, water from the funnel is collected in the beaker for a measured time (e.g., 10 minutes). The mass of the beaker and water m_1 is then measured. The beaker is then emptied for the next stage.

2. With the heater on, the procedure is repeated for exactly the same time. The joulemeter readings before and after the heater is switched on are recorded. After the heater is switched off, the mass of the beaker and the water m_2 is measured once more.

To calculate the specific latent heat of fusion of ice, note that:

- the mass of ice melted because of the heater is $m = m_2 - m_1$
- the energy supplied E to the heater = the difference between the joulemeter readings
- the specific latent heat of fusion of ice is $L_F = \dfrac{E}{m} = \dfrac{E}{m_2 - m_1}$.

Safety: Take care with a hot immersion heater, and wear eye protection.

Latent heat of vaporisation

When a liquid substance is heated, at its boiling point, the substance boils and turns into vapour. The energy supplied is called latent heat of vaporisation. It is the energy needed by the particles to break away from their neighbouring particles in the liquid.

If the substance in its gas state is cooled, it will condense at the same temperature as its boiling point. Latent heat is transferred to the surroundings as the substance condenses into a liquid and its particles form new bonds.

The **specific latent heat of vaporisation** L_v of a substance is the energy needed to change the state of 1 kg of the substance from liquid to vapour, at its boiling point (i.e., without changing its temperature).

The unit of specific latent heat of vaporisation is the joule per kilogram (J/kg).

If energy E is transferred to a liquid at its boiling point, and mass m of the substance boils away without change in temperature:

$$\text{specific latent heat of vaporisation, } L_v = \frac{\text{energy, } E \text{ (joules, J)}}{\text{mass, } m \text{ (kilograms, kg)}}$$
(joules per kilogram, J/Kg)

1 In the experiment shown in Figure 1, 0.024 kg of water was collected in the beaker in 300 s with the heater turned off. The beaker was then emptied and placed under the funnel again. With the heater on for exactly 300 s, the joulemeter reading increased from zero to 15 000 J, and 0.068 kg of water was collected in the beaker.
   a Calculate the mass of ice melted because of the heater being on. [1 mark]
   b Use the data to calculate the specific latent heat of fusion of water. [2 marks]

2 In the experiment shown in Figure 2, the balance reading decreased from 0.152 kg to 0.144 kg in the time taken to supply 18 400 J of energy to the boiling water. Use the data to calculate the specific latent heat of vaporisation of water. [3 marks]

3 An ice cube of mass 0.008 kg at 0 °C was placed in water at 15 °C in an insulated plastic beaker. The mass of water in the beaker was 0.120 kg. After the ice cube had melted, the water was stirred, and its temperature was found to have fallen to 9 °C. The specific heat capacity of water is 4200 J/kg °C.
   a Calculate the energy transferred from the water. [2 marks]
   b Show that when the melted ice warmed from 0 °C to 9 °C, it gained 300 J of energy. [2 marks]
   c Use this data to calculate the specific latent heat of fusion of water. [3 marks]

4 Estimate how long a 3000 W electric kettle would take to boil away 100 g of water. The specific latent heat of vapourisation of water is 2.25 MJkg. [3 marks]

Specific latent heat of vaporisation of water

Use a low-voltage heater (Figure 2) to bring water in an insulated beaker to the boil. The joulemeter reading and the top pan balance reading are then measured and then remeasured after a certain time (e.g., 5 minutes).

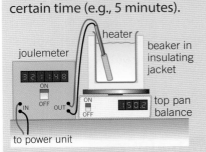

Figure 2 *Measuring the specific latent heat of vaporisation of water*

In this time:
● the energy supplied E = the difference between the joulemeter readings.
● the mass of water boiled away, m = the difference between the readings of the top pan balance.
● the specific latent heat of vaporisation of water is $L_v = \frac{E}{m}$.

Safety: This experiment will be demonstrated by your teacher. You should wear eye protection and stand behind a safety screen. Your teacher will need to take care with the hot immersion heater.

Key points

● Latent heat is the energy needed for a substance to change its state without changing its temperature.
● Specific latent heat of fusion (or of vaporisation) is the energy needed to melt (or to boil) 1 kg of a substance without changing its temperature.
● In latent heat calculations, use the equation $E = m L$.
● The specific latent heat of ice (or of water) can be measured using a low-voltage heater to melt the ice (or to boil the water).

P6.6 Gas pressure and temperature

Learning objectives

After this topic, you should know:

- how a gas exerts pressure on a surface
- how the pressure of a gas in a sealed container is affected by changing the temperature of the gas
- how to see evidence of gas molecules moving around at random.

Synoptic links

You will learn more about how pressure is measured in pascals (Pa) in Topic P11.1.

In the kitchen

Never heat food in a sealed can. The can will probably explode because the pressure of gas inside it increases as the temperature increases. This is because the molecules of gas in the can collide repeatedly with each other and with the surface inside their container, rebounding after each collision. Each impact with the surface exerts a tiny force on the surface. Millions of millions of these impacts happen every second, and together the total force causes a steady pressure on the surface inside the container. The pressure of a gas on a surface is the total force exerted on a unit area of the surface.

Increasing the temperature of any sealed gas container increases the pressure of the gas inside it. This is because:

- the energy transferred to the gas when it's heated increases the kinetic energy of its molecules. So the average kinetic energy of the gas molecules increases when the temperature of the gas is increased.

- the average speed of the molecules increases when the kinetic energy increases, and the molecules on average hit the container surfaces with more force and more often. So the pressure of the gas increases.

Gas pressure and temperature

Figure 1 shows dry air in a sealed flask connected to a pressure gauge. The flask is in a big beaker of water, which is heated to raise the temperature of the gas. The water is heated in stages to raise the temperature in stages. At each stage, the water is stirred to make sure that its temperature is the same throughout. The temperature of the water is measured using the thermometer. The pressure is read off the pressure gauge.

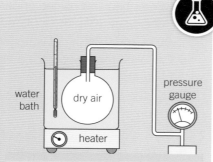

Figure 1 *Measuring gas pressure at different temperatures*

- If the measurements are plotted on a graph of pressure against temperature in °C, the results give you a straight-line graph as shown in Figure 2. This shows that the increase of pressure is the same for equal increases of temperature.

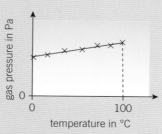

Figure 2 *Pressure–temperature graph for a gas*

Safety: This experiment should be carried out by your teacher and behind a safety screen. You should wear safety goggles.

Observing random motion

Individual molecules are too small for you to see directly. But you can see the effects of them by observing the motion of smoke particles in air. Figure 3 shows how you can do this using a smoke cell and a microscope. The smoke particles move about haphazardly and follow unpredictable paths.

1 A small glass cell is filled with smoke

2 Light is shone through the cell

3 The smoke is viewed through a microscope

4 You see the smoke particles constantly moving and changing direction. The path taken by one smoke particle will look something like this

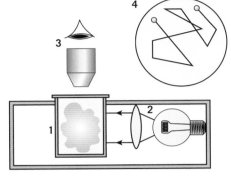

Figure 3 *A smoke cell*

Figure 4 shows how the random motion of smoke particles in air happens. Air molecules repeatedly collide at random with each smoke particle. The air molecules must be moving very fast to make this happen, because they are much too small to see, and the smoke particles are much, much bigger than the air molecules. What you see is the random motion of the smoke particles caused by the random impacts that the gas (air) molecules make on each smoke particle.

The smoke particle is much larger than the air molecules

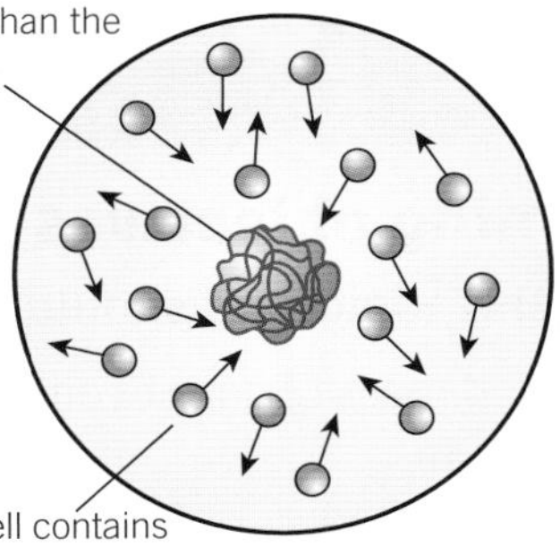

The glass cell contains air molecules that are in constant erratic motion. As they collide with the smoke particle they give it a push. The direction of the push changes at random

Figure 4 *The random motion of smoke particles*

Go further!

The random motion of tiny particles in a fluid is called **Brownian motion,** after the botanist Robert Brown who first observed it in 1785. He used a microscope to observe pollen grains floating on water. He was amazed to see that the pollen grains were constantly moving about and changing direction haphazardly as if they had a life of their own. Brown couldn't explain what he saw. Brownian motion puzzled scientists until the kinetic theory of matter provided an explanation.

1 When a gas is heated in a sealed container, write how, if at all, each of the following properties of the gas changes:
   a The pressure of the gas [1 mark]
   b The average separation of the molecules [1 mark]
   c The number of impacts the molecules make on the surface of the container each second. [1 mark]

2 Explain why smoke particles in air move about faster if the temperature of the air is increased. [3 marks]

3 A gas cylinder is fitted with a valve that opens and lets gas out if the gas becomes too hot. Explain how the gas pressure changes if the gas becomes too hot and the valve opens. 🖊 [3 marks]

4 Look back at the practical on the previous page.
   a Explain why the water must be stirred before its temperature is measured. [2 marks]
   b Explain why the pressure gauge does not read zero before the water is heated. [2 marks]

Key points

- The pressure of a gas is caused by the random impacts of gas molecules on surfaces that are in contact with the gas.
- If the temperature of a gas in a sealed container is increased, the pressure of the gas increases because:
  - the molecules move faster so they hit the surfaces with more force
  - the number of impacts per second of gas molecules on the surfaces of a sealed container increases, so the total force of the impacts increases.
- The unpredictable motion of smoke particles is evidence of the random motion of gas molecules.

P6.7 Gas pressure and volume

Learning objectives

After this topic, you should know:

For a fixed mass of gas at constant temperature:

- how pressure (or volume) changes affect the volume (or pressure) of the gas,
- why the pressure of a gas changes when its volume is changed at constant temperature
- when to use the equation $pV = \text{constant}$
- **H** why the temperature of a gas increases when it is compressed quickly enough.

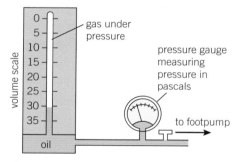

Figure 1 *Testing the variation of pressure and volume of a fixed mass of air*

Table 1

Pressure in kPa	Volume in cm³	Pressure × volume in kPa cm³
100	36	3600
120	30	3600
150	24	3600
180	20	3600

The volume of a fixed mass of gas depends on its pressure and on its temperature. A gas can be compressed or expanded by pressure changes. The pressure produces a net force at right angles to the wall of the container of the gas.

When gas is stored in a tube and a piston is pushed into the tube, the volume of the air in the tube decreases as the air is compressed. Because of this, the pressure of the air in the tube increases. As long as the compression happens slowly, the temperature of the air in the tube does not change.

Higher

To push the piston into the tube, work must be done (i.e., energy transferred) by applying a force to the piston. The applied force has to overcome the force that is caused by the pressure of the gas (air) enclosed in the tube. If the compression did *not* happen slowly, the work done on the gas would increase its internal energy store and its temperature. By compressing the gas slowly, the gas loses energy by heating its surroundings at the same rate as energy is transferred into it. So the internal energy store and the temperature of the gas do not change.

Investigating pressure and volume

Figure 1 shows a gas trapped by oil in an inverted glass tube. The pressure of the air in the tube is measured in pascals (Pa) using a pressure gauge. The volume of the air is measured in metres cubed (m³) using the vertical scale alongside the tube. A foot pump is used to increase the pressure of the gas. The tube has a thick wall so that it can withstand the high internal pressure of the gas.

The volume of the gas in the tube is measured at different pressures as the pressure is increased slowly from atmospheric pressure in stages. At each stage, the tap is closed so that the gas in the tube does not leak out, and the pressure and volume of the gas are measured when they have stopped changing. The measurements are recorded in Table 1. The measurements show that:

- the pressure increases as the volume decreases
- the pressure decreases as the volume increases.

Safety: This experiment should be carried out by your teacher and behind a safety screen. You should wear safety goggles.

Explanation of the variation of pressure with volume

For a fixed mass of gas, the number of gas molecules is constant. If the temperature is constant, the average speed of the molecules is constant.

If the volume of a fixed mass of gas at constant temperature is reduced, the gas pressure increases because:

- the space the molecules move in is smaller, so they don't travel as far between each impact with the surface of their container

- the molecules hit the surfaces more often, so the number of impacts per second increases. So the total force of the impacts per square metre of surface area (i.e., the pressure of the gas) increases.

Boyle's Law

The measurements in Table 1 show that the pressure × volume is constant. This is called Boyle's Law. So for a fixed mass of gas at constant temperature:

$$\underset{\text{(pressure, Pa)}}{\text{pressure, } p} \times \underset{\text{(metres cubed, m}^3\text{)}}{\text{volume, } V} = \text{constant}$$

Worked example

In a chemistry experiment, $0.000\,20\,\text{m}^3$ (= $200\,\text{cm}^3$) of gas was collected in a flask at a pressure of $125\,\text{kPa}$. Calculate the volume of this mass of gas at a pressure of $100\,\text{kPa}$ and the same temperature.

Solution

Let $p_1 = 125\,\text{kPa} = 125\,000\,\text{Pa}$ and $V_1 = 0.000\,20\,\text{m}^3$

$p_1 V_1 = 125\,000\,\text{Pa} \times 0.000\,20\,\text{m}^3 = 25\,\text{Pa}\,\text{m}^3$

Let $p_2 = 100\,\text{kPa} = 100\,000\,\text{Pa}$, where V_2 is the volume to be calculated.

Applying $p_2 V_2 = p_1 V_1$ therefore gives $100\,000\,\text{Pa} \times V_2 = 25\,\text{Pa}\,\text{m}^3$

So, $V_2 = \dfrac{25\,\text{pa}\,\text{m}^2}{100\,000\,\text{pa}} = 0.000\,25\,\text{m}^3 = 250\,\text{cm}^3$

Rearranging equations

The equation for Boyle's law is $pV =$ constant, and it can be rearranged to:

$$p = \frac{\text{constant}}{V} \quad \text{or} \quad V = \frac{\text{constant}}{p}$$

The rearranged equation $p = \dfrac{\text{constant}}{V}$ shows that the pressure is inversely proportional to the volume of the gas. For example, if the volume is doubled, the pressure is halved (Figure 2).

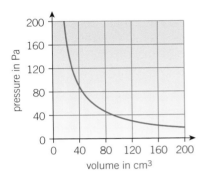

Figure 2 *The inversely proportional relationship between volume and pressure*

1 When a fixed mass of gas expands slowly at constant temperature, write how the volume and the pressure of the gas changes. [1 mark]

2 Calculate the unknown quantities in the table below. The values are for a gas of fixed mass and constant pressure.

	Initial pressure in Pa	Initial volume in m³	Final pressure in Pa	Final volume in m³	
a	100 000	0.000 20	50 000		[2 marks]
b	100 000	0.000 30		0.000 15	[2 marks]
c	120 000		100 000	0.000 60	[2 marks]
d		0.000 15	60 000	0.000 45	[2 marks]

3 A bicycle pump contains $20\,\text{cm}^3$ of air at a pressure of $100\,\text{kPa}$. The air is then pumped in a single stroke through a valve into a tyre of volume $100\,\text{cm}^3$, which contains air at the same pressure. Calculate the pressure of the air in the tyre after the stroke. Assume the tyre volume does not change. [3 marks]

4 Ⓗ A cylinder contains air trapped by a piston. Explain why the temperature of the air in the cylinder increases if the piston is used to compress the air suddenly. [4 marks]

Key points

- For a fixed mass of gas at constant temperature:
  - its pressure is increased if its volume is decreased
  - reducing the volume of a gas increases the number of molecular impacts per second on the surfaces that are in contact with the gas.
- Use the equation $pV =$ constant if the mass and the temperature of the gas do not change.
- Ⓗ The temperature of a gas can increase if it is compressed rapidly because work is done on it and the energy isn't transferred quickly enough to its surroundings.

P6 Molecules and matter

Summary questions

1 In a paint factory, empty steel tins of mass 0.320 kg and volume 0.001 m³ are filled with paint of density 2500 kg/m³.

   a Calculate the mass of paint in each filled paint tin. [2 marks]

   b Calculate the total weight in newtons of each filled paint tin. [2 marks]

2 This question is about A4 paper for use in a photocopier.

   a Use a millimetre ruler to measure the length and the width of a sheet of this paper. [1 mark]

   b The paper has a mass per unit area of 80 g/m². Calculate the mass of a single sheet of the paper. [2 marks]

   c A packet of 500 sheets of this paper has a thickness of 50 mm. Calculate the thickness of a single sheet of the paper. [1 mark]

   d Use your answers to **b** and **c** to calculate the density of the paper in **i** g/cm³, [3 marks]
     ii kg/m³ [1 mark]

3 A test tube containing a solid substance is heated in a beaker of water. The temperature–time graph shows how the temperature of the substance changed with time as it was heated.

Figure 1

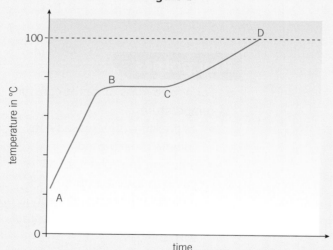

   a State why the temperature of the substance:

    i increased from **A** to **B** [1 mark]

    ii stayed the same from **B** to **C** [1 mark]

    iii increased from **C** to **D**. [1 mark]

   b Use the graph to estimate the melting point of the solid. [1 mark]

   c Describe how the arrangement and motion of the particles changes as the temperature increases from **A** to **D**. [3 marks]

4 a A plastic beaker containing 0.10 kg of water at 18 °C was placed in a refrigerator for 450 seconds. After this time, the temperature of the water was found to be 3 °C. The specific heat capacity of water is 4200 J/kg °C.

    i Calculate the energy transferred from the water. [2 marks]

    ii Calculate the rate of transfer of energy from the water. [2 marks]

   b **i** Calculate how much more energy would need to be removed from the water to cool it from 3 °C and to freeze it. The specific latent heat of fusion of water is 340 kJ/kg. [4 marks]

    ii Estimate how long it would take to cool it from 3 °C and freeze it. [2 marks]

5 A 3.0 kW electric kettle is fitted with a safety cut-out designed to switch it off as soon as the water boils. Unfortunately, the cut-out does not operate correctly and allows the water to boil for 30 seconds longer than it is supposed to.

   a Calculate how much energy is supplied to the kettle in this time. [2 marks]

   b The specific latent heat of vaporisation of water is 2.3 MJ/kg. Estimate the mass of water boiled away in this time. [2 marks]

6 In a chemistry experiment, 25 cm³ of a gas is collected in a syringe at a pressure of 120 kPa.

   a Calculate the volume of this amount of gas if its pressure was changed to 100 kPa without changing its temperature. [2 marks]

   b If the gas in **a** was then cooled without changing its volume, write and explain how its pressure would change. 🛇 [4 marks]

7 a Explain in terms of particles why gases have a much lower density than solids and liquids. 🛇 [3 marks]

   b Describe how the arrangement and motion of the particles of a substance change when the substance changes its state from liquid to solid. [4 marks]

Practice questions

01 A large statue stands on a square base at the side of a garden pond. Both the base and the statue are made of granite.

01.1 Describe how the volume of the base can be calculated without moving it. [1 mark]

01.2 The statue falls into the pond. A gardener states that the volume of the statue can be estimated by measuring the new water level. Explain if this statement is correct. [2 marks]

01.3 Calculate whether a mechanical hoist with a maximum lift force of 750 N can be used to lift the statue out of the pond. The volume of the statue is 0.027 m³ and the density of granite is 2800 kg/m³. [3 marks]

02 A teacher demonstrates heating some naphthalene in a fume cupboard. The temperature of the naphthalene is measured every 2 minutes.

Figure 1

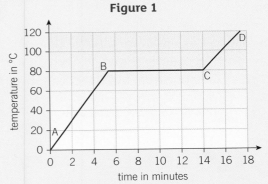

02.1 State the melting point of naphthalene. [1 mark]

02.2 Describe what is happening to the naphthalene between **A** to **B**, **B** to **C**, and **C** to **D**. [3 marks]

02.3 Suggest a reason why the teacher heated the naphthalene in a fume cupboard. [1 mark]

03.1 Particles in a solid are arranged in a regular fixed pattern. Draw diagrams to show the arrangement of particles in a liquid and a gas. [2 marks]

03.2 When a substance changes state, which property of the substance changes?
Choose the correct word from the box. [1 mark]

> mass   temperature   volume

03.3 Calculate the energy required to melt an ice cube that has a mass of 0.075 kg.
The specific latent heat of fusion of ice is 3.34 × 10⁵ J/kg. [3 marks]

03.4 A bar tender prefers to use artificial, non-melting ice cubes to cool drinks. Estimate the final temperature of 200 g of water when the energy transferred from the artificial ice cube is 3360 joules. The water is at a temperature of 20 °C Specific latent heat capacity of water is 4200 J/kg °C. [3 marks]

04 The apparatus in Figure 2 was used to measure the specific latent heat of vaporisation of water.

Figure 2

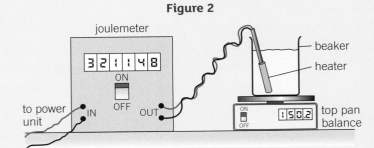

This is the method that was used.

1 Fill the beaker with water and switch on the heater.
2 When the water boils set the joulemeter to zero and take the reading on the top pan balance.
3 Allow the water to boil for 10 minutes.
4 Take the new readings on the joulemeter and top pan balance.
These are the results of the investigation.

Table 1

Time (mins)	Mass (g)	Energy (Joules)
0	184	0
10	168	37 800

04.1 Calculate the mass of water changed into steam in 10 minutes. [1 mark]

04.2 Calculate the latent heat of vaporisation of water using the results in **Table 1**.
Use the equation E = mL [3 marks]

04.3 Give **one** reason why the value calculated in **04.2** is greater than the actual value. [1 mark]

04.4 Suggest **two** ways the investigation could be improved. [2 marks]

Learning objectives

After this topic, you should know:

- what a radioactive substance is
- the types of radiation given out from a radioactive substance
- when a radioactive source emits radiation (radioactive decay)
- there are different types of radiation emitted by radioactive sources.

A key discovery

If your photos showed a mysterious image, what would you think? In 1896 a French physicist, Henri Becquerel, discovered the image of a key on a photographic film he developed. He remembered the film had been in a drawer under a key. On top of that there had been a packet of uranium salts. The uranium salts must have sent out some form of radiation that passed through paper (the film wrapper) but not through metal (the key).

Becquerel asked a young physicist, Marie Curie, to investigate. She found that the salts emitted radiation all the time. She used the word radioactivity to describe this strange new property of uranium.

She and her husband, Pierre, did more research into this new branch of science. They discovered new radioactive elements. They named one of the elements polonium, after Marie's native country, Poland.

Investigating radioactivity

You can use a Geiger counter to detect radioactivity. This is made up of a detector called a Geiger–Müller tube (or Geiger tube) connected to an electronic counter (Figure 2). The counter clicks each time a particle of radiation from a radioactive substance enters the Geiger tube.

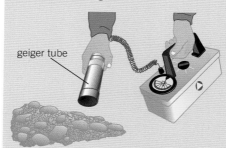

geiger tube

Figure 2 *Using a Geiger counter*

Safety: Avoid touching and inhaling radioactive material.

URANIUM SALTS

photographic plate

Figure 1 *Becquerel's key*

Inside the atom

What stops the radiation? The physicist Ernest Rutherford carried out tests to answer this question about a century ago. He put different materials between the radioactive substance and a detector.

He discovered two types of radiation:

- One type (**alpha radiation** α) was stopped by paper.
- The other type (**beta radiation** β) went through the paper.

Scientists later discovered a third type, **gamma radiation** γ, which is even more penetrating than beta radiation.

Rutherford carried out more investigations and discovered that alpha radiation is made up of positively charged particles. He realised that these particles could be used to probe the atom. His research students included Hans Geiger, who invented what was later called the Geiger counter. They carried out investigations in which a narrow beam of alpha particles was directed at a thin metal foil. Rutherford was astonished that some of the alpha particles rebounded from the foil. He proved that this happens because every atom has at its centre a positively charged nucleus containing most of the mass of the atom. He went on to propose that the nucleus contains two types of particle – protons and neutrons.

A radioactive puzzle

Why are some substances radioactive? Every atom has a nucleus made up of protons and neutrons. Electrons move about in energy levels (or shells) surrounding the nucleus.

Most atoms each have a stable nucleus that doesn't change. But the atoms of a radioactive substance each have a nucleus that is unstable. An unstable nucleus becomes stable or less unstable by emitting alpha, beta, or gamma radiation.

An unstable nucleus is described as decaying when it emits radiation. No one can tell exactly when an unstable nucleus will decay. It is a **random** event that happens without anything being done to the nucleus.

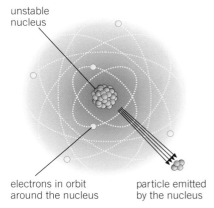

unstable nucleus

electrons in orbit around the nucleus

particle emitted by the nucleus

Figure 3 *Radioactive decay*

Synoptic link

For more about X-rays, look at Topic P13.5

1 **a** Write two differences between the radiation from uranium and the radiation from a lamp. [2 marks]
   b Write two differences between radioactive atoms compared with the atoms in a lamp filament. [2 marks]

2 **a** **i** The radiation from a radioactive source is stopped by paper. Name the type of radiation the source emits. [1 mark]
      ii The radiation from a different source goes through paper. Name the type of radiation this source emits. [1 mark]
   b Name the type of radiation from radioactive sources that is the most penetrating. [1 mark]

3 Explain why some substances are radioactive. [2 marks]

4 A Geiger counter clicks very rapidly when a certain substance is brought near it.
   a Describe the substance that made the Geiger counter click. [2 marks]
   b When the Geiger tube was near the substance, the counter clicked much less when a sheet of paper was placed between the substance and the tube. Explain why the counter clicked much less. [3 marks]

Key points

- A radioactive substance contains unstable nuclei that become stable by emitting radiation.
- There are three main types of radiation from radioactive substances – α, β, and γ.
- Radioactive decay is a random event – you can't predict or influence when it will happen.
- Radioactive sources emit α, β, and γ radiation.

P7.2 The discovery of the nucleus

Learning objectives

After this topic, you should know:

- how the nuclear model of the atom was established
- why the 'plum pudding' model of the atom was rejected
- what conclusions were made about the atom from experimental evidence
- why the nuclear model was accepted.

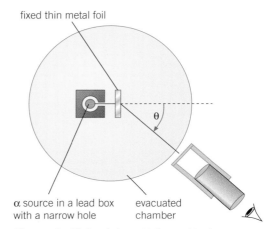

fixed thin metal foil

θ

α source in a lead box with a narrow hole

evacuated chamber

Figure 1 *Alpha (α) particle scattering*

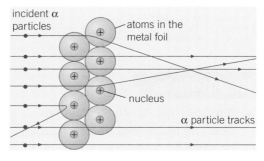

incident α particles

atoms in the metal foil

nucleus

α particle tracks

Figure 2 *Alpha (α) particle tracks*

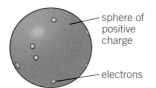

sphere of positive charge

electrons

Figure 3 *The plum pudding atom*

The physicist Ernest Rutherford discovered that alpha and beta radiation is made up of different types of particles. He realised that alpha (α) particles could be used to probe the atom. He asked two of his research workers, Hans Geiger and Ernest Marsden, to investigate how a thin metal foil scatters a beam of alpha particles. Figure 1 shows the arrangement they used.

The apparatus was in a vacuum chamber to prevent air molecules absorbing the alpha (α) particles. The detector consisted of a microscope focused on a small glass plate. Each time an alpha particle hit the plate, a spot of light was observed. The detector was moved to different positions. At each position, the number of spots of light observed in a certain time was counted.

Their results showed that:

- most of the alpha particles passed straight through the metal foil
- the number of alpha particles deflected per minute decreased as the angle of deflection increased
- about 1 in 10 000 alpha particles were deflected by more than 90°.

Rutherford was astonished by the results. He said it was like firing naval shells at tissue paper and discovering that a small number of the shells rebound. He knew that α particles are positively charged and that the radius of an atom is about 10^{-10} m. He deduced from the results that there is a positively charged nucleus at the centre of every atom that is:

- much smaller than the atom because most α particles pass through the atom without deflection
- where most of the mass of the atom is located.

Rutherford's nuclear model of the atom was quickly accepted because it:

- agreed exactly with the measurements Geiger and Marsden made in their experiments
- explained radioactivity in terms of changes that happen to an unstable nucleus when it emits radiation
- predicted the existence of the neutron, which was later discovered.

The plum pudding model

Before the nucleus was discovered in 1914, scientists didn't know what the structure of the atom was. They did know that atoms contain tiny negatively charged particles (which they called electrons).

Some scientists thought the atom was like a 'plum pudding' with positively charged matter evenly spread about (as in a pudding), and electrons buried inside (like plums in the pudding). But Rutherford's discovery meant that the plum pudding model of the atom was no longer accepted by scientists.

Bohr's model of the atom

After Rutherford's discovery, scientists knew that every atom has a positively charged nucleus that negatively charged electrons move around. The physicist Niels Bohr put forward the theory that the electrons in an atom orbit the nucleus at specific distances and specific energy values or energy levels. His model of the atom showed that the electrons in an orbit can move to another orbit by absorbing electromagnetic radiation to move away from the nucleus or by emitting electromagnetic radiation to move closer to the nucleus. His calculations based on his atomic model agreed with experimental observations of the light emitted by atoms.

A nuclear puzzle

More α-scattering experiments showed that:

● the hydrogen nucleus has the least amount of charge

● the charge of any nucleus is shared equally between a whole number of smaller particles, each with the same amount of positive charge.

The name proton was given to the hydrogen nucleus because scientists reckoned that every other nucleus contained hydrogen nuclei. But they also knew that the mass of every nucleus except for the hydrogen nucleus is bigger than the total mass of its protons. So there must be an uncharged type of particle with about the same mass as a proton in every nucleus except the hydrogen nucleus. They called this uncharged particle the neutron. The proton–neutron model of the nucleus explains all the mass and charge values of every nucleus. Direct experimental evidence for its existence was found by the physicist James Chadwick about 20 years after Rutherford's discovery of the nucleus.

1 Write four features of every nucleus of every atom. [4 marks]

2 a Figure 4 shows four possible paths, labelled A, B, C, and D, of an alpha particle deflected by a nucleus. Choose the path the alpha particle would travel along. [1 mark]

b Explain why each of the other paths in part **a** is not possible. [3 marks]

3 a i Write the conclusions that scientists made about the atom as a result of the discovery of electrons. [2 marks]

ii Describe two differences between the nuclear model of the atom and the plum pudding model. [2 marks]

b Explain why the alpha-scattering experiment led to the acceptance of the nuclear model of the atom and the rejection of the plum pudding model. [2 marks]

4 a Write one difference and one similarity between a proton and a neutron. 🖊 [2 marks]

b Explain why the mass of a helium nucleus is four times the mass of a hydrogen nucleus and its charge is only twice as much as the charge of a hydrogen nucleus. [3 marks]

Go further!

An atom emits electromagnetic radiation when an electron moving around the nucleus jumps from one energy level to a lower energy level. The radiation is emitted as a **photon**, which is a packet of waves emitted in a short burst. The energy of the emitted photon is equal to the energy change of the electron. Einstein put forward the photon theory and Bohr used it in his calculations.

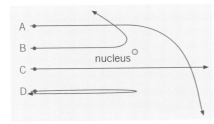

Figure 4 *See Summary question 2*

Key points

● Rutherford used α particles to probe inside atoms. He found that some of the α particles were scattered through large angles.

● The 'plum pudding' model could not explain why some α particles were scattered through large angles.

● An atom has a small positively charged central nucleus where most of the atom's mass is located.

● The nuclear model of the atom correctly explained why some α particles scattered through large angles.

P7.3 Changes in the nucleus

Learning objectives

After this topic, you should know:

- what an isotope is
- how the nucleus of an atom changes when it emits an alpha particle or a beta particle
- how to represent the emission of an alpha particle from a nucleus
- how to represent the emission of a beta particle from a nucleus.

In alpha (α) or beta (β) decay, the number of protons in a nucleus changes. In α decay, the number of neutrons also changes. How do the changes happen in α and β decay, and how can you represent these changes?

Table 1 gives the relative masses and the relative electric charges of a proton, a neutron, and an electron.

The **atomic number** (or proton number) of a nucleus is the number of protons in it. It has the symbol Z. Atoms of the same element each have the same number of protons.

The **mass number** of a nucleus is the number of protons plus neutrons in it. It has the symbol A.

Isotopes are atoms of the same element with different numbers of neutrons. The isotopes of an element have nuclei with the same number of protons but a different number of neutrons. Figure 1 shows how to represent an isotope of an element X, which has Z protons and A protons plus neutrons.

Radioactive decay

An unstable nucleus becomes more stable by emitting an α (alpha) or a β (beta) particle or by emitting a γ (gamma) ray.

α emission

An α particle is made up of two protons plus two neutrons. Its relative mass is 4, and its relative charge is +2. So it is usually represented by the symbol $_{2}^{4}\alpha$. It is identical to a helium nucleus, so in nuclear equations you might see it represented by the symbol $_{2}^{4}\text{He}$.

When an unstable nucleus emits an α particle:

- its atomic number goes down by 2, and its mass number goes down by 4
- the mass and the charge of the nucleus are both reduced.

For example, the thorium isotope $_{90}^{228}\text{Th}$ decays by emitting an α particle. So, it forms the radium isotope $_{88}^{224}\text{Ra}$.

Figure 2 shows an equation to represent this decay.

- The numbers along the top show that the total number of protons and neutrons after the change (= 224 + 4) is equal to the total number of neutrons and protons before the change (= 228).
- The numbers along the bottom show that the total number of protons after the change (= 88 + 2) is equal to the total number of protons before the change (= 90).

β emission

A β particle is an electron created and emitted by a nucleus that has too many neutrons compared with its protons. A neutron in the nucleus changes into a proton and a β particle (i.e. an electron), which is instantly emitted. The relative mass of a β particle is effectively zero, and its relative charge is −1. So a β particle can be represented by the symbol $_{-1}^{0}\beta$.

Table 1 *Relative mass and charge of subatomic particles*

	Relative mass	Relative charge
proton	1	+1
neutron	1	0
electron	$\sim\frac{1}{2000}$	−1

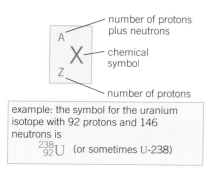

example: the symbol for the uranium isotope with 92 protons and 146 neutrons is
$_{92}^{238}\text{U}$ (or sometimes U-238)

Figure 1 *Representing an isotope*

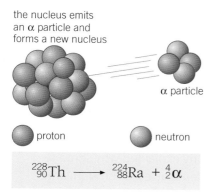

the nucleus emits an α particle and forms a new nucleus

α particle

proton   neutron

$_{90}^{228}\text{Th} \longrightarrow {}_{88}^{224}\text{Ra} + {}_{2}^{4}\alpha$

Figure 2 *α emission*

When an unstable nucleus emits a β particle:

- the atomic number of the nucleus goes up by 1, and its mass number is unchanged (because a neutron changes into a proton)
- the charge of the nucleus is increased, and the mass of the nucleus is unchanged.

For example, the potassium isotope $^{40}_{19}K$ decays by emitting a β particle. So it forms a nucleus of the calcium isotope $^{40}_{20}Ca$. Figure 3 shows an equation to represent this decay.

- The numbers along the top show that the total number of protons and neutrons after the change (= 40 + 0) is equal to the total number of neutrons and protons before the change (= 40).
- The numbers along the bottom show that the total charge (in relative units) after the change (= 20 − 1) is equal to the total charge before the change (= 19).

γ emission

A γ-ray is electromagnetic radiation from the nucleus of an atom. It is uncharged and has no mass. So its emission does not change the number of protons or neutrons in a nucleus. So the mass and the charge of the nucleus are both unchanged.

Neutron emission

Neutrons are emitted by some radioactive substances as a result of α particles colliding with unstable nuclei in the substance. Such a collision causes the unstable nuclei to become even more unstable and emit a neutron. Because the emitted neutrons are uncharged, they can pass through substances more easily than an α particle or a β particle can.

1 How many protons and how many neutrons are there in the nucleus of each of the following isotopes:
   a $^{12}_{6}C$ [1 mark]   b $^{60}_{27}Co$ [1 mark]   c $^{235}_{92}U$? [1 mark]
   d How many more protons and how many more neutrons are in $^{238}_{92}U$ compared with $^{224}_{88}Ra$? [2 marks]

2 A substance contains the radioactive isotope $^{238}_{92}U$, which emits alpha radiation. The product nucleus X emits beta radiation and forms a nucleus Y. Determine how many protons and how many neutrons are present in:
   a a nucleus of $^{238}_{92}U$ [1 mark]   b a nucleus of X [2 marks]
   c a nucleus of Y [2 marks]

3 Copy and complete the following equations for α and β decay.
   a $^{238}_{92}U \rightarrow ^{?}_{?}Th + ^{4}_{2}\alpha$ b $^{64}_{29}Cu \rightarrow ^{?}_{?}Zn + ^{0}_{-1}\beta$ [4 marks]

4 A radioactive isotope of polonium (Po) has 84 protons and 126 neutrons. The isotope is formed from the decay of a radioactive isotope of bismuth, which emits a β particle in the process. Copy and complete the equation below to represent this decay.

$Bi \rightarrow Po + \beta$ [3 marks]

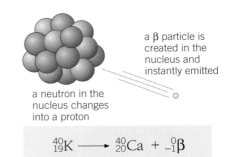

$$^{40}_{19}K \longrightarrow ^{40}_{20}Ca + ^{0}_{-1}\beta$$

Figure 3 β emission

Go further!

Most nuclei are stable because the protons and neutrons inside a nucleus are held together by a strong attractive force called the strong nuclear force. This force is strong enough in stable nuclei to overcome the electrostatic repulsion between protons, and to stop the neutrons moving away from the nucleus.

Study tip

Make sure you know the changes to mass number and to atomic number in alpha decay and in beta decay.

Key points

- Isotopes of an element are atoms with the same number of protons but different numbers of neutrons. So they have the same atomic number but different mass numbers.

α decay	β decay
Change in the nucleus	
Nucleus loses 2 protons and 2 neutrons	A neutron in the nucleus changes into a proton
Particle emitted	
2 protons and 2 neutrons emitted as an α particle	An electron is created in the nucleus and instantly emitted
Equation	
$^{A}_{Z}X \rightarrow ^{A-4}_{Z-2}Y + ^{4}_{2}\alpha$	$^{A}_{Z}X \rightarrow ^{A}_{Z+1}Y + ^{0}_{-1}\beta$

P7.4 More about alpha, beta, and gamma radiation

Learning objectives

After this topic, you should know:

- how far each type of radiation can travel in air
- how different materials absorb alpha, beta, and gamma radiation
- the ionising power of alpha, beta, and gamma radiation
- why alpha, beta, and gamma radiation is dangerous.

Table 1 *The results of the two tests*

Type of radiation	Absorber materials	Range in air
alpha α	Thin sheet of paper	about 5 cm
beta β	Aluminium sheet (about 5 mm thick) Lead sheet (2–3 mm thick)	about 1 m
gamma γ	Thick lead sheet (several cm thick) Concrete (more than 1 m thick)	unlimited – spreads out in air without being absorbed

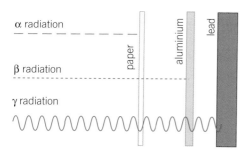

Figure 2 *The penetrating power of alpha, beta, and gamma radiation*

Penetrating power

Alpha radiation can't penetrate paper. But what stops beta and gamma radiation? And how far can each type of radiation travel through air? You can use a Geiger counter to find out, but you must take account of background radiation, which is radiation from unstable nuclei in materials around us and in the atmosphere. To do this you need to:

1. measure the count rate (which is the number of counts per second) without the radioactive source present. This is the background count rate.

2. measure the count rate with the source in place. Subtracting the background count rate from this gives you the count rate from the source alone.

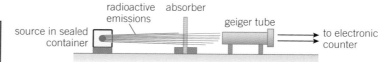

Figure 1 *Absorption tests*

You can then test absorber materials and the range that each type of radiation travels in air.

To test different materials, you need to place each material between the tube and the radioactive source (Figure 1). Then you measure the count rate. You can add more layers of material until the count rate from the source is zero. The radiation from the source has then been stopped by the absorber material.

To test the range that each type of radiation travels in air, you need to move the tube away from the source. When the tube is beyond the range of the radiation, the count rate from the source is zero.

Radioactivity dangers

The radiation from a radioactive substance can knock electrons out of atoms. The atoms become charged because they lose electrons. The process is called **ionisation**. When an object is exposed to ionising radiation, it is said to be **irradiated**, but it does not become radioactive.

Radioactive substances can contaminate other materials that they come into contact with. Radioactive contamination is the unwanted presence of materials containing radioactive atoms on other materials. The hazard from contamination is due to the decay of the nuclei of the contaminating atoms. The type of radiation emitted affects the level of hazard.

X-rays, fast-moving protons, and fast-moving neutrons also cause ionisation. Ionisation in a living cell can damage or kill the cell. Damage to the genes in a cell can be passed on if the cell generates more cells. Strict safety rules must always be followed when radioactive substances are used.

Alpha radiation is more dangerous in the body than beta or gamma radiation. This is because the ionising power of alpha radiation is much greater than the ionising power of beta or gamma radiation.

Workers who use ionising radiation reduce their exposure by:

● keeping as far away as possible from the source of radiation (e.g., by using special handling tools with long handles)

● spending as little time as possible in at-risk areas

● shielding themselves by staying behind thick concrete barriers and/or using thick lead plates.

Peer review

Scientists have studied the effects of radiation on humans, including the survivors of the atom bombs dropped on Japan in 1945. Their findings are published and shared with other scientists so that the findings can be checked by them. This process is called peer review.

Radiation in use

When a radioactive substance is used, the substance must emit the appropriate type of radiation for that use.

Smoke alarms contain a radioactive isotope that sends out alpha particles into a gap in a circuit in the alarm. The alpha particles ionise the air in the gap so there is a current across the gap. In a fire, smoke absorbs the ions so the current across the gap drops and the alarm sounds. Beta or gamma radiation could not be used because they do not create enough ions to make the air in the gap conduct electricity.

Automatic thickness monitoring in metal foil production uses a radioactive source that sends out β radiation (Figure 3). The amount of β radiation passing through the foil depends on the thickness of the foil. The detector measures the amount of radiation passing through the foil. If the foil is too thick, the detector reading drops and the detector sends a signal to increase the pressure of the rollers on the metal sheet. This makes the foil thinner again. Gamma radiation isn't used because it would all pass through the foil unaffected. Alpha radiation isn't used as it would all be stopped by the foil.

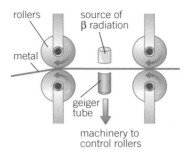

Figure 3 *Thickness monitoring using a radioactive source*

Key points

● α radiation is stopped by paper and has a range of a few centimetres in air. It consists of particles, each composed of two protons and two neutrons. It has the greatest ionising power.

● β radiation is stopped by a thin sheet of metal and has a range of about one metre in air. It consists of fast-moving electrons emitted from the nucleus. It is less ionising than alpha radiation and more ionising than gamma radiation.

● γ radiation is stopped by thick lead and has an unlimited range in air. It consists of electromagnetic radiation.

● Alpha, beta, and gamma radiation ionise substances they pass through. Ionisation in a living cell can damage or kill the cell.

---

1 a State why a radioactive source is stored in a lead-lined box. [1 mark]

b Name the type of ionising radiation from radioactive substances that is most easily absorbed. [1 mark]

c Name the type or types of radiation from a radioactive source that are stopped by a thick aluminium plate. [1 mark]

2 a Name the type of radiation from a radioactive source that is:

i uncharged [1 mark]

ii positively charged [1 mark]

iii negatively charged. [1 mark]

b Name the type of radiation from a radioactive source that:

i has the longest range in air [1 mark]

ii has the greatest ionising power. [1 mark]

3 a Explain why ionising radiation is dangerous. [2 marks]

b Explain how you would use a Geiger counter to find the range of the radiation from a source of α radiation. [3 marks]

4 Explain why γ radiation is not suitable for monitoring the thickness of metal foil. [2 marks]

P7.5 Activity and half-life

Learning objectives

After this topic, you should know:

- what is meant by the half-life of a radioactive source
- what is meant by the count rate from a radioactive source
- what happens to the count rate from a radioactive isotope as it decays
- how to calculate count rates after a given number of half-lives. **H**

Every atom of an element always has the same number of protons in its nucleus. But the number of neutrons in the nucleus can differ. An atom of a specific element with a certain number of neutrons is called an isotope of that element.

The **activity** of a radioactive source is the number of unstable atoms in the source that decay per second. The unit of activity is the Becquerel (Bq), which is 1 decay per second. As the nucleus of each unstable atom (the parent atom) decays, the number of parent atoms decreases. So the activity of the sample decreases.

You can use a Geiger counter to monitor the activity of a radioactive sample. To do this, you need to measure the **count rate** from the sample. The count rate is the number of counts per second. This is proportional to the activity of the source, as long as the distance between the tube and the source stays the same. The graph in Figure 1 shows that the count rate of a sample decreases with time.

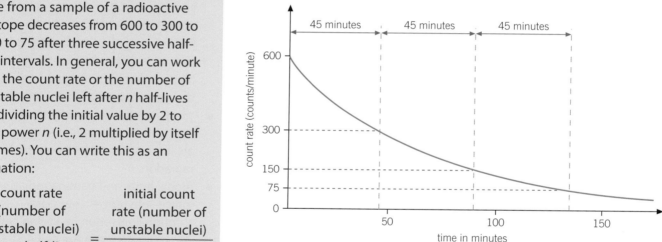

Figure 1 *A graph of count rate against time. The count rate here is measured in counts per minute*

The average time taken for the count rate (and so the number of parent atoms) to fall by half is always the same. This time is called the **half-life**. The half-life shown on the graph is 45 minutes.

The half-life of a radioactive isotope is the average time it takes:

- for the number of nuclei of the isotope in a sample (and so the mass of parent atoms) to halve
- for the count rate from the isotope in a sample to fall to half its initial value.

Half-life calculations

Figure 1 shows that the count rate from a sample of a radioactive isotope decreases from 600 to 300 to 150 to 75 after three successive half-life intervals. In general, you can work out the count rate or the number of unstable nuclei left after n half-lives by dividing the initial value by 2 to the power n (i.e., 2 multiplied by itself n times). You can write this as an equation:

$$\text{count rate (number of unstable nuclei) after } n \text{ half-lives} = \frac{\text{initial count rate (number of unstable nuclei)}}{2^n}$$

Worked example

A particular radioactive isotope has a half-life of 6.0 hours. A sample of this isotope contains 60 000 radioactive nuclei. Calculate the number of radioactive nuclei of this isotope remaining after 24 hours.

Solution

$n = 4$ because 24 hours equals 4 half-lives for this isotope.

$2^4 = 16$, so the number of radioactive nuclei of the isotope remaining after 24 hours = $60\,000 \div 2^4 = 60\,000 \div 16 = \textbf{3750}$

The random nature of radioactive decay

Radioactive decay is a random process. This means that no one can predict exactly *when* an individual atom will suddenly decay. But you *can* predict how many atoms will decay in a given time – because there are so many of them. This is a bit like throwing dice. You can't predict what number you will get with a single throw. But if you threw 1000 dice, you would expect one-sixth to come up with a particular number.

Suppose you start with 1000 unstable atoms. Look at the graph in Figure 2.

If 10% decay every hour:

- 100 atoms will decay in the first hour, leaving 900 atoms
- 90 atoms (= 10% of 900) will decay in the second hour, leaving 810 atoms.

Table 1 shows what you get if you continue the above calculations. The results are plotted as a graph in Figure 2. The graph is like Figure 1, except the half life is just over 6 hours. The similarity is because radioactive decay, like throwing dice, is a random process.

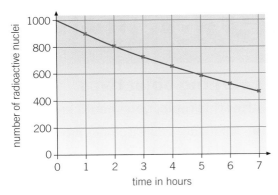

Figure 2 *Half-life*

Table 1 *What you get if you continue the calculations. The results are plotted as a graph in Figure 2*

Time from start (in hours)	No. of unstable atoms present	No. of unstable atoms that decay in the next hour
0	1000	100
1	900	90
2	810	81
3	729	73
4	656	66
5	590	59
6	531	53
7	478	48

1 a Define the half-life of a radioactive isotope. [1 mark]
  b Determine what the count rate in Figure 1 will be after 75 minutes from the start. [1 mark]

2 A radioactive isotope has a half-life of 15 hours. A sealed tube contains 8 milligrams of this isotope.
  a Calculate what mass of the isotope is in the tube:
    i 15 hours later [1 mark]
    ii 45 hours later. [1 mark]
  b Estimate how long it would take for the mass of the isotope to decrease to less than 5% of the initial mass. [3 marks]

3 a Ⓗ A sample of a radioactive isotope contains 320 million atoms of the isotope. [1 mark]
    i Calculate how many atoms of the isotope are present after one half-life. [1 mark]
    ii Calculate the ratio of the number of atoms of the isotope left after five half lives to the initial number of atoms. [1 mark]
    iii Calculate the number of atoms of the isotope left after five half-lives. [2 marks]
  b Estimate how long it would take for the count rate in Figure 1 to decrease to less than 40 counts per minute. [2 marks]

4 A sample of old wood was carbon dated and found to have 25% of the count rate measured in an equal mass of living wood. The half-life of the radioactive carbon is 5600 years. Calculate the age of the sample of wood. [2 marks]

Key points

- The half-life of a radioactive isotope is the average time it takes for the number of nuclei of the isotope in a sample to halve.
- The count rate of a Geiger counter caused by a radioactive source decreases as the activity of the source decreases.
- The number of atoms of a radioactive isotope and the count rate both decrease by half every half-life.
- Ⓗ The count rate after n half-lives = the initial count rate $\div 2^n$.

P7.6 Nuclear radiation in medicine

Learning objectives

After this topic, you should know:

- what radioactive isotopes are used for in medicine
- how to choose a radioactive isotope for a particular job
- what type of nuclear radiation can be used for medical imaging
- how to use radioactivity to destroy cancer cells.

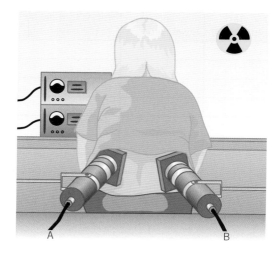

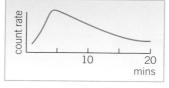

a chart recorder A

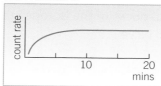

b chart recorder B

Figure 1 *Using a tracer to monitor a patient's kidneys*

Radioactivity has lots of uses. Nuclear radiation is used in medicine to help doctors diagnose internal disorders in patients and to treat disorders to make patients well again. For each use, a radioactive isotope is needed that emits a specific type of radiation and has a suitable half-life.

The examples below describe some of the ways nuclear radiation is used in medicine.

1 Radioactive tracers are used to trace the flow of a substance through an organ. The tracer contains a radioactive isotope that emits gamma radiation as it can be detected outside the system. For example, doctors use radioactive iodine to find out if a patient's kidney is blocked.

Before the test, the patient drinks water containing a tiny amount of the radioactive substance. A detector is then placed against each kidney. Each detector is connected to a chart recorder.

- The radioactive substance flows in and out of a normal kidney. So the detector reading goes up then down (Figure 1a).
- For a blocked kidney, the reading goes up and stays up. This is because the radioactive substance goes into the kidney but doesn't flow out again (Figure 1b).

Radioactive iodine is used for this test because:

- Its half-life is eight days, so it lasts long enough for the test to be done, but decays almost completely after a few weeks.
- It emits gamma radiation, so it can be detected outside the body.
- It decays into a stable product.

2 Gamma cameras are used to take images of internal body organs. Before an image is taken, the patient is injected with a solution that contains a gamma-emitting radioactive isotope. The solution is then absorbed by the organ, and a nearby gamma camera detects the gamma radiation emitted by the solution. The gamma rays pass through the holes in the thick lead grid in front of the detector. The detector only detects gamma rays from nuclei directly in front of it. The detector signals are used to build up an image of where the radioactive isotope is located in the organ (Figure 2).

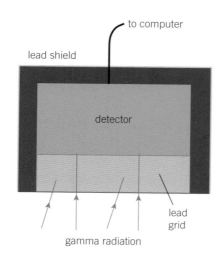

Figure 2 *The gamma camera*

The radioactive isotope must be a gamma emitter with a half-life long enough to give a useful image, but short enough so that its nuclei have mostly decayed after the image has been taken.

3   Gamma radiation in a narrow beam is used to destroy cancerous tumours. The radiation is emitted from a radioactive isotope of cobalt. It has a half-life of five years. Gamma radiation is used because it can penetrate deeper into the body than beta radiation and alpha radiation.

4   Radioactive implants are used to destroy cancer cells in some tumours. Beta- or gamma-emitting isotopes are used in the form of small seeds or tiny rods. Permanent implants use isotopes with half-lives long enough to irradiate the tumour over a given time, but short enough so that most of the unstable nuclei will have decayed soon afterwards.

Reducing risk

Everybody is exposed to background radiation, which is ionising radiation from radioactive substances in the environment such as radon gas or other sources. Figure 3 shows the percentage contributions of the different origins of background radiation.

The risk to the general public is very small but workers who use ionising radiation need to reduce their exposure to the radiation by following certain rules (Topic P7.4). Each worker must also wear a personal radiation monitor, such as a film badge, while he or she is in at-risk areas (Topic P13.4).

Synoptic link

Medical imaging systems use different types of radiation to see inside the body. For more about ultrasound imaging, look back at Topic P12.6. For more about X-ray imaging, look back at Topic P13.5. See Topic P13.4 for more about the use of gamma radiation to destroy tumours.

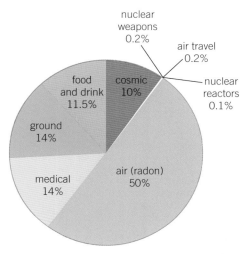

Figure 3 *The origins of background radiation*

Study tip

Make sure you can decide on the appropriate radioactive isotope to use for a particular task.

1   Radiation from radioactive sources is used for different purposes. Identify the type of radiation you would use and give a reason for your choice:
    a obtaining an image of an internal organ   [2 marks]
    b finding out whether a kidney in a patient is blocked.   [3 marks]

2   Give two sources of background radiation that originate in your environment.   [1 mark]

3   **a** Describe how nuclear radiation is used to destroy a tumour using a radioactive implant. 🖉   [4 marks]
    b State one type of radiation emitted by a radioactive implant.   [1 mark]

4   **a** Explain why a radioactive isotope used in a kidney scan should have a half-life that is not too short and not too long.   [2 marks]
    b Evaluate whether this consideration is important for a temporary radioactive implant.   [2 marks]

5   **a** Write the ideal properties of a radioactive isotope used as a medical tracer.   [3 marks]
    b When a radioactive tracer is used, explain why it is best to use a radioactive isotope that decays into a stable isotope.   [3 marks]

Key points

- Radioactive isotopes are used in medicine for medical imaging, treatment of cancer, and as tracers to monitor organs.
- How useful a radioactive isotope is depends on:
  - its half-life
  - the type of radiation it gives out.
- For medical imaging with a radioactive isotope and for medical tracers, the half-life should be not too short and not too long.
- A gamma beam or a radioactive implant can destroy cancer cells in a tumour.

P7.7 Nuclear fission

Learning objectives

After this topic, you should know:

- what nuclear fission is
- the difference between spontaneous fission and induced fission
- what a chain reaction is
- how a chain reaction in a nuclear reactor is controlled.

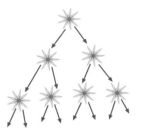

In a chain reaction, each reaction causes more reactions which cause more reactions, etc.

Figure 1 A chain reaction

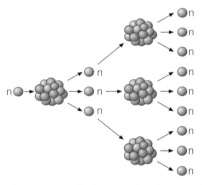

Figure 2 A chain reaction in a nuclear reactor

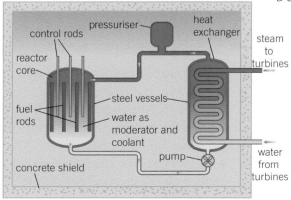

Figure 3 A nuclear reactor

Energy is released in a nuclear reactor because of **nuclear fission**. In induced fission, the nucleus of an atom of a fissionable substance is struck by a neutron, causing the nucleus to split into two smaller fragment nuclei of roughly equal size and to release several neutrons. Fission can also happen very rarely *without* a neutron being absorbed. This process is called spontaneous fission. The nucleus splits and several neutrons are released, just as in induced fission.

Fission neutrons

When a nucleus undergoes fission, it releases:

- two or three neutrons (called fission neutrons) at high speeds
- energy, in the form of gamma radiation, plus the kinetic energy stored in the fission neutrons and the *fragment* nuclei.

Chain reactions

A fission event can release several neutrons, which can cause other fissionable nuclei to split. This then produces a **chain reaction** of fission events. The neutrons released from each fission event can cause more fission events, to maintain the chain reaction.

In a nuclear fission reactor, on average, exactly one fission neutron from each fission event goes on to produce more fission. This makes sure that energy is released at a steady rate in the reactor.

Fissionable isotopes

The fuel in a nuclear reactor must contain fissionable isotopes.

Most reactors today are designed to use enriched uranium as the fuel. This is mostly made up of the non-fissionable uranium isotope $^{238}_{92}U$ (U-238) and about 2–3% of the uranium isotope $^{235}_{92}U$ (U-235), which *is* fissionable. In comparison, natural uranium is more than 99% U-238.

The U-238 nuclei in a nuclear reactor do not undergo fission, but they do change into other heavy nuclei, including plutonium-239. The isotope $^{239}_{94}Pu$ is fissionable. It can be used in a different type of reactor, but not in a uranium-235 reactor, which is the most common type of reactor.

Inside a nuclear reactor

A nuclear reactor has uranium fuel rods spaced out evenly in the reactor core. Figure 3 shows a cross-section of a pressurised nuclear reactor.

The reactor core contains the fuel rods, control rods, and water at high pressure. The fission neutrons are slowed down by collisions with the atoms in the water molecules. This is needed because fast neutrons don't cause further fission of U-235. The water is said to act as a **moderator** because it slows down the fission neutrons.

Control rods in the core absorb surplus neutrons. This keeps the chain reaction under control. The depth of the rods in the core is adjusted to maintain a steady chain reaction.

The fuel rods become very hot and the water acts as a coolant. Its molecules' kinetic energy stores increase as energy is transferred from the neutrons and fuel rods. The water is pumped through the core. Then it goes through sealed pipes to and from a heat exchanger outside the core. The water transfers energy from the core to the heat exchanger.

The **reactor core** is in a vessel made of thick steel to withstand the very high temperature and pressure in the core. The vessel is enclosed by thick concrete walls. These absorb ionising radiation that escapes through the walls of the steel vessel.

1 Natural uranium consists mainly of uranium-238. Uranium fuel is produced from natural uranium by increasing the proportion of uranium-235 in it.
   a Describe what happens to a uranium-235 nucleus when a neutron collides with it and causes it to undergo fission. [2 marks]
   b Describe what happens to a uranium-238 nucleus when a neutron collides with it. [2 marks]

2 a Put statements **A** to **D** in the list below into the correct sequence, starting with **B**, to describe a steady chain reaction in a nuclear reactor.
   A a U-235 nucleus splits   **C** neutrons are released
   B a neutron hits a U-235 nucleus   **D** energy is released [1 mark]

3 a In a nuclear reactor, describe the purpose of the control rods. [2 marks]
   b If the control rods in a nuclear reactor are pushed further into the reactor core, write and explain what would happen to the number of fission neutrons in the reactor. [2 marks]
   c Explain why the core of a nuclear reactor is in a container made of thick steel surrounded by thick concrete walls. [3 marks]

4 Look at the chain reaction shown in Figure 4.

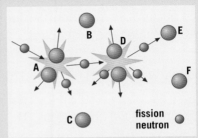

Figure 4

   a Write which of the nuclei **A** to **F** have been hit by a neutron. [1 mark]
   b Describe what has happened to these nuclei. [1 mark]
   c Write which two of the other nuclei **B** to **F** could undergo fission from a fission neutron shown in the figure. [1 mark]

Study tip

'Fission' means splitting. Do not confuse nuclear fission and nuclear fusion. (You will learn about nuclear fusion in Topic P7.8)

Key points

- Nuclear fission is the splitting of an atom's nucleus into two smaller nuclei and the release of two or three neutrons and energy.
- Induced fission occurs when a neutron is absorbed by a uranium-235 nucleus or a plutonium-239 nucleus and the nucleus splits. Spontaneous fission occurs without a neutron being absorbed.
- A chain reaction occurs in a nuclear reactor when each fission event causes further fission events.
- In a nuclear reactor, control rods absorb fission neutrons to ensure that, on average, only one neutron per fission goes on to produce further fission.

P7.8 Nuclear fusion

Learning objectives

After this topic, you should know:

- what nuclear fusion is
- how nuclei can be made to fuse together
- where the Sun's energy comes from
- why it is difficult to make a nuclear fusion reactor.

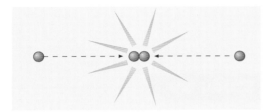

Figure 1 *A nuclear fusion reaction*

Figure 3 *The effects of a hydrogen bomb. A hydrogen bomb is a uranium bomb surrounded by the hydrogen isotope, $_1^2$H. When the uranium bomb explodes, it makes the surrounding hydrogen fuse together and release even more energy*

Imagine if you could get energy from water. Stars release energy by fusing together small nuclei such as hydrogen to form larger nuclei. Water contains lots of hydrogen atoms. A glass of water could provide the same amount of energy as a tanker full of petrol. But only if a fusion reactor could be made here on Earth.

Fusion reactions

Two small nuclei release energy when they are fused together to form a single larger nucleus. This process is called **nuclear fusion**. Some of the mass of the small nuclei is converted to energy. Some of this energy is transferred as nuclear radiation from the large nucleus that's formed. Nuclear fusion happens only if the relative mass of the nucleus that's formed is no more than about 55 (about the same as an iron nucleus). To form bigger nuclei, energy must be supplied.

The Sun is about 75% hydrogen and 25% helium. The core is so hot that it's made up of a plasma of bare nuclei with no electrons. These nuclei move about and fuse together when they collide. When they fuse together, they release energy. Figure 2 shows how protons fuse together to form a $_2^4$He nucleus. Energy is released at each stage.

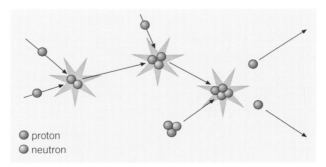

- proton
- neutron

Figure 2 *Fusion reactions in the Sun*

- When two protons (i.e. hydrogen nuclei) fuse together, they form a 'heavy hydrogen' nucleus, $_1^2$H. At the same time, other particles are created and emitted.
- Two more protons collide separately with two $_1^2$H nuclei and turn them into heavier nuclei.
- The two heavier nuclei collide to form the helium nucleus $_2^4$He.
- The energy released at each stage is carried away as the kinetic energy of the product nucleus and other particles emitted.

Fusion reactors

There are very big technical difficulties with making fusion a useful source of energy. The plasma of light nuclei has to be heated to very high temperatures and very high pressures before the nuclei will fuse. This is because two nuclei approaching each other will repel each other because of their positive charges. But if the nuclei are moving fast enough, they can overcome this force of repulsion and fuse together.

In a fusion reactor:

- the plasma is heated by passing a very big electric current through it
- the plasma is contained by a magnetic field so that it doesn't touch the reactor walls. If it did, it would go cold, and fusion would stop.

Scientists have been working on these problems since the 1950s. For a fusion reactor to be successful, it would have to release more energy than it uses to heat the plasma. Today, scientists working on experimental fusion reactors are able to do this by fusing heavy hydrogen nuclei to form helium nuclei – but only for a few minutes!

A promising future

Practical fusion reactors could meet all the energy needs of people.

- The fuel for fusion reactors is easily available as heavy hydrogen and is naturally present in sea water.
- The reaction product, helium, is a non-radioactive gas, so it is harmless.
- The energy released could be used to generate electricity.

In comparison, fission reactors mostly use uranium, which is found in only some parts of the world. Also, fission reactors produce nuclear waste that has to be stored securely for many years. But fission reactors have been used for over 50 years, unlike fusion reactors, which are still being developed.

Go further!

Two nuclei will fuse together if they get close enough for the strong nuclear force to pull them together. This can only happen if the nuclei approach each other at high speed, because they need to have enough kinetic energy to overcome the electrostatic repulsion between them.

---

1 a Explain what is meant by nuclear fusion.   [1 mark]
   b Look at Figure 2 and work out what is formed when a proton collides with a $_1^2$H nucleus.   [1 mark]

2 a Explain why the plasma of light nuclei in a fusion reactor needs to be very hot.   [1 mark]
   b Explain why a fusion reactor that needs more energy than it produces would not be much use.   [1 mark]

3 Give two advantages and two disadvantages that a fusion reactor has compared with a fission reactor.   [4 marks]

4 a Write how many protons and how many neutrons are present in a $_1^2$H nucleus.   [1 mark]
   b Write an equation to represent the fusion of a proton and a $_1^2$H nucleus when they form a helium $_2^3$He nucleus. The symbol for a proton is $_1^1$p (because its proton number (the lower number) is 1 and its mass number (the top number) is 1).   [1 mark]
   c Copy and complete the equation below to show the reaction in Figure 2 that takes place when two $_2^3$He nuclei fuse together to form a $_2^4$He nucleus.

   $_2^3$He + $_2^3$He → $_2^4$He + ... + ...   [2 marks]

Key points

- Nuclear fusion is the process of forcing the nuclei of two atoms close enough together so that they form a single larger nucleus.
- Nuclear fusion can be brought about by making two light nuclei collide at very high speed.
- Energy is released when two light nuclei are fused together. Nuclear fusion in the Sun's core releases energy.
- A fusion reactor needs to be at a very high temperature before nuclear fusion can take place. The nuclei to be fused are difficult to contain.

P7.9 Nuclear issues

Learning objectives

After this topic, you should know:

- what radon gas is and why it is dangerous
- how safe nuclear reactors are
- why nuclear waste is dangerous
- what happens to nuclear waste.

Table 1 *Sources of background radiation in the UK. 1 μSv = 1 millionth of 1 Sv*

Source	Radiation dose in μSv
cosmic rays	238
ground & buildings	332
food & drink	274
natural radioactivity in the air	1190
medical applications	332
nuclear weapons tests	5
air travel	5
nuclear power	2

Figure 1 *Storage of nuclear waste*

Background radiation

A Geiger counter clicks even without a radioactive source near it. This is because of background radiation. Radioactive substances are found naturally all around you.

Table 1 shows the sources of background radiation. The radiation from radioactive substances is hazardous because it ionises substances it passes through. The numbers in Table 1 are the radiation dose measured in sieverts (Sv). This tells you how much radiation on average each person gets in a year from each source. These measurements can be used to compare the effect on human health of the radiation from different sources of background radioactivity. Medical sources include X-rays as well as radioactive substances, because X-rays have an ionising effect. See Topic P13.4.

Background radiation in the air is caused mostly by radon gas that seeps through the ground from radioactive substances in rocks deep underground. Radon gas emits alpha particles, so radon is a health hazard if it is breathed in. It can seep into homes and other buildings in some locations. In affected homes, pipes under the building can be installed and fitted to a suction pump to draw the gas out of the ground before it seeps into the building.

Nuclear waste

Used fuel rods are very hot and very radioactive. After they are removed from a reactor, they are stored in big tanks of water for up to a year. The water cools the rods down.

Remote-control machines are then used to open the fuel rods. The unused uranium and plutonium are removed chemically from the used fuel. These are stored in sealed containers so that they can be used again.

The material that's left contains lots of radioactive isotopes with long half-lives. This radioactive waste has to be stored in secure conditions for many years to prevent radioactive contamination of the environment

Chernobyl and Fukushima

In 1986, a nuclear reactor in Ukraine exploded. A cloud of radioactive material from the fire drifted over many parts of Europe, including Britain. More than 100 000 people were evacuated and over 30 people died in the accident. More have developed leukaemia or other cancer types since then.

The Chernobyl reactor did not have a high-speed shutdown system like most reactors have. The operators at Chernobyl ignored safety instructions.

The lessons learned from Chernobyl were put into practice at Fukushima in Japan after three nuclear reactors were crippled in March 2011 by an earthquake and a tsunami. The entire population within 20 km were evacuated from their homes. Radiation levels, food and milk production, and health effects over a much wider area will need to be monitored for many years. Nearby reactors with greater protection from tsunamis were much less affected than the three crippled reactors.

New nuclear reactors

Most of the world's nuclear reactors in use at the present will need to be replaced in the next 20 years with new third-generation nuclear reactors. The new types of reactors could have:

● a standard design to reduce costs and construction time

● a longer operating life – typically 60 years

● more safety features, such as convection of outside air through cooling panels along the reactor walls

● much less effect on the environment.

Half-lives and instability

Radioactive isotopes have a wide range of half-lives. Some radioactive isotopes have half-lives of a fraction of a second whilst others have half-lives of more than a billion years. Isotopes with the shortest half-lives have the most unstable nuclei so they emit a lot of radiation in a short time.

The half-life of a radioactive source tells you how quickly its activity decreases. As its activity decreases, the rate it gives out radiation decreases. So the hazards caused by the ionising effect of the radiation from radioactive materials decrease with time according to the half-lives of their isotopes.

Radioactive risks

The effect of radiation on living cells depends on the type and amount of radiation received (the dose), whether the source of radiation is inside or outside the body, and how long the living cells are exposed to the radiation. The bigger the dose of radiation someone gets, the higher the risk of cancer. High doses kill living cells.

Smaller doses pose less risk but the risk is never zero. There is a very low level of risk to every person because of background radiation.

Table 2 *Risks of α, β, and γ radiation*

	α radiation	β and γ radiation
inside body	**very dangerous** – affects all the surrounding tissue	**dangerous** – reaches cells throughout the body
outside body	**some danger** – absorbed by skin, damages skin cells, retinal cells	

1 **a** Explain why radioactive waste needs to be stored:
   i securely [1 mark]   **ii** for many years. [1 mark]
   b Explain why a source of alpha radiation is very dangerous inside the human body but not as dangerous outside it. [3 marks]

2 In some locations, the biggest radiation hazard comes from radon gas that seeps up through the ground and into buildings.
   a Explain why radon gas is dangerous in a house. [3 marks]
   b Describe one way of making an existing house safe from radon gas. [2 marks]

3 Suggest whether the UK government should replace existing nuclear reactors with new reactors, either fission, fusion, or both. Answer this question by discussing the benefits and drawbacks of new fission and fusion reactors. [4 marks]

4 The risk to the average person in the UK from background radiation is about the same as the risk from road traffic, which causes about 10 deaths per 100 000 people each year. Use Table 1 to evaluate whether measures such as avoiding air travel would reduce the risk from background radiation. [5 marks]

Key points

● Radon gas is an α-emitting isotope that seeps into houses through the ground in some areas.

● There are hundreds of fission reactors safely in use in the world. None of them is of the same type as the Chernobyl reactors that exploded.

● Nuclear waste contains many different radioactive isotopes that emit nuclear radiation for many years. The radiation is dangerous because it can cause cancer.

● Nuclear waste is stored in safe and secure conditions for many years after unused uranium and plutonium (to be used in the future) are removed from it.

P7 Radioactivity

Summary questions

1 **a** Calculate how many protons and how many neutrons are in a nucleus of each of the following isotopes:

   i $^{14}_{6}C$ **ii** $^{228}_{90}Th$.   [2 marks]

  b $^{14}_{6}C$ emits a β particle and becomes an isotope of nitrogen (N).

    i Write how many protons and how many neutrons are in this nitrogen isotope. [2 marks]

    ii Write the symbol for this isotope. [1 mark]

  c $^{228}_{90}Th$ emits an α particle and becomes an isotope of radium (Ra).

    i Write how many protons and how many neutrons are in this isotope of radium. [2 marks]

    ii Write the symbol for this isotope. [1 mark]

2 Copy and complete the following table about the properties of alpha, beta, and gamma radiation. [4 marks]

	α	β	γ
Identity		electrons	
Stopped by			thick lead
Range in air		about 1 m	
Relative ionisation			weak

3 The following measurements were made of the count rate from a radioactive source.

Time in hours	0	0.5	1.0	1.5	2.0	2.5
Count rate due to the source in counts per minute	510	414	337	276	227	188

  a Plot a graph of the count rate (on the vertical axis) against time. [3 marks]

  b Use your graph to find the half-life of the source. [1 mark]

4 In a radioactive carbon dating experiment of ancient wood, a sample of the wood had an activity of 40 Bq. The same mass of living wood had an activity of 320 Bq.

  a i State what is meant by the activity of a radioactive source. [1 mark]

    ii **H** Calculate how many half-lives the activity took to decrease from 320 to 40 Bq. [2 marks]

  b **H** The half-life of the radioactive carbon in the wood is 5600 years. Calculate the age of the sample. [1 mark]

5 In an investigation to find out what type of radiation was emitted from a given source, the following measurements were made with a Geiger counter.

Source at 20 mm from tube	Average count rate in counts per minute
no source present	29
no absorber present	385
sheet of metal foil between S and T	384
thick aluminium plate between S and T	32

  a Write what caused the count rate when no source was present. [1 mark]

  b Write the count rate from the source with no absorbers present. [1 mark]

  c Write what type of radiation was emitted by the source. Explain how you arrived at your answer. [4 marks]

6 Figure 1 shows the path of two α particles labelled A and B that are deflected by the nucleus of an atom.

Figure 1

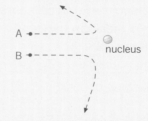

  a Explain why they are deflected by the nucleus. [2 marks]

  b Explain why B is deflected less than A. [2 marks]

  c Explain why most α particles directed at a thin metal foil pass straight through it. [2 marks]

7 **a** Explain what is meant by a nuclear chain reaction. [3 marks]

  b Explain what would happen in a nuclear reactor if:

    i the coolant fluid leaked out of the core [2 marks]

    ii the control rods were pushed further into the reactor core. [3 marks]

8 **a i** Explain what is meant by nuclear fusion. [1 mark]

    ii Explain why two nuclei repel each other when they get close. [3 marks]

    iii Explain why they need to collide at high speed to fuse together. [2 marks]

  b Give three reasons why nuclear fusion is difficult to achieve in a reactor. [3 marks]

Practice questions

01 A group of students investigated the nature of radioactive decay. They used 80 one-penny coins and a stopwatch to perform the investigation.

  1 Place the 80 one-penny coins in a container with lid.

  2 Shake container for 10 seconds and tip the coins onto the bench.

  3 Count the number of coins with the heads side showing and record result.

  4 Replace only the coins showing the tails side into container and repeat the tests.

Table 1

Time in seconds	Number of heads shown
0	80
10	39
20	20
30	11
40	5
50	3
60	2
70	1

01.1 Draw a graph of number of heads (*y*-axis) against time (*x*-axis). [3 marks]

01.2 Give some conclusions about the results. [2 marks]

01.3 Give a reason why only one-penny coins were used and not a mix of coins. [1 mark]

01.4 Suggest one reason for a possible human error in the investigation. [1 mark]

01.5 State what the head coins are meant represent in nuclear decay. [1 mark]

02 Background radiation depends on many sources. **Figure 1** shows some of the most common sources of radiation.

Figure 1

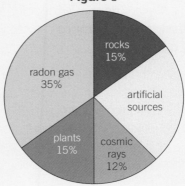

02.1 Calculate the percentage of radiation from artificial sources. [1 mark]

02.2 Give the names of two artificial sources of radiation. [1 mark]

02.3 In Australia adults receive on average 1.5–2.0 mSv each year of radiation. The maximum recommended dose for workers is 20.0 mSv each year. Passengers on each long haul flight from Sidney to London will receive an additional 0.3 mSv of radiation. Explain why airline pilots are only allowed to make a certain number of long haul flights. Your answer should include a calculation. [3 marks]

03.1 Complete the table of information for a radon atom.

Number of protons	86
Number of electrons	
Number of neutrons	132

[1 mark]

03.2 Calculate the mass number of this radon atom. [1 mark]

03.3 Radon can exist in the form of many different isotopes. State the differences in the nuclei of these isotopes. [1 mark]

03.4 The half-life of radon-222 is 3.8 days. Calculate how long a 48 g sample of radon will last before it contains only 3 g of radon and stops being effective. [2 marks]

04.1 Match each type of radiation to the correct description.

alpha radiation   electromagnetic radiation
beta radiation   same as a helium nucleus
gamma radiation   high-speed electron [3 marks]

04.2 When nuclear isotopes decay nuclear radiation is emitted. Complete the nuclear decay diagram.

$$^{230}_{90}\text{Th} \xrightarrow{\alpha} {}^{226}\text{Ra} \longrightarrow {}^{}_{86}\text{Rn}$$ [3 marks]

A sample of radioactive material is tested to find out whether it passes through different materials. The results are shown in **Figure 2**.

Figure 2

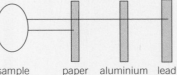

sample   paper aluminium lead

04.3 Name the types of radiation in the sample. [1 mark]

3 Forces in action

An astronaut in a space station can float around and perform acrobatic tricks. Many people think this is because there is no gravity in space. However, this is wrong. The force of gravity due to the Earth stretches far into space and keeps the space station orbiting the Earth.

You and all of the objects around you are acted on by the force of gravity. You are also acted on by other forces, such as friction, which acts between objects when they touch each other, and non-contact forces like magnetic and electrostatic forces.

In this section, you will learn about what forces do, how we measure them and their effects, and how we calculate the effect forces have on objects.

Key questions

- How do we represent a force and what do we mean by a resultant force?
- How do we work out the effect of a resultant force acting on an object?
- What do we mean by momentum and how is it linked to force?
- What do we mean by elasticity?

Making connections

- When a force does work on an object to make an it move faster, energy is transferred to the object to increase its kinetic energy store. As you work through this section you will meet content that describes what happens when forces do work on objects. Make sure that you look back at **P1 Conservation and dissipation of energy** to recall the different types of energy transfers between objects.

- The planets of the solar system orbit the Sun because there is a gravitational force of attraction between each planet and the Sun. There is a force of gravitational attraction between any two objects, but it's usually too small to notice unless one or both of the objects is very, very big. You will learn more about the forces on objects in space in **P16 Space**.

I already know...

Force is measured in newtons (N) using a newton-meter.

An object is in equilibrium when the forces acting on it are balanced.

Speed is measured in metres per second.

Drag forces and friction resist the motion of moving objects.

When objects interact, each one exerts a force on the other.

The force in a stretched object is called tension and it increases if the object is stretched more.

The weight of an object is due to the force of gravity on it.

I will learn...

The difference between a vector and a scalar and how to represent a vector.

How to find the resultant of two forces and to resolve a force into perpendicular components.

The difference between speed and velocity and what we mean by acceleration.

What is meant by terminal velocity and why objects fall through water at a constant velocity.

What is meant by conservation of momentum and when we can use this rule.

How to measure the stiffness of a spring and what is meant by elasticity.

How to calculate the weight of an object from its mass and the gravitational field strength of where it is.

Required Practicals

Practical		Topic
6	Stretch tests	P10.8
7	Investigating force and acceleration	P10.1

Distance and displacement

When you travel to school, the distance you travel may be much greater than the direct distance from your home to your school. The map in Figure 1 shows the route from home to school for a student who has to catch a bus to school. The bus has to pick up lots of other students along a route that has quite a few changes of direction. So the distance travelled by the student is much greater than the direct distance from the student's home to the school.

Distance without change of direction is called **displacement**. In other words, displacement is distance in a certain direction.

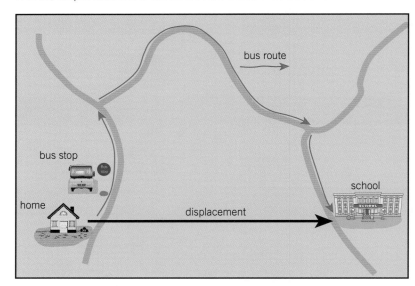

Figure 1 *A school journey*

Vectors and scalars

As well as velocity and displacement, many other physical quantities have both a *size* and a *direction*. Physical quantities that have direction are called **vectors**. Other examples of vectors include acceleration, force, momentum, weight, and gravitational field strength.

Physical quantities that have size, but no specific direction are called **scalars**. Examples include speed, distance, time, mass, energy, and power.

The size of a quantity is its **magnitude**. A vector has magnitude (i.e., size) as well as a direction. A scalar has magnitude only. For example, in Figure 1 the displacement from home to school is 5 km due East. So the magnitude of the displacement is 5 km, and the direction is East.

- A vector quantity has a magnitude and a direction.
- A scalar quantity has magnitude only.

Representing a vector quantity

Any vector quantity can be represented by an arrow, like the displacement arrow in Figure 1.

- The direction of the arrow shows the direction of the vector quantity.
- The length of the arrow represents the magnitude of the vector quantity.

Because a force has a magnitude and a direction, it is a vector quantity and can be represented on a diagram by an arrow. Figure 2 shows the force acting on a nail when it is struck by a hammer. The force of the hammer on the nail is represented by the red arrow (Figure 2). If the magnitude of the force is known, this can also be represented by the length of the arrow.

Scale diagrams

When more than one force acts on an object, the forces on the object sometimes need to be shown on a scale diagram. For example, suppose two forces of 3.0 N and 4.0 N act at right angles to each other on a small object. To show both forces on a scale diagram, we could choose a scale in which 1 unit of distance (for example 10 mm) represents a force of 1.0 N. On this scale, the length of the two arrows would need to be 30 mm and 40 mm respectively (Figure 3).

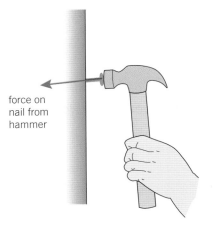

force on nail from hammer

Figure 2 *Representing a force*

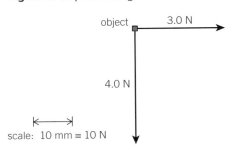

scale: 10 mm ≡ 10 N

Figure 3 *A scale diagram*

1 **a** State what is meant by the magnitude of a vector quantity.
[1 mark]

b State the difference between a scalar quantity and a vector quantity.
[1 mark]

2 Look at the journey shown on the map in Figure 1. If the displacement arrow represents a displacement of 10 km, use the map to estimate the approximate distance travelled by the student on their journey to school.
[1 mark]

3 A small object is acted on by a horizontal force, **A**, of magnitude 15 N, and another horizontal force, **B**, of magnitude 12 N, which acts in the opposite direction to A. Draw a to-scale vector diagram showing forces **A** and **B** acting on the object.
[2 marks]

4 A force **A** of 48 N acts on a small object as shown in Figure 4.

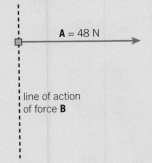

A = 48 N

line of action of force **B**

Figure 4

a Copy Figure 4 and state the scale you have used on your diagram.
[1 mark]

b Add a further arrow to your diagram to represent a force **B** of 36 N acting on the object in a direction at right angles to the direction of the 48 N force.
[1 mark]

> ### Study tip
>
> Force diagrams usually show more than one force. For this reason, the forces on a diagram should always be clearly labelled to identify the force. Force diagrams do not always need to be scale diagrams, but if a diagram is a scale diagram, the scale should be shown (e.g., 10 mm ≡ 1.0 N, where ≡ means 'represents').

> ### Key points
>
> - Displacement is distance in a given direction.
> - A vector quantity is a physical quantity which has magnitude and direction.
> - A scalar quantity has magnitude but no direction.
> - A vector quantity can be represented by an arrow in the direction of the vector and of length in proportion to the magnitude of the vector.

P8.2 Forces between objects

Learning objectives

After this topic, you should know:

- what forces can do
- the unit of force
- what is meant by a contact force
- the forces being exerted when two objects interact.

When you apply **forces** to a tube of toothpaste, the forces you apply squeeze the tube and change its shape and push toothpaste out of the tube. Be careful not to apply too much force, if you do the toothpaste might come out too fast. Forces can change the shape of an object or change its state of rest or its motion.

A force is a push or pull that acts on an object because of its interaction with another object. If two objects must touch each other to interact, the forces are called contact forces. Examples include friction, air resistance, stretching forces (or tension), and normal contact forces. Contact forces occur when an object is supported by or strikes another object. Non-contact forces include magnetic force, electrostatic force, and the force of gravity.

Synoptic link

You will meet Newton's Second Law when you study forces and acceleration in Topic P10.1.

Equal and opposite forces

Newton's third law states that when two objects interact with each other, they exert equal and opposite forces on each other. The unit of force is the newton (abbreviated as N).

- A boxer who punches a bag with a force of 100 N experiences an equal and opposite force of 100 N from the bag.
- The weight of an object is the force of gravity on the object due to the Earth. The object exerts an equal and opposite force on the Earth.
- Two roller skaters pull on opposite ends of a rope (Figure 1). The skaters move towards each other because they pull on each other with equal and opposite forces. Two newton-meters could be used to show this.

Study tip

Remember that when two objects interact, although they exert equal and opposite forces on each other, the effects of these forces on each object will depend on the masses of the objects – the larger the mass, the smaller the effect.

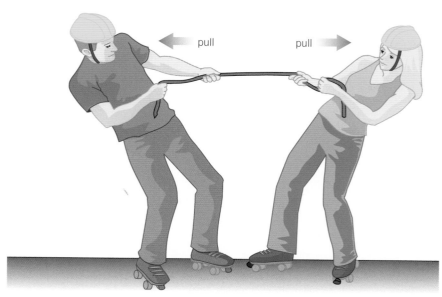

Figure 1 *Equal and opposite forces*

In the mud

A car stuck in mud can be difficult to move. Figure 2 shows how a tractor can be very useful here. At any stage, the force of the rope on the car is equal and opposite to the force of the car on the rope.

To pull the car out of the mud, the force of the mud on the tractor needs to be greater than the force of the mud on the car. These two forces are not equal and opposite to each other. The 'equal and opposite force' to the force of the mud on the tractor is the force of the tractor on the mud. The 'equal and opposite force' to the force of the mud on the car is the force of the car on the mud.

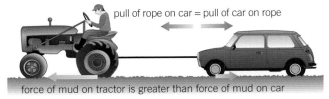

pull of rope on car = pull of car on rope

force of mud on tractor is greater than force of mud on car

Figure 2 *In the mud*

Friction in action

The driving force on a car is the force that makes it move. This is sometimes called the engine force or the motive force. This force is caused by the **friction** between the ground and the tyre of each drive wheel. Friction acts where the tyre is in contact with the ground.

When the car moves forward:

● the force of the friction of the road on the tyre is in the forward direction

● the force of the friction of the tyre on the road is in the reverse direction.

These two forces are equal and opposite to each other (Figure 3).

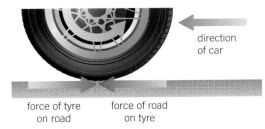

direction of car

force of tyre on road   force of road on tyre

Figure 3 *Driving force*

1 **a** The brakes of a moving car are applied. Describe the effect of the braking force on the car. [1 mark]
   b When a car brakes, the road exerts a force on each tyre to slow the car down. Describe the force that each tyre exerts on the road. [1 mark]

2 **a** A hammer hits a nail with a downward force of 50 N. What is the size and direction of the force of the nail on the hammer? [1 mark]
   b A lorry tows a broken-down car. The force of the lorry on the tow rope is 200 N. How much force is exerted on the tow rope by the lorry? [1 mark]

3 A book is at rest on a table. Compare the force of the book on the table with each of the following forces.
   a the force of the table on the book [2 marks]
   b the force of the table on the floor. [2 marks]

4 When a student is standing at rest on bathroom scales, the scales read 500 N.
   a What is the size and direction of the force of the student on the scales? [1 mark]
   b What is the size and direction of the force of the scales on the student? [1 mark]
   c What is the size and direction of the force of the floor on the scales? [1 mark]

Key points

● Forces can change the shape of an object, or change its motion or its state of rest.
● The unit of force is the newton (N).
● A contact force is a force that acts on objects only when the objects touch each other.
● When two objects interact, they always exert equal and opposite forces on each other.

P8.3 Resultant forces

Learning objectives

After this topic, you should know:

- what a resultant force is
- what happens if the resultant force on an object is:
  - zero
  - greater than zero
- how to calculate the resultant force when an object is acted on by two forces acting along the same line
- what a free-body force diagram is. **H**

Figure 1 *The linear air track*

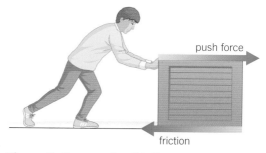

Figure 2 *Overcoming friction*

Wherever you are right now, at least two forces are acting on you. These are the gravitational force on you and a force supporting you. Most objects around you are acted on by more than one force. You can work out the effect of the forces on an object by replacing them with a single force, the **resultant force**. This is a single force that has the same effect as all the forces acting on the object. If the resultant force is zero, we say that the forces acting on the object are balanced.

Balanced forces

Newton's First Law of motion states that if the forces acting on an object are balanced, the resultant force on the object is zero, and:

- if the object is at rest, it stays stationary
- if the object is moving, it keeps moving with the same speed and in the same direction.

If only two forces act on an object with zero resultant force, the forces must be equal to each other and act in opposite directions.

1 A glider on a linear air track floats on a cushion of air (Figure 1). As long as the track stays level, the glider moves at the same speed and direction along the track. That is because friction is absent. Newton's First law tells you that the glider will continue moving with the same speed in the same direction.

2 When a heavy crate is pushed across a rough floor at a constant speed without changing its direction, the push force on it is equal in size, and acting in the opposite direction, to the friction of the floor on the crate (Figure 2). Newton's First Law states that the crate will continue moving with the same speed, and in the same direction.

Unbalanced forces

When the resultant force on an object is not zero, the forces acting on the object are not balanced. The movement of the object depends on the size and direction of the resultant force.

1 When a jet plane is taking off, the thrust force of its engines is greater than the force of air resistance (or drag) on it. The resultant force on the plane is the difference between the thrust force and the force of air resistance acting on it. The resultant force is therefore greater than zero.

2 When a car driver applies the brakes, the braking force is greater than the force from the engine. The resultant force is the difference between the braking force and the engine force. It acts in the opposite direction to the car's direction, so it slows the car down.

The examples show that if an object is acted on by two unequal forces acting in opposite directions, the resultant force is:

- equal to the difference between the two forces
- in the direction of the larger force.

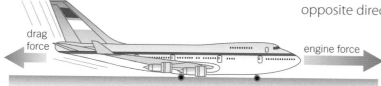

Figure 3 *A passenger jet on take-off*

For example, Figure 4 shows two forces, **A** and **B**, acting on an object in opposite directions. If **A** = 5 N and **B** = 9 N, the resultant force on the object is 4 N (= 9 N − 5 N) in the direction of B. If the two forces act in the same direction, the resultant force is equal to the sum of the two forces and is in the same direction.

Figure 5 shows a tug-of-war in which the pull force of each team is represented by a vector. A scale of 10 mm to 200 N is used. Team A pulls with a force of 1000 N, and team B pulls with a force of 800 N. So the resultant force is 200 N in team A's direction.

Figure 4 *Forces in opposite directions*

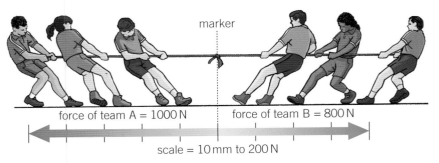

Figure 5 *A tug-of-war*

Force diagrams

When an object is acted on by more than one force, you can draw a free-body force diagram to work out the resultant force on the object. A free-body force diagram shows the forces acting on an object without any other objects or other forces shown. Each force is shown on the diagram by a vector, which is an arrow pointing in the direction of the force. Figure 4 is a simple example of a free-body force diagram. Figure 5 is not a free-body force diagram because it shows more than one object.

1 Describe and explain what happens to the glider in Figure 1 if the air track blower is switched off. [2 marks]

2 A jet plane lands on a runway and stops.
   a State the direction of the resultant force on the plane as it lands. [1 mark]
   b State the resultant force on the plane when it has stopped. [1 mark]

3 A car is stuck in the mud. A tractor tries to pull it out.
   a The tractor pulls the car with a force of 250 N. Give the reason why the car does not move. [1 mark]
   b Increasing the tractor force to 300 N pulls the car steadily out of the mud at a constant speed and direction. Calculate the force of the mud on the car now. [1 mark]

4 a Copy the car in Figure 6 and show the weight of the car as a vector arrow midway between the wheels. [1 mark]
   b Ⓗ Show the support forces of the road on each wheel as vector arrows. Explain your reasoning for their direction. [2 marks]

Figure 6 *Braking*

Key points

- The resultant force is a single force that has the same effect as all the forces acting on an object.
- If the resultant force on an object is:
  - zero, the object stays at rest or at the same speed and direction
  - greater than zero, the speed or direction of the object will change.
- If two forces act on an object along the same line, the resultant force is:
  - their sum, if the forces act in the same direction
  - their difference, if the forces act in opposite directions.
- Ⓗ A free-body force diagram of an object shows the forces acting on it.

P8.4 Moments at work

Learning objectives

After this topic, you should know:

- what the moment of a force measures
- how to calculate the moment of a force
- how the moment of a force can be increased
- why levers are force multipliers.

Figure 1 *A turning effect*

To undo a very tight wheel-nut on your bicycle, you need a spanner. Figure 1 shows the idea. The force you apply to the spanner has a turning effect on the nut. You can not undo a tight nut with your fingers, but you can with the spanner. The spanner exerts a much larger turning effect on the nut than the force you apply to the spanner.

If you had a choice between a long-handled spanner and a short-handled one, which would you choose? The longer the spanner handle, the less force you need to exert on it to loosen the nut.

In this example, the turning effect of the force, called the **moment** of the force, can be increased by:

- increasing the size of the force
- using a spanner with a longer handle.

Levers

A crowbar is a lever that can be used to raise one edge of a heavy object. Look at Figure 2.

The weight of the object is called the **load**, and the force the person applies to the crowbar is called the effort. The point about which the crowbar turns is called the pivot. Using the crowbar, the effort needed to lift the object is only a small fraction of its weight. The lever used in this way is an example of a force multiplier because the effort moves a much bigger load.

The line that a force acts along is called its line of action.

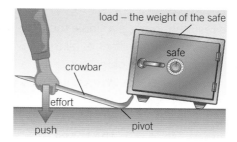

Figure 2 *Using a crowbar*

Investigating the turning effect of a force

The diagram in Figure 3 shows one way to investigate the turning effect of a force. You can move the weight W along the metre ruler.

You should find that the newton-meter reading (i.e., the force needed to support the ruler) increases if the weight is increased or when you move the weight further away from the pivot.

- Explain why this happens.

Safety: Clamp the stand to your bench and protect your feet and bench top from falling weights. Wear eye protection.

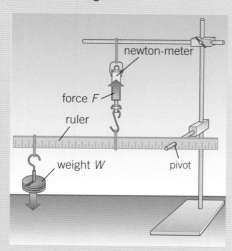

Figure 3 *Investigating turning forces*

You can work out the moment of a force by using this equation:

moment, M = **force, F** × **perpendicular distance from**
(newton   (newtons, N)   **the line of action of the force**
metres, N m)   **to the pivot, d** (metres, m)

Calculating moments

The word equation can be written using symbols as:

$$M = F \times d$$

where:

M = moment of the force in newton metres, N m

F = force in newtons, N

d = perpendicular distance from the line of action of the force to the pivot, in metres, m.

Worked example

A force of 50 N is exerted on a claw hammer of length 0.30 m, as shown in Figure 4. Calculate the moment of the force.

Solution

Moment of the force = 50 N × 0.30 m = **15 N m**

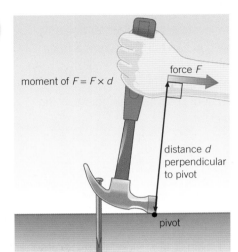

moment of $F = F \times d$

force F

distance d perpendicular to pivot

pivot

Figure 4 *Using a claw hammer. The claw hammer is being used to remove a nail from a wooden beam*

---

1 **a** A force acts on an object and makes it turn about a fixed point. State the effect on the moment of the force if:

   i the force is increased without changing its line of action [1 mark]

   ii the force is doubled and the perpendicular distance from its line of action to the pivot is halved [1 mark]

   iii the force is halved and the perpendicular distance from its line of action to the pivot is halved. [1 mark]

   b A force of 72 N is exerted on a claw hammer of length 0.25 m, as shown in Figure 4. Calculate the moment of the force. [2 marks]

2 In Figure 1, a force is applied to a spanner to undo a nut. Determine whether the moment of the force is:

   a clockwise or anticlockwise [1 mark]

   b increased or decreased by:

   i increasing the force [1 mark]

   ii exerting the force nearer to the nut. [1 mark]

3 Explain each of the following statements:

   a it is easier to remove a nail with a claw hammer if the hammer has a long handle [1 mark]

   b a door with rusty hinges is more difficult to open than a door of the same size with lubricated hinges. [2 marks]

4 A spanner of length 0.25 m is used to turn a nut as in Figure 1. Calculate the force that needs to be applied to the end of the spanner if the moment of the force that it should exert is 18 N m. [1 mark]

Study tip

Remember that when calculating moments, it is the perpendicular distance from the pivot that is needed. The perpendicular distance is along a line from the pivot that is at a right angle (90°) to the line of action of the force.

Key points

- The moment of a force is a measure of the turning effect of the force on an object.
- The moment of a force about a pivot is $M = F d$, where d is the perpendicular distance from the line of action of the force F to the pivot.
- To increase the moment of a force, increase F or increase d.
- Levers can used to exert a force that is greater than the effort.

P8.5 More about levers and gears

Learning objectives

After this topic, you should know:

- how levers act as force multipliers
- how you can tell if a lever is a force multiplier
- what gears do
- how gears can give a bigger turning effect.

More about force multipliers

You can use levers to increase the size of a force acting on an object or to make the object turn more easily. When a lever is used to increase or multiply a force, the force applied to the lever must act further from the pivot (the point about which the lever turns) than the force it has to overcome. For example:

- A bottle opener (Figure 1) is a simple device you can use to force the cap off a bottle. The force on the cap is much bigger than the force you apply to the bottle opener. This is because the line of action of the force you apply is further from the pivot than the edge of the cap. So the bottle opener acts as a force multiplier and exerts a larger force on the cap.
- A pair of scissors with sharp edges cuts string easily. This is because the force of the scissors on the string is much bigger than the force you can apply to the scissors. As shown in Figure 2, this is because the line of action of the force you apply is much further from the pivot than the point where the scissors cut the string. So the scissors act as a force multiplier.

Gears

Gears are like levers because they can multiply the effect of a turning force. For example, when a car is in low gear, a small gear wheel turns a larger gear wheel (Figure 3). The gears multiply the turning effect of the engine force, so a bigger turning effect is exerted on the wheels when in a low gear.

The gear wheels exert equal and opposite forces on each other where their edges are in contact. The force on each gear wheel acts tangentially at this point (i.e., at right angles to the line to the centre of the wheel). But the turning force on the larger gear wheel acts further from its shaft than the turning force of the smaller gear wheel acting on its shaft. So the turning effect of the wheel shaft is bigger than the turning effect of the engine shaft.

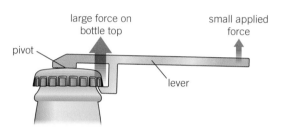

Figure 1 *Using a bottle opener*

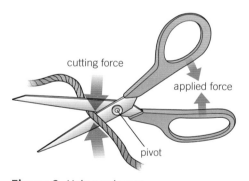

Figure 2 *Using scissors*

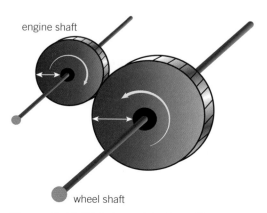

Figure 3 *Multiplying a turning effect*

Wheels and axles

The wheel and axle shown in Figure 4 is like a simple gear system. Use the arrangement shown in Figure 4 to investigate how much effort (the force pulling on the rim of the wheel) is needed to raise a load of known weight (the force pulling on the axle).

Increase the effort by adding known weights to it until the load just starts to rise.

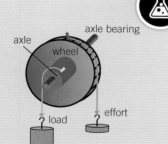

Figure 4 *Testing a wheel and axle*

- From your results, work out the ratio of the load to the effort.
- Repeat the test for different loads.
- Give some conclusions from your results about how the ratio of load to the effort changes as the load is increased.

Safety: Wear eye protection. Make sure the equipment is secure and protect your feet from falling masses.

Changing gears

Figure 5 shows a simplified version of some of the gear wheels in a car.

When a low gear is chosen:

- a small gear wheel driven by the engine shaft is used to turn a large gear wheel on the output shaft. So the output shaft turns slower than the engine shaft.
- The turning effect of the output shaft is greater than the turning effect of the engine shaft.

Low gear gives low speed and a high turning effect

When a high gear is chosen:

- a large gear wheel driven by the engine shaft is used to turn a small gear wheel on the output shaft. So the output shaft turns faster than the engine shaft, so the car can move at a higher speed.
- But the force of the smaller gear wheel acts nearer to its shaft than the force of the larger gear wheel acting on its shaft. So the turning effect of the output shaft is less than the turning effect of the engine shaft.

High gear gives high speed and a low turning effect

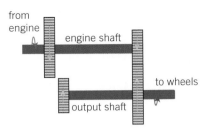

a) **Low gear** – low speed

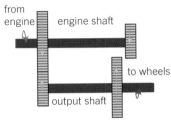

b) **High gear**– high speed

Figure 5 *Changing gears*

Worked example

A gear wheel of radius 20 mm is used to turn another gear wheel of radius 10 mm with a force of 80 N.

Calculate the moment of the force on:

a the 10 mm wheel

b the 20 mm wheel.

Solution

The perpendicular distance of the line of action of the force on each gear wheel is equal to the radius of the gear wheel.

a Moment = force × gear wheel radius = 80 N × 0.010 m = 0.80 N m.

b An equal and opposite force acts on the 20 mm wheel. The moment of this force = 80 N × 0.020 m = 1.6 N m.

---

1 Figure 6 shows a T bar used to undo a wheel nut on a car wheel. Explain why the turning force on the wheel nut is much greater than the turning forces applied to the T bar. [2 marks]

2 The moment of the force on a gear wheel of radius 40 mm should always be less than 4.8 N m. Calculate the maximum force on the gear wheel to make it turn. [1 mark]

3 The bicycle chain of a bicycle is used to turn the rear wheel of the bicycle. The chain pulls on a gear wheel on the axle of the wheel. The gear wheel is one of a set of gear wheels of different diameters. Explain why the largest gear wheel should be used in an uphill climb.
[3 marks]

4 The handbrake of a car is used to apply the car brakes when the car is parked. Design a handbrake that will pull on a horizontal cable with a force eight times greater than the force applied to it. [3 marks]

Study tip

When you calculate the moment of a force, make sure all your distance measurements are in metres.

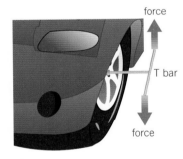

Figure 6

Key points

- A lever used as a force multiplier exerts a greater force than the force applied to the lever by the effort.
- The pivot of a force multiplier is nearer to the line of action of the force it exerts than to the force applied to it.
- Gears are used to change the moment of a turning effect.
- To increase the moment of a turning effect, a small gear wheel needs to drive a larger gear wheel.

P8.6 Centre of mass

The design of racing cars has changed a lot since the first models. Look at Figure 1, which shows examples of past and modern racing car designs. One thing that has not changed is the need to keep the car near to the ground. The weight of the car must be as low as possible. Otherwise, the car would overturn when going round corners at high speeds.

Figure 1 *Racing cars from the 1920s to the modern day*

You can think of the weight of an object as if it acts at a single point. This point is called the centre of mass (or the centre of gravity) of the object. The idea of centre of mass is very useful to designers and engineers. For example, the designer of a chair needs to use a force diagram showing its weight as well as all the other forces acting on it, to make sure the finished chair will not tip over when someone sitting on it leans back.

The centre of mass of an object is the point at which its mass can be thought of as being concentrated.

Suspended equilibrium

If you suspend an object and then release it, it will sooner or later come to rest with its centre of mass directly below the point of suspension, as shown in Figure 2a. The object is then in equilibrium, which means it is at rest. Its weight does not exert a turning effect on the object, because its centre of mass is directly below the point of suspension.

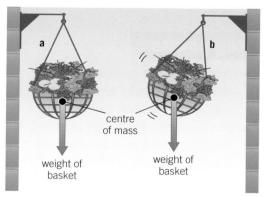

Figure 2 *Suspension* **a** *In equilibrium* **b** *Non-equilibrium*

If the object is turned from this position and then released, it will swing back to its equilibrium position. This is because its weight has a turning effect that returns the object to equilibrium, as shown in Figure 2b. You say that the object is *freely suspended* if it returns to its equilibrium position after the turning force is taken away.

The centre of mass of a symmetrical object

For a flat object that is symmetrical, its centre of mass is along the axis of symmetry. You can see this in Figure 3.

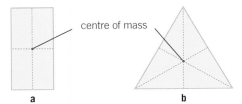

Figure 3 *Symmetrical objects*

If the object has more than one axis of symmetry, its centre of mass is where the axes of symmetry meet.

- A rectangle or a uniform ruler has two axes of symmetry, as shown Figure 3a. The centre of mass is where the axes meet.
- The equilateral triangle in Figure 3b has three axes of symmetry, each bisecting one of the angles of the triangle. The three axes meet at the same point. This is where the centre of mass of the triangle is.

Centre of mass

Figure 4 shows how to find the centre of mass of an irregular-shaped card.

1. Put a hole in one corner of the card and suspend the card from a rod.
2. Use a plumb line to draw a vertical line on the card from the rod.
3. Repeat the procedure, hanging the card from a different corner.

The point where the two lines meet is the centre of mass.

Use this method to find the centre of mass of a semicircular card of a radius 100 mm.

- Evaluate the accuracy of your experiment. For example, the card should balance on the flat end of a pencil placed directly under the card's centre of mass.

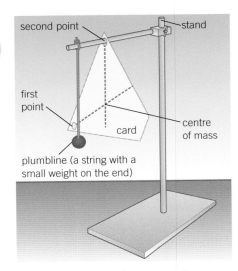

Figure 4 *Finding the centre of mass of a card*

1. Sketch each of the objects shown and mark its centre of mass.

   a 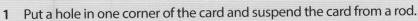 [1 mark]

   b [1 mark]   c [1 mark]

2. Explain why a child sitting on a swing comes to rest directly below the top of the swing. [2 marks]

3. Describe how you would find the centre of mass of a flat semicircular card. ✏ [4 marks]

4. Look again at the flower baskets in Figure 2. Describe the resultant force on each basket. Give a reason for your answer in each case. [2 marks]

Key points

- The centre of mass of an object is the point where its mass can be thought of as being concentrated.
- The centre of mass of a uniform ruler is at its midpoint.
- When an object is freely suspended, it comes to rest with its centre of mass directly underneath the point of suspension.
- The centre of mass of a symmetrical object is along the axis of symmetry.

P8.7 Moments and equilibrium

Learning objectives

After this topic, you should know:

- how to use your knowledge of forces and moments to explain why objects at rest don't turn
- identify the forces that can turn an object about a fixed point
- identify whether a turning force that can turn an object turns it clockwise or anticlockwise
- how to calculate the size of a force (or its perpendicular distance from a pivot) acting on an object that is balanced.

A seesaw is an example in which clockwise and anticlockwise moments might balance each other out. The girl in Figure 1 sits near the pivot to balance her younger brother at the far end of the seesaw. Her brother is not as heavy as she is. She sits nearer the pivot than he does. That means her anticlockwise moment about the pivot balances his clockwise moment.

A model seesaw

Look at the model seesaw in Figure 2. The ruler is balanced horizontally by adjusting the position of the two weights. When it is balanced:

- the anticlockwise moment due to W_1 about the pivot $= W_1 d_1$, and
- the clockwise moment due to W_2 about the pivot $= W_2 d_2$.

The anticlockwise moment due to W_1 = the clockwise moment due to W_2 and therefore:

$$W_1 d_1 = W_2 d_2$$

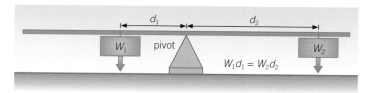

Figure 1 *The seesaw*

Figure 2 *The principle of moments*

The seesaw is an example of the **principle of moments**. This states that, for an object that is not turning:

the sum of all the clockwise moments about any point $=$ **the sum of all the anticlockwise moments about that point**

> **Study tip**
>
> Make sure the units in your calculations are consistent.

> **Study tip**
>
> You can use the arrangement in Figure 2 to find an unknown weight, W_1, if we know the other weight, W_2, and we measure the distances d_1 and d_2. You can then calculate the unknown weight using the equation $W_1 d_1 = W_2 d_2$

Worked example

Calculate W_1 in Figure 2, if $W_2 = 4.0\,\text{N}$, $d_1 = 0.25\,\text{m}$ and $d_2 = 0.20\,\text{m}$.

Solution

Rearranging $W_1 d_1 = W_2 d_2$ gives

$$W_1 = \frac{W_2 d_2}{d_1} = 4.0\,\text{N} \times \frac{0.20}{0.25\,\text{m}} = \textbf{3.2\,N}$$

Measuring the weight of a beam

Figure 3 shows how we can measure the weight of a beam by balancing it off-centre using a known weight. The weight of the beam acts at its centre of mass, which is at distance d_0 from the pivot.

- The moment of the beam about the pivot = $W_0 d_0$ clockwise, where W_0 is the weight of the beam.
- The moment of W_1 about the pivot = $W_1 d_1$ anticlockwise, where d_1 is the perpendicular distance from the pivot to the line of action of W_1.

Applying the principle of moments gives $W_1 d_1 = W_0 d_0$.

So we can calculate W_0 if we know W_1 and distances d_1 and d_0.

Safety: Take care with falling weights.

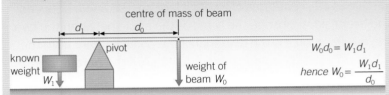

$$W_0 d_0 = W_1 d_1$$

$$\text{hence } W_0 = \frac{W_1 d_1}{d_0}$$

Figure 3 *Finding the weight of a beam*

If you have to move a heavy load, think beforehand about how to make the job easier. Figure 4 shows a wheelbarrow and a trolley being used to move a load. The load (weight W_0) is lifted using a much smaller effort (force F_1). Once the load has been lifted, it can easily be moved by pushing the wheelbarrow or the trolley forwards.

1 **a i** In Figure 2, calculate W_1 if $W_2 = 6.0\,N$, $d_1 = 0.30\,m$, and $d_2 = 0.15\,m$. [1 mark]
  ii In Figure 3, calculate the weight of the beam if $W_1 = 2.0\,N$, $d_1 = 0.15\,m$, and $d_0 = 0.25\,m$. [1 mark]
  b In Figure 4 explain why the effort is smaller than the load. [2 marks]

2 Dawn sits on a seesaw 2.50 m from the pivot. Jasmin balances the seesaw by sitting 2.00 m on the other side of the pivot.
  a Who is lighter, Dawn or Jasmin? [1 mark]
  b Jasmin weighs 425 N. What is Dawn's weight? [1 mark]
  c John now sits on the seesaw on the same side as Dawn at a distance of 0.50 m from the pivot. Jasmin stays in the same position as before. John's weight is 450 N. How far and in which direction should Dawn move to rebalance the seesaw? [2 marks]

3 For the balanced beam in the figure, work out its weight, W.

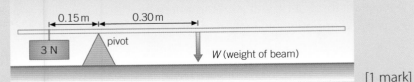

[1 mark]

4 In Figure 4, a two-wheeled trolley is used to carry a box of weight 120 N. Given that $d_1 = 6d_0$, calculate the effort force needed to raise the box off the ground. [1 mark]

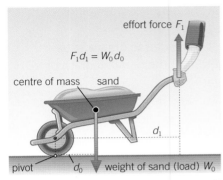

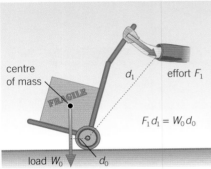

Figure 4 *Using moments*

Key points

- If an object at rest doesn't turn, the sum of the anticlockwise moments about any point = the sum of the clockwise moments about that point.
- All the forces acting on an object that don't pass through a fixed point can turn an object about that point
- The direction of the force and the position of the fixed point determines whether the moment acts clockwise or anticlockwise
- To calculate the force needed to stop an object turning we use the equation above. We need to know all the forces that don't act through the pivot and their perpendicular distances from the line of action to the pivot.

P8.8 The parallelogram of forces

Learning objectives

After this topic, you should know:

- what the parallelogram of forces is
- what the parallelogram of forces is used for
- what is needed to draw a scale diagram of the parallelogram of forces
- how to use the parallelogram of forces to find the resultant of two forces.

In Topic P8.2, you learnt how to find the resultant of two forces that act along the same line. What if the two forces do not act along the same line? Figure 2 shows a ship being towed by cables from two tugboats. The tension force in each cable pulls on the ship. The combined effect of these tension forces is to pull the vessel forwards. This is the resultant force.

Figure 2 shows how the two tension forces T_1 and T_2, represented as vectors, combine to produce the resultant force. The tension forces are drawn to scale as adjacent sides of a parallelogram. The angle between the two adjacent sides must be the same as the angle between the two forces. The resultant force is the diagonal of the parallelogram from the origin of T_1 and T_2. This geometrical method is called the **parallelogram of forces**.

Figure 1 *In tow*

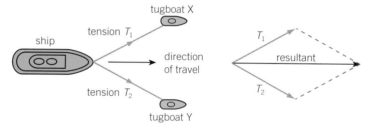

Figure 2 *Combining forces*

Investigating the parallelogram of forces

You can use weights and pulleys to demonstrate the parallelogram of forces (Figure 3). The tension in each string is equal to the weight it supports, either directly or over a pulley.

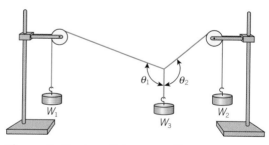

Figure 3 *The parallelogram of forces*

The point where the three strings meet is at rest. The string supporting the middle weight (W_3) is vertical. Using a protractor, you can measure angles θ_1 and θ_2 and note the values of the three known weights. You can then draw a scale diagram of a parallelogram so that:

- the line down the centre of the diagram represents the vertical line through the point where the three strings meet
- adjacent sides of the parallelogram at angles θ_1 and θ_2 to the vertical line represent the tensions in the strings supporting W_1 and W_2.

The resultant force of W_1 and W_2 represented by the diagonal line should be equal and opposite in direction to the vector representing W_3.

Study tip

Remember that you cannot always use arithmetic to add and subtract forces. When the two forces act at an angle, you will need to use geometry (the parallelogram of forces).

Make a model zip wire

Use a length of thin string and a weight hanger (or other suitable object) to make and test a model zip wire. Figure 4 shows the idea.

Release the weight hanger on the string at the top end and observe where it comes to rest.

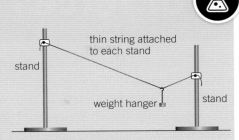

Figure 4 *A model zip wire*

Investigate how the height difference between the ends of the string affects the horizontal distance from the rest position of the hanger to one of the stands.

- Plot your results on a graph and discuss the effect of the height difference on the rest position of the hanger.

Safety: Make sure stands are clamped to the bench.

1 Figures 6 and 7 show examples where two forces act on an object X. In each case, work out the magnitude and direction of the resultant force on X.

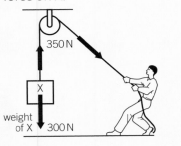

Figure 6

Figure 7

[2 marks]

2 A force of 3.0 N and a force of 4.0 N act on a point. Determine the magnitude and direction of the resultant of these two forces if the angle between their lines of action is:

   a 90°   [2 marks]   **b** 60°   [2 marks]   **c** 45°.   [2 marks]

3 In Figure 5, suppose the angle between the two sections of rope joined to the car had been 50° instead of 30°. Use the parallelogram of forces to find the maximum tension in the main tow rope. [2 marks]

4 In a model zip wire like the one shown in Figure 4, the two sections of the string are both at an angle of 20° to a horizontal line through the lowest point P of the string.

   a Draw a diagram to show the line of action of the forces due to each string acting on point P.   [1 mark]

   b i The weight hanger has a weight of 2.0 N. Using a suitable scale, draw a vector arrow on your diagram to show the weight of the weight hanger. Label the scale on your diagram. [1 mark]

   ii Use your diagram to find the tension in each section of the string. [3 marks]

Worked example

A tow rope is attached to a car at two points 0.80 m apart. The two sections of rope joined to the car are the same length and are at 30° to each other (Figure 5). The pull on each attachment should not exceed 3000 N. Use the parallelogram of forces to determine the maximum tension in the main tow rope.

Solution

The maximum tension T in the main tow rope is the resultant force of the two 3000 N forces at 30° to each other.

Drawing the parallelogram of forces as shown in Figure 5 gives:

$$T = 5800\,N$$

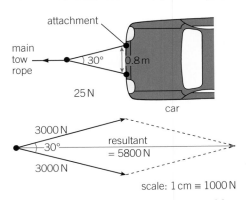

Figure 5 *Using the parallelogram of forces*

Key points

- The parallelogram of forces is a scale diagram of two force vectors.
- The parallelogram of forces is used to find the resultant of two forces that do not act along the same line.
- You will need a protractor, a ruler, a sharp pencil, and a blank sheet of paper.
- The resultant is the diagonal of the parallelogram that starts at the origin of the two forces.

P8.9 Resolution of forces

Learning objectives

After this topic, you should know:

- what is meant by resolution of a force
- how to resolve a force
- about the forces on an object in equilibrium
- how to use a force diagram to work out whether or not an object is in equilibrium.

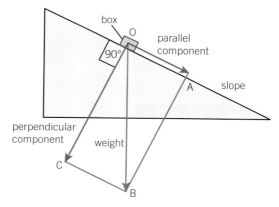

Figure 1 *Cyclists on an uphill road*

Figure 2 *Resolving a force*

Cyclists know that it is more difficult to travel uphill than it is to travel on a flat road (Figure 1). The reason is that the weight of the cyclist and the bicycle have a downhill effect. To understand this effect, consider a small box on a slope as shown in Figure 2. The weight of the box as a force vector is shown by the line labelled OB. You can think of the force vector as two parts or *components* – one force component acting down the slope, and the other force component acting perpendicular or normal to the slope. The process of looking at force in this way is called resolving a force into two components and is carried out as follows:

- A rectangle OABC is drawn on Figure 2. The weight of the box, OB, lies along the diagonal of the rectangle, and the rectangle's sides are parallel and perpendicular to the slope.
- Side OA of the rectangle lies along the slope, and represents the component of the weight acting down the slope. The box stays at rest on the slope and does not slide down, because friction acts on it. The component of weight acting down the slope is too small to overcome this frictional force. For the cyclist on an uphill road, the component of weight acting down the slope is the amount of force that the cyclist has to match in order to keep moving up the hill.
- Side OC of the rectangle OABC gives the component of the weight acting normal to the slope. This is the force pressing on the slope due to the box.

Test an incline

Use the arrangement in Figure 3 to measure the force *F* needed to keep a trolley in the same position on the inclined board.

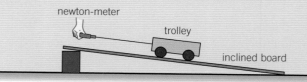

Figure 3 *Testing an incline*

1 Use a newton-meter to measure the weight *W* of the trolley and the force *F*.

Add weights to the trolley to find out how *F* changes as the weight of the trolley is increased.

2 Repeat the test with the board at a different angle to the laboratory bench.

Record all your measurements in a table, and plot a graph of your results.

- Write down the conclusions you make from your graph about the relationship between force *F* and weight *W*.

Safety: Beware of falling objects.

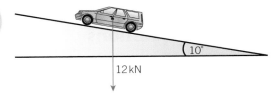

Figure 4 *Car parked on an uphill road*

Worked example

A car of weight 12 kN is parked on an uphill road. The road is inclined at an angle of 10° to the horizontal as shown in Figure 4.

a Use a geometrical method to find the component of the car's weight acting down the slope.

b Describe the force of friction of the road on the car tyres.

Solution

a See Figure 5. The ratio of the rectangle's small side to the diagonal = 1 : 6.0. Therefore, the parallel component of the weight = 12/6 kN = 2.0 kN.

b The frictional force = 2.0 kN acting up the slope.

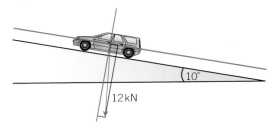

Figure 5

Equilibrium

An object at rest is in equilibrium. The key conditions for an object to be in equilibrium are:

● The resultant force on the object is zero

● The forces acting on the object have no overall turning effect. See Topic P8.7.

To work out whether or not an object is in equilibrium:

● If the lines of action of the forces are parallel, the sum of the forces in one direction must be equal to the sum of the forces in the opposite direction. This means that the resultant force on the object is zero.

● If the lines of action of the forces are not all parallel, the forces can be resolved into two components along the same perpendicular lines. The components along each line must balance out if the resultant force is zero.

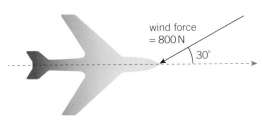

Figure 6 *Aircraft in level flight*

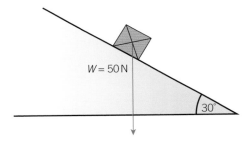

Figure 7 *Box resting on a slope*

1 An aircraft in level flight is travelling due east at a constant speed when it is acted on by a horizontal wind force of 800 N as shown in Figure 6. By resolving the wind force into two perpendicular components, determine the component of the wind force along the line in which the aircraft is moving. [2 marks]

2 A student pushed a trolley of weight 510 N up a slope that is inclined at 15° to the horizontal.
   a Determine the component of the trolley's weight down the slope. [2 marks]
   b The force exerted by the student was greater than the answer to part **a**. Give one possible reason for this difference. [1 mark]

3 Explain why a ladder placed against a wall would slide down the wall if the floor was too slippery. [2 marks]

4 A box of weight 50 N is at rest on a slope. The slope is at an angle of 30° to the horizontal as shown in Figure 7.
   a Copy the diagram and determine the components of the box's weight parallel and perpendicular to the slope. [3 marks]
   b Describe the friction between the box and the slope. [1 mark]

Key points

● Resolving a force means finding perpendicular components that have a resultant force that is equal to the force.

● To resolve a force in two perpendicular directions, draw a rectangle with adjacent sides along the two directions so that the diagonal represents the force vector.

● For an object in equilibrium, the resultant force is zero.

● An object at rest is in equilibrium because the resultant force on it is zero.

P8 Forces in balance

Summary questions

1 Figure 1 shows an iron bar suspended at rest from a spring balance that reads 1.6 N.

Figure 1

  a i Calculate the magnitude and the direction of the force on the spring balance due to the iron bar. [1 mark]

   ii Calculate the weight of the bar in newtons. [1 mark]

  b When a magnet is held under the iron bar, the spring balance reading increases to 2.0 N. Calculate the magnitude and the direction of:

   i the force on the iron bar due to the magnet. [1 mark]

   ii the force on the magnet due to the iron bar. [1 mark]

2 The bottle opener in Figure 2 is being used to force the cap off a bottle. Explain why the force of the bottle opener on the cap is much larger than the force applied to the bottle opener by the person opening the bottle. [2 marks]

Figure 2

3 Figure 3 shows a toy suspended from a ceiling.

  a The star on the toy has a weight of 0.04 N and is a distance of 0.30 m from the point **P** where the thread is attached to the toy. Calculate the moment of the star about point P. [2 marks]

  b The crescent moon attached to the toy is at a distance of 0.20 m from **P**. The star and the crescent moon balance each other because their moments about **P** are equal and opposite. Calculate the weight of the crescent moon. [2 marks]

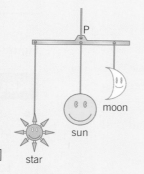
Figure 3

4 Figure 4 shows a wheelbarrow being used to move a bag of sand.

  a Explain why the vertical force F needed to lift the wheelbarrow's legs off the ground is much less than the combined weight of the sand and the wheelbarrow. [3 marks]

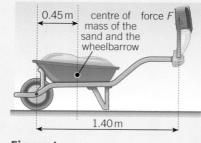
Figure 4

  b A vertical force of 48 N was needed to lift the wheelbarrow's legs off the ground. The force was applied to the handles at a horizontal distance of 1.40 m from the wheel axle. The centre of mass of the bag of sand and the wheelbarrow was a horizontal distance of 0.45 m from the wheel axle.

   i Calculate the combined weight of the bag of sand and the wheelbarrow. [2 marks]

   ii The weight of the wheelbarrow was 65 N. Calculate the weight of the sand. [1 mark]

5 **H** a Look at Figure 5. Two tugboats are used to pull a ship. Each tugboat exerts a force of 7200 N on the ship at an angle of 45° between their cables, as shown in the figure. Use the parallelogram of forces to find the magnitude of the resultant of the tugboat forces on the ship. [2 marks]

  b The ship moves at a constant speed and direction because it is acted on by a drag force. Explain why the drag force has this effect on the ship. [2 marks]

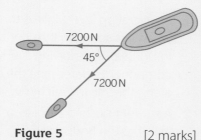
Figure 5

6 Describe how you would investigate how the force needed to make a box slide across a flat level surface depends on the weight of the box. [5 marks]

7 A gear wheel of radius 20 mm is used to turn a bigger gear wheel of radius 50 mm with a force of 360 N.

  a Calculate the moment of the force acting on the bigger gear wheel. [2 marks]

  b Using moments, explain why the turning effect about the bigger gear wheel is greater than the turning effect about the smaller gear wheel. [3 marks]

Practice questions

01.1 Forces can be either contact forces or non-contact forces. Give the name of one contact force and one non-contact force. [2 marks]

01.2 **Figure 1** shows a water skier being pulled by a speed boat.

Figure 1

The motive force of the speed boat is 20 000 N. Copy and complete the sentence using the correct words from the box.

less than	equal to	greater than

When the water skier accelerates through the water, the resistive force of the water is _____ 20 000 N. [1 mark]

01.3 Describe what happens to the speed and resistive force on the water skier as she accelerates through the water. [3 marks]

02.1 An angler is fishing off the beach. His hook has caught in some rocks and he is trying to pull it free.

Figure 2

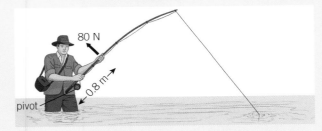

The angler pulls on the fishing road with a force of 80 N. Calculate the turning effect of the pulling force about the pivot. Write down the equation you use. [2 marks]

02.2 The angler uses a hydro-bike to travel to some rocks further out to sea. The centre of mass of a hydro-bike is close to the water to make it stable. State what is meant by the centre of mass of an object. [1 mark]

02.3 The propeller at the rear of the hydro-bike is attached to a small sprocket. A much larger sprocket at the front of the bike is attached to the pedals. A chain connects the two sprockets.

Figure 3

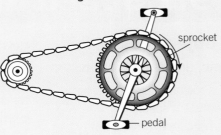

Explain why the front sprocket attached to the pedals must have a larger radius than the rear sprocket attached to the propeller. [3 marks]

03 Ⓗ A drone is a radio-controlled flying device.

Figure 4

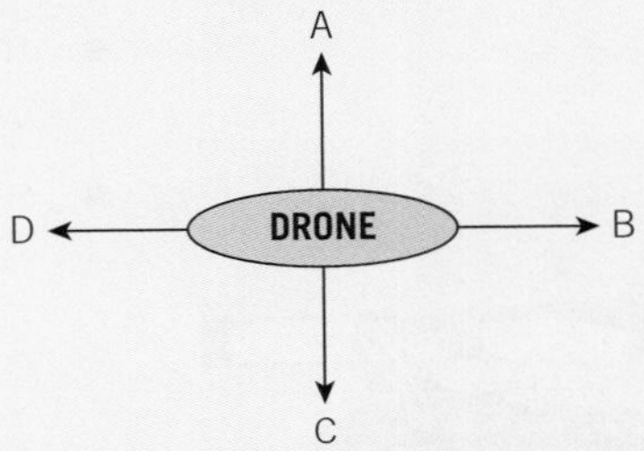

```
           A
           ↑
           |
D ←——— DRONE ———→ B
           |
           ↓
           C
```

03.1 Copy and complete the sentences using correct answers from the box.

force A	force B	force C	force D

When the drone is flying at a constant height _____ and _____ are equal and opposite. When the drone is flying at constant speed _____ and _____ are equal and opposite. [2 marks]

03.2 A video camera is attached to the drone. The drone is used to film alligators in swampland. Give two advantages of using a drone for this purpose. [2 marks]

03.3 Some people object to the use of drones in public places. Suggest two reasons why the use of a drone may be a problem. [2 marks]

04 A tractor pulls a trailer across a field. There is a resultant horizontal force of 400 N acting on the trailer, and a vertical force of 100 N from the weight of the trailer.
Draw a vector diagram to determine the magnitude and direction of the resultant force on the trailer. Use graph paper. [4 marks]

Learning objectives

After this topic, you should know:

- how speed is calculated for an object moving at constant speed
- how to use a distance–time graph to determine whether an object is stationary or moving at constant speed
- what the gradient of the line on a distance–time graph can tell you
- how to use the equation for constant speed to calculate distance moved or time taken.

Figure 1 *The Budweiser Rocket attempting the land speed record in 1979. Thust SSC driven by Andy Green in 1997 achieved a speed of 341 m/s, and still held the world land speed record in 2015*

Study tip

Be careful when plotting points on a graph grid. Make a clear point or a small cross, *not* a large 'blob'.

Rearranging the speed equation

The equation for speed can be written as:

$$v = \frac{s}{t}$$

where *v* is the speed, *s* is the distance travelled, and *t* is the time taken. If you know two of these three quantities, you can find the third by using $v = \frac{s}{t}$ or by rearranging it to give:

$$s = v\,t \quad \text{or} \quad t = \frac{s}{v}.$$

Next time you are travelling on a motorway, look for the marker posts positioned every kilometre. If you are a passenger in a car on a motorway, you can use these posts to check the speed of the car. You need to time the car as it passes each post. Table 1 shows some measurements made on a car journey.

Table 1 *Measurements made on a car journey*

Distance in metres	0	1000	2000	3000	4000	5000	6000
Time in seconds	0	40	80	120	160	200	240

The measurements in Table 1 are plotted on a graph of distance against time in Figure 2.

- the car took 40 s to go from each marker post to the next. So its speed was constant (or uniform).
- the car went a distance of 25 metres every second (= 1000 metres ÷ 40 seconds). So its speed was 25 metres per second.

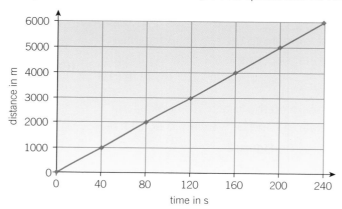

Figure 2 *A distance–time graph*

If the car had travelled faster, it would have gone further than 1000 metres every 40 seconds. So the line on the graph would have been steeper. In other words, the **gradient** of the line would have been greater.

The gradient of a line on a distance–time graph represents speed.

Equation for constant speed

For an object moving at constant speed, you can calculate its speed using the equation:

speed, *v* (metres per second, m/s) = $\dfrac{\text{distance travelled, } s \text{ (metres, m)}}{\text{time taken, } t \text{ (seconds, s)}}$

The scientific unit of speed is the metre per second, usually written as metre/second or m/s.

This equation can also be used to calculate the average speed of an object whose speed varies. For example, if a motorist in a traffic queue took 50 s to travel a distance of 300 m, the car's average speed was 6.0 m/s (= 300 m ÷ 50 s).

Speed in action

Long-distance vehicles are fitted with recorders called tachographs. These can check that their drivers do not drive for too long. Look at the distance–time graph in Figure 3 for three lorries, X, Y, and Z, on the same motorway.

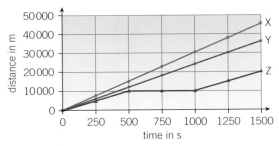

Figure 3 *Comparing distance–time graphs*

- X went fastest because it travelled further than Y or Z in the same time.
- Y travelled more slowly than X. You can see this from the graph because the line for Y has a smaller gradient than the line for X. Also you can see that Y travelled 30 000 metres in 1250 seconds. So its speed was:

  distance ÷ time = 30 000 m ÷ 1250 s = 24 m/s.

- Z travelled the least distance. It stopped for some of the time. This is shown by the flat section of the graph (from 500 s to 1000 s). The gradient is zero in this section because Z was stationary. Its speed was zero during this time. When it was moving, its speed was also less than that of X or Y. You can see this from the graph because the gradient of the line for Z when it was moving is less than the gradient of the lines for X and Y.

1 **a** For an object travelling at constant speed:
   i describe the distance it travels every second. [1 mark]
   ii describe the gradient of its distance–time graph. [1 mark]
   b Look at the distance–time graphs in Figure 3.
   i Calculate the speed of X. [2 marks]
   ii Calculate how long Z stopped for. [1 mark]
   iii Calculate the *average* speed of Z for the 1500 s journey. [2 marks]

2 A vehicle on a motorway travels 1800 m in 60 s. Calculate:
   a the average speed of the vehicle in m/s. [2 marks]
   b how far it would travel if it travelled at this speed for 300 s. [2 marks]
   c how long it would take to travel a distance of 3300 m at this speed. [2 marks]

3 A car on a motorway travels a certain distance in six minutes at a speed of 21 m/s. A coach takes seven minutes to travel the same distance. Calculate the distance and the speed of the coach. [2 marks]

4 A train takes 2 hours and 40 minutes to travel a distance of 360 km.
   a Calculate the average speed of the train in metres per second on this journey. [2 marks]
   b The train travelled at a constant speed of 40 m/s for a distance of 180 km. Calculate the time taken in minutes for this section of the journey. [2 marks]

Synoptic links

For more information on rearranging equations, see Maths skills for Physics.

Study tip

If time is given in minutes or hours in these calculations, always convert it into seconds.

Key points

- The speed of an object is: $v = \frac{s}{t}$.
- The distance–time graph for any object that is:
  - stationary, is a horizontal line
  - moving at constant speed, is a straight line that slopes upwards.
- The gradient of a distance–time graph for an object represents the object's speed.
- The speed equation $v = \frac{s}{t}$ can be rearranged to give:
  $$s = v\,t \text{ or } t = \frac{s}{v}$$

P9.2 Velocity and acceleration

Learning objectives

After this topic, you should know:

- the difference between speed and velocity
- how to calculate the acceleration of an object
- the difference between acceleration and deceleration.

Figure 1 *You experience plenty of changes in velocity on a corkscrew ride*

Synoptic link

Velocity and displacement are vector quantities because they have magnitude (size) and direction. Speed is a scalar quantity because it has magnitude only. See Topic P8.1.

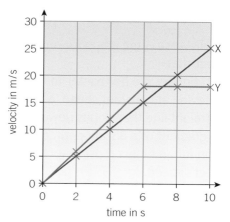

Figure 2 *Velocity–time graph for cars X and Y*

In fairground rides that throw you round and round, your speed and direction of motion keep changing. The word **velocity** is used for speed in a given direction.

Velocity is speed in a given direction.

- Two moving objects can have the same speed but different velocities. For example, a car travelling north at 30 m/s on a motorway has the same speed as a car travelling south at 30 m/s. But their velocities are not the same, because they are moving in opposite directions.

- An object moving round in a circle has a direction of motion that changes continuously as it goes round. So its velocity is not constant even if its speed is constant. For example, a car travelling round a roundabout at constant speed has a continually changing velocity. This is because the direction in which it is moving is continually changing.

An object that travels at constant velocity travels at a constant speed without changing its direction. So it travels in a straight line in a given direction. The word **displacement** is used for the distance travelled in a given direction.

Acceleration

A car maker claims that their new car accelerates more quickly than any other new car. A rival car maker is not pleased by this claim and issues a challenge. Each car in turn is tested on a straight track with a velocity recorder fitted. The results are shown in Table 1.

Table 1 *Results of the car velocity tests*

Time from a standing start in seconds (s)	0	2	4	6	8	10
Velocity of car X in metres per second (m/s)	0	5	10	15	20	25
Velocity of car Y in metres per second (m/s)	0	6	12	18	18	18

Which car has a greater **acceleration**? The results are plotted on the velocity–time graph in Figure 2. You can see that the velocity of Y goes up from zero faster than the velocity of X does. So Y has a greater acceleration in the first six seconds. After that, the velocity of Y is constant so its acceleration is zero after six seconds.

The acceleration of an object is its change of velocity per second. The unit of acceleration is the metre per second squared, or m/s².

Any object with a changing velocity is accelerating. You can work out its average acceleration a for a change of velocity Δv in time t using the equation:

acceleration, *a* (metres per second squared, m/s²)

$$= \frac{\textbf{change in velocity, } \Delta \textbf{\textit{v}} \text{ (metres per second, m/s)}}{\textbf{time taken for the change, } \textbf{\textit{t}} \text{ (seconds, s)}}$$

For an object that accelerates steadily from an initial velocity u to a final velocity v, in time t:

its change of velocity Δv = final velocity v – initial velocity u.

So you can write the equation for acceleration as:

$$a = \frac{v - u}{t}$$

Deceleration

A car decelerates when the driver brakes. The term **deceleration** or negative acceleration is used for any situation where an object slows down.

Worked example

A car moving at a velocity of 28 m/s brakes and stops in 8.0 s.

Calculate its deceleration.

Solution

initial velocity u = 28 m/s, final velocity v = 0, time taken t = 8.0 s

$$acceleration\ a = \frac{change\ in\ velocity}{time\ taken} = \frac{v - u}{t}$$
$$= \frac{0\ m/s - 28\ m/s}{8.0\ s} = -3.5\ m/s^2$$

The deceleration is therefore 3.5 m/s².

Worked example

In Figure 2, the velocity of Y increases from 0 to 18 m/s in 6.0 s.

Calculate its acceleration.

Solution

Change of velocity = $v - u$
$$= 18\ m/s - 0\ m/s$$
$$= 18\ m/s$$

Time taken t = 6.0 s

$$Acceleration\ a = \frac{change\ in\ velocity}{time\ taken}$$
$$= \frac{v - u}{t}$$
$$= \frac{18\ m/s}{6.0\ s}$$
$$= 3.0\ m/s^2$$

Study tip

Be careful with units, especially the unit of acceleration. The unit is m/s², that is, the change in speed measured in m/s that occurs every second.

Study tip

A minus value for acceleration means that the object is slowing down.

If you are talking about deceleration, you do not need to include a minus sign.

1 **a** Compare speed and velocity. [1 mark]
   b A car on a motorway is travelling at a constant speed of 30 m/s when it overtakes a lorry travelling at a speed of 22 m/s. If both vehicles maintain their speeds, calculate how far ahead of the lorry the car will be after 300 s. [2 marks]

2 The velocity of a car increased from 8 m/s to 28 m/s in 16 s without change of direction. Calculate its acceleration. [2 marks]

3 The driver of a car increased the speed of the car as it joined the motorway. It then travelled at constant velocity before slowing down as it left the motorway at the next junction.
   a i State when the car decelerated. [1 mark]
   ii State when the acceleration of the car was zero. [1 mark]
   b When the car joined the motorway, it accelerated from a speed of 7.0 m/s for 10 s at an acceleration of 2.0 m/s². Calculate its speed at the end of this time. [2 marks]

4 A sprinter in a 100 m race accelerated from rest and reached a speed of 9.2 m/s in the first 3.1 s.
   a Calculate the acceleration of the sprinter in this time. [2 marks]
   b The sprinter continued to accelerate to top speed and completed the race in 10.4 s. Calculate the sprinter's average speed. [2 marks]

Key points

- Velocity is speed in a given direction.
- A vector is a physical quantity that has a direction as well as a magnitude. A scalar is a physical quantity that has a magnitude only and does not have a direction.
- The acceleration of an object is $a = \frac{\Delta v}{t}$
- Deceleration is the change of velocity per second when an object slows down.

Learning objectives

After this topic, you should know:

- how to measure velocity changes
- what a horizontal line on a velocity–time graph tells you
- how to use a velocity–time graph to work out whether an object is accelerating or decelerating
- **H** what the area under a velocity–time graph tells you.

Investigating acceleration

Use a motion sensor and a computer to find out how the steepness of a runway affects a trolley's acceleration.

- In this investigation, name:
  - **i** the independent variable
  - **ii** the dependent variable.
- Describe the relationship you find between the two variables.

Safety: Use foam or an empty cardboard box to stop the trolley falling off the bench. Mind your feet!

Investigating acceleration

You can use a motion sensor linked to a computer to record how the velocity of an object changes. Figure 1 shows how you can do this using a trolley as the moving object. The computer can also be used to display the measurements as a velocity–time graph.

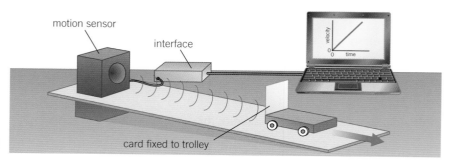

Figure 1 *Investigating velocity–time graphs using a computer*

Test A: If you let the trolley accelerate down the runway, its velocity increases with time. Look at the velocity–time graph from a test run on the laptop screen in Figure 1.

- The line goes up because the velocity increases with time. So it shows that the trolley was accelerating as it ran down the runway.
- The line is straight, which tells you that the increase in velocity was the same every second. In other words, the acceleration of the trolley was constant.

Test B: If you make the runway steeper, the trolley accelerates faster. This would make the line on the graph in Figure 1 steeper than for test A. So the acceleration in test B is greater.

These tests show that:

the gradient of the line on a velocity–time graph represents acceleration.

Braking

Braking reduces the velocity of a vehicle. Look at the graph in Figure 2. It is the velocity–time graph for a vehicle that brakes and stops at a set of traffic lights. The velocity is constant until the driver applies the brakes.

- The section of the graph with the horizontal line shows constant velocity. The gradient of the line is zero, so the acceleration in this section is zero.
- When the brakes are applied, the vehicle decelerates, and its velocity decreases to zero. The gradient of the line is negative in this section. So the acceleration is negative.

Higher

The area under the line on a velocity–time graph represents distance travelled in a given direction (or displacement).

Look at Figure 2 again.

- Before the brakes are applied, the vehicle moves at a velocity of 20 m/s for 10 s. So it travels 200 m in this time (= 20 m/s × 10 s). This distance is represented on the graph by the area under the line from 0 s to 10 s. This is the rectangle shaded yellow on the graph.

- When the vehicle decelerates in Figure 2, its velocity drops from 20 m/s to 0 m/s in 5 s. You can work out the distance travelled in this time from the area of the purple triangle in Figure 2. This area is:

$$\frac{1}{2} \times \text{the height of the triangle} \times \text{the base of the triangle}$$

$$= \frac{1}{2} \times 20 \,\text{m/s} \times 5 \,\text{s} = 50 \,\text{m}.$$

So the vehicle must have travelled a distance of 50 m when it was decelerating.

Speed and velocity can also be described in kilometres per hour (km/h).

Because 1000 m = 1 km, and 3600 s = 1 hour, then a speed of 1 km/h is equal to 1000 m ÷ 3600 s = 0.278 m/s.

If speed and velocity are given in kilometres per hour (km/h), their values must be converted to metres per second before plotting them on a graph or using them in an equation.

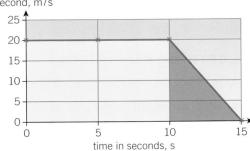

velocity in metres per second, m/s

Figure 2 *Braking*

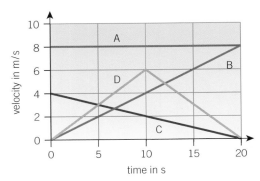

velocity in m/s

time in s

Figure 3

1 Match each of the following descriptions (**i** to **iv**) to one of the lines, labelled **A**, **B**, **C**, and **D** on the velocity–time graph (Figure 3).
   i Accelerated motion throughout.
   ii Zero acceleration.
   iii Accelerated motion, then decelerated motion.
   iv Deceleration throughout. [2 marks]

2 Ⓗ Look at Figure 3.
   a Identify the line representing the object that travelled:
      i the furthest distance [1 mark]
      ii the least distance. [1 mark]
   b Which object, **B** or **D**, travelled further? [1 mark]

3 Ⓗ Look again at Figure 3. Show that the object that produced the data for line A (the horizontal line) travelled a distance of 160 m. [1 mark]
   b Determine the distance travelled by object B. [1 mark]

4 Ⓗ Look again at the graph in Figure 3.
   a Calculate the distance travelled by object **C**. [2 marks]
   b Calculate the difference in the distances travelled by **A** and **D**. [3 marks]

Study tip

If you are drawing a straight line graph, always use a ruler.

Key points

- A motion sensor linked to a computer can be used to measure velocity changes.
- The gradient of the line on a velocity–time graph represents acceleration.
- If a velocity–time graph is a horizontal line, the acceleration is zero.
- A positive gradient on a velocity–time graph represents positive acceleration, a negative gradient represents deceleration.
- Ⓗ The area under the line on a velocity–time graph represents distance travelled.

P9.4 Analysing motion graphs

Learning objectives

After this topic, you should know:

- how to calculate acceleration from a velocity–time graph
- **H** how to calculate distance from a velocity–time graph
- **H** how to calculate speed from a distance–time graph:
  - where the speed is constant
  - where the speed is changing.

Using distance–time graphs

For an object moving at constant speed, the distance–time graph is a straight line sloping upwards (Figure 1).

The speed of the object is represented by the gradient of the line. To find the gradient, you need to draw a triangle under the line, as shown in Figure 2. The height of the triangle represents the distance travelled, and the base of the triangle represents the time taken. So:

$$\text{the gradient of the line} = \frac{\text{the height of the triangle}}{\text{the base of the triangle}}$$

and this represents the object's speed.

For a moving object with changing speed, the distance–time graph is not a straight line. The red line in Figure 2 shows an example.

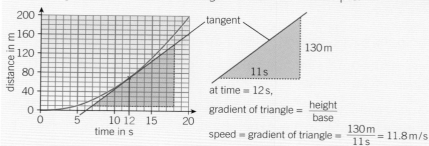

at time = 12 s,

$$\text{gradient of triangle} = \frac{\text{height}}{\text{base}}$$

$$\text{speed} = \text{gradient of triangle} = \frac{130\,\text{m}}{11\,\text{s}} = 11.8\,\text{m/s}$$

Figure 2 *A distance–time graph for changing speed*

In Figure 2, the gradient of the line increases, so the object's speed must have increased. You can find the speed at any point on the line by drawing a tangent to the line at that point, as shown in Figure 2. The **tangent** to the curve is a straight line that touches the curve without cutting through it (in this case at 12 s). The gradient of the tangent is equal to the speed of the object at that instant in time.

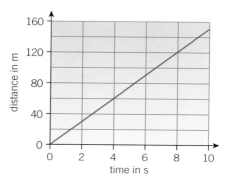

Figure 1 *A distance–time graph for constant speed*

Acceleration

Use the graph in Figure 3 to find the acceleration of the object.

Solution

The height of the triangle represents an increase of velocity of 8 m/s (= 12 m/s – 4 m/s).

The base of the triangle represents a time of 10 s.

So the acceleration = $\dfrac{\text{change of velocity}}{\text{time taken}}$

$$= \frac{8\,\text{m/s}}{10\,\text{s}} = 0.8\,\text{m/s}^2$$

Using velocity–time graphs

Figure 3 shows the velocity–time graph of an object moving with a constant acceleration. Its velocity increases at a steady rate. So the graph shows a straight line that has a constant gradient.

To find the acceleration from the graph, remember that the gradient of the line on a velocity–time graph represents the acceleration.

In Figure 3, the gradient is given by the height divided by the base of the triangle under the line. The height of the triangle represents the change of velocity, and the base of the triangle represents the time taken.

So the gradient represents the acceleration, because:

$$\text{acceleration} = \frac{\text{change of velocity}}{\text{time taken}}$$

To find the distance travelled from the graph, remember that the area under a line on a velocity–time graph represents the distance travelled. The shape under the line in Figure 3 is a triangle on top of a rectangle. So the distance travelled is represented by the area of the triangle plus the area of the rectangle under it. Prove for yourself that the triangle represents a distance travelled of 40 m and that the rectangle also represents a distance of 40 m. So the total distance travelled is 80 m (= 40 m + 40 m).

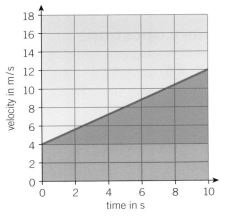

Figure 3 *A velocity–time graph for constant acceleration*

Distance, velocity, and acceleration

In Figure 3, the total distance travelled is 80 m, and the time taken t is 10 s. So the average velocity is 8 m/s, which is equal to $\frac{1}{2}(u + v)$, where the initial velocity $u = 4$ m/s, and the final velocity $v = 12$ m/s. So the distance travelled $s = \frac{1}{2}(u + v) \times t$.

Because the acceleration $a = \frac{v - u}{t}$, then

$$a \times s = \frac{v - u}{t} \times \frac{1}{2}(u + v) \times t = \frac{1}{2}(v^2 - u^2)$$

Rearranging this equation gives $v^2 - u^2 = 2as$

This equation is useful for calculations where the time taken is not given and the acceleration is constant. You *do not* need to know how to prove this equation.

Study tip

Make sure that you know whether you are dealing with a distance–time graph or a velocity–time graph. The gradients of these two types of graph represent different quantities.

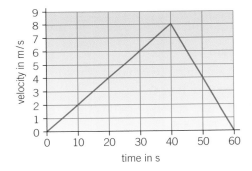

Figure 4

1 **a** Determine the speed of the object shown on the graph in Figure 1.   [2 marks]
   b Describe how the speed of the object shown in Figure 2 changes.   [2 marks]

2 The graph in Figure 4 shows how the velocity of a cyclist on a straight road changes with time.
   a Describe the motion of the cyclist.   [2 marks]
   b Use the graph in Figure 4 to determine the acceleration of the cyclist and the distance travelled in:
      i the first 40 s   [4 marks]   **ii** the following 20 s.   [4 marks]
   c Ⓗ Calculate the average speed of the cyclist over the journey.   [2 marks]

3 In a motorcycle test, the speed from rest was recorded at intervals in Table 1.

   Table 1 *Motorcycle test results*

Time in s	0	5	10	15	20	25	30
Velocity in m/s	0	10	20	30	40	40	40

   a Plot a velocity–time graph of these results.   [3 marks]
   b Calculate the initial acceleration of the motorcycle.   [2 marks]
   c Ⓗ Calculate how far the motorcycle moved in:
      i the first 20 s   **ii** the following 10 s.   [4 marks]

4 Use Table 1 and the equation $v^2 = u^2 + 2as$ to calculate the velocity of the motorcycle after 1.0 km from the start if it had kept the same acceleration as it had during the first 20 s.   [2 marks]

Key points

- The speed of an object moving at constant speed is given by the gradient of the line on its distance–time graph.
- The acceleration of an object is given by the gradient of the line on its velocity–time graph.
- Ⓗ The distance travelled by an object is given by the area under the line on its velocity–time graph.
- Ⓗ The speed, at any instant in time, of an object moving at changing speed is given by the gradient of the tangent to the line on its distance–time graph.

P9 Motion

Summary questions

1 A model car travels round a circular track at constant speed.
   If you were given a stopwatch, a marker, and a tape measure, discuss how you would measure the speed of the car. [5 marks]

2 Figure 1 shows the distance–time graph for a car on a motorway.

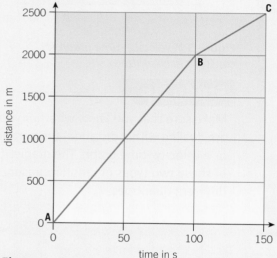

Figure 1

   a Determine which part of the journey was faster – **A** to **B**, or **B** to **C**. [1 mark]
   b i Calculate the speed of the car between **A** and **B**. [2 marks]
     ii Calculate the speed of the car between **B** and **C**. [2 marks]
   c If the car had travelled the whole distance of 2500 m at the same speed as it travelled between **A** and **B**, calculate how long the journey would have taken. [2 marks]

3 Figure 2 shows a distance–time graph for a motorcycle approaching a speed limit sign.

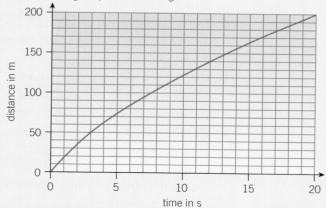

Figure 2

   a Describe how the speed of the motorcycle changed with time. [2 marks]
   b 🅗 Use the graph to determine the speed of the motorcycle:
     i initially   ii 10 s later. [5 marks]

4 a A car took 10 s to increase its velocity from 5 m/s to 30 m/s. Calculate its acceleration. [2 marks]
   b 🅗 The graph (Figure 3) shows how the velocity of the car changed with time during the 10 s.

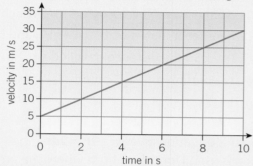

Figure 3

   i Calculate how far the car travelled in this time. [3 marks]
   ii Calculate the average speed of the car in this time. [2 marks]

5 Table 1 shows how the velocity of a train changed as it travelled from one station to the next.

Table 1

Time in s	0	20	40	60	80	100	120	140	160
Velocity in m/s	0	5	10	15	20	20	20	10	0

   a Plot a velocity–time graph using this data. [3 marks]
   b Calculate the train's acceleration in each of the three parts of the journey. [5 marks]
   c 🅗 Calculate the total distance travelled by the train. [4 marks]
   d Show that the average speed for the train's journey was 12.5 m/s. [2 marks]

6 A water skier started from rest and accelerated steadily to 12 m/s in 15 s, then travelled at constant speed for 45 s, before slowing down steadily and coming to a halt 90 s after she started.
   a Draw a velocity–time graph for this journey. [3 marks]
   b Calculate the acceleration of the water skier in the first 15 s. [2 marks]
   c Calculate the deceleration of the water skier in the final 30 s. [2 marks]
   d 🅗 Calculate the total distance travelled by the water skier. [4 marks]

Practice questions

01 An electric truck is used to deliver parcels to different locations in a warehouse.
The distance-time graph shows the journey taken by the truck.

Figure 1

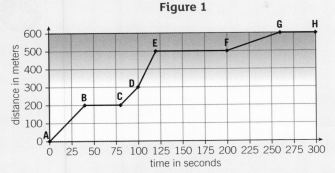

01.1 Determine the total distance travelled by the truck. [1 mark]

01.2 Determine the total time taken to offload all the parcels. [2 marks]

01.3 Between which two points was the truck travelling at the greatest speed?
Give a reason for your answer. [2 marks]

01.4 Between which two points did the truck change speed without stopping. [2 marks]

01.5 Calculate the average speed of the truck over the complete journey.
Write down the equation you use. [3 marks]

01.6 State one advantage of using an electric truck rather than a diesel truck in a warehouse. [1 mark]

02.1 Describe the difference between the terms speed and velocity. [2 marks]

02.2 A firework rocket is sent up into the air. The velocity–time graph of the journey is shown.

Figure 2

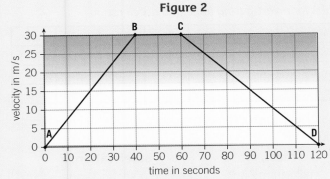

Calculate the average acceleration of the rocket between 0 s and 40 s.
Give the unit. [3 marks]

02.3 Ⓗ Use the graph in Figure 2 to calculate the total distance the rocket travels. [2 marks]

02.4 Copy and complete the sentences using the correct words from the box.

| stationary   decelerating   accelerating |
| travelling at a constant speed |

The rocket is _____ between points A and B.
The rocket is _____ between points B and C.
The rocket is _____ between points C and D. [3 marks]

03 Engineers are testing a new water slide in an aqua park. They are using a plastic dummy and measuring the time it takes for the dummy to travel from top to bottom. The rate of running water through the slide is being changed.

03.1 Name the independent variable. [1 mark]

03.2 Name the dependent variable. [1 mark]

03.3 Suggest one reason why this is not a suitable test. [1 mark]

03.4 In one test the dummy is sliding at 10.0 m/s at the end of the slide. The total slide length is 20 m.
Calculate the acceleration of the dummy through the slide. Use the equation:
$v^2 - u^2 = 2as$ [2 marks]

04 Some physical quantities are called scalars and some physical quantities are called vectors.
Copy **Table 1**. Tick **one** box in each line.

Table 1

Physical Quantity	Scalar	Vector
speed		
mass		
acceleration		
weight		
energy		

[5 marks]

05 Competitors are taking part in a triathlon race which involves running, swimming and cycling in immediate succession over a 50 km distance.
Describe the factors that may affect the eventual outcome of the race.
Your answer should include typical speed values at various stages and what other factors may affect the time of all the competitors in the race. [4 marks]

Learning objectives

After this topic, you should know:

- how the acceleration of an object depends on the size of the resultant force acting upon it
- the effect that the mass of an object has on its acceleration
- how to calculate the resultant force on an object from its acceleration and its mass
- (H) what is meant by the inertia of an object.

Synoptic links

Calculating acceleration from a velocity-time graph was covered in Topic P9.4.

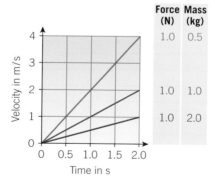

Force (N)	Mass (kg)
1.0	0.5
1.0	1.0
1.0	2.0

Figure 2 *Velocity–time graph for different combinations of force and mass*

Rearranging resultant force equation

You can rearrange the equation $F = m \times a$ to give $a = \dfrac{F}{m}$ or $m = \dfrac{F}{a}$

You can use the apparatus shown in Figure 1 to accelerate a trolley with a constant force.

Investigating force and acceleration

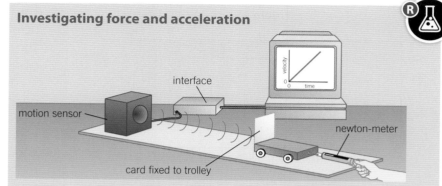

Figure 1 *Investigating the link between force and motion*

Use the newton-meter to pull the trolley along a known distance with a constant force.

A motion sensor and a computer can be used to record the velocity of the trolley as it accelerates.

- Predict what will happen to the acceleration of the trolley if the force is increased or decreased.

You can double or treble the total moving mass of the trolley by using double-deck and triple-deck trolleys.

- Predict what will happen to the acceleration of the trolley if the mass of the trolley is increased.

Safety: Protect your bench and feet from falling trolleys.

You can display the results of this investigation as a velocity–time graph on the computer screen. Figure 2 shows velocity–time graphs for different trolley masses. You can work out the acceleration from the gradient of the line. Some typical results using different forces and masses are given in Table 1.

Table 1 *Typical results for investigating force and acceleration*

Resultant force in N	0.5	1.0	1.5	2.0	4.0	6.0
Mass in kg	1.0	1.0	1.0	2.0	2.0	2.0
Acceleration in m/s²	0.5	1.0	1.5	1.0	2.0	3.0
Mass × acceleration in kg m/s²	0.5	1.0	1.5	2.0	4.0	6.0

The results show that the resultant force, the mass, and the acceleration are linked by the equation:

resultant force, F = mass, m × acceleration, a
(newtons, N)   (kilograms, kg)   (metres per second squared, m/s²)

Newton's Second Law

Newton's Second Law of motion says that the acceleration of an object is:

- proportional to the resultant force on the object
- inversely proportional to the mass of the object.

So the resultant force is proportional to the object's mass multiplied by its acceleration. You can write this as $F \propto ma$, where the symbol $\propto$ means 'is proportional to'. You can see from the equation for resultant force that 1 N is the force that gives a 1 kg mass an acceleration of 1 m/s².

Inertia

A resultant force is needed to change the velocity of an object. The tendency of an object to stay at rest or to continue in uniform motion (i.e., moving at constant velocity) is called its **inertia**. The inertial mass of an object is a measure of the difficulty of changing the object's velocity.

Inertial mass can be defined as $\dfrac{\text{force}}{\text{acceleration}}$.

Speeding up or slowing down

If the velocity of an object changes, it must be acted on by a resultant force. Its acceleration is always in the same direction as the resultant force.

● The velocity of the object increases (i.e., it accelerates) if the resultant force is in the *same* direction as the velocity. You say that its acceleration is positive because it is in the same direction as its velocity.

● The velocity of the object decreases (i.e., it decelerates) if the resultant force is in the *opposite* direction to its velocity. You say that its acceleration is negative because it is in the opposite direction to its velocity.

1 a Calculate the resultant force on a sprinter of mass 80 kg who accelerates at 8 m/s². [1 mark]
   b Calculate the acceleration of a car of mass 800 kg acted on by a resultant force of 3200 N. [1 mark]
2 Copy and complete the table. [5 marks]

	Force in N	Mass in kg	Acceleration in m/s²
a		20	0.80
b	200		5.0
c	840	70	
d		0.40	6.0
e	5000		0.20

3 A car and a trailer have a total mass of 1500 kg.
   a Calculate the force needed to accelerate the car and the trailer at 2.0 m/s². [1 mark]
   b The mass of the trailer is 300 kg. Determine:
      i   the force of the tow bar on the trailer [1 mark]
      ii the resultant force on the car. [2 marks]
4 A constant force was used as in Figure 1 to accelerate a trolley from rest. The acceleration of the trolley was 0.60 m/s². A mass of 0.5 kg was then fixed onto the trolley, and the same force as before gave it an acceleration of 0.48 m/s².
   a i Ⓗ Explain what is meant by the inertia of an object. [1 mark]
      ii Explain why the acceleration in the second case was less than before. [2 marks]
   b Use the data above to calculate the mass of the trolley. [2 marks]

Worked example

Calculate the resultant force on an object of mass 6.0 kg when it has an acceleration of 3.0 m/s².

Solution

resultant force = mass × acceleration
= 6.0 kg × 3.0 m/s²
= 18.0 N

Worked example

Calculate the acceleration of an object of mass 5.0 kg acted on by a resultant force of 40 N.

Solution

Rearranging $F = m \times a$ gives
$a = \dfrac{F}{m} = \dfrac{40\,\text{N}}{5.0\,\text{kg}} = 8.0\,\text{m/s}^2$

Key points

● The greater the resultant force on an object, the greater the object's acceleration.
● The greater the mass of an object, the smaller its acceleration for a given force.
● The resultant force acting on an object is $F = ma$
● Ⓗ The inertia of an object is its tendency to stay at rest or in uniform motion.

Learning objectives

After this topic, you should know:

- the difference between mass and weight
- about the motion of a falling object acted on only by gravity
- what terminal velocity means
- what can be said about the resultant force acting on an object that is falling at terminal velocity.

If you release an object above the ground, it falls because of its weight (i.e., the force acting on the object due to gravity).

If no other forces act on it, the resultant force on it is its weight. The object is said to be falling freely. It accelerates downwards at a constant acceleration of $10 \, \text{m/s}^2$. This is the acceleration due to gravity (or the acceleration of free fall) and is represented by the symbol g. For example, if you release a 1 kg object above the ground:

- the gravitational force on it is 10 N, and
- its acceleration $\left(= \dfrac{\text{force}}{\text{mass}} = \dfrac{10 \, \text{N}}{1 \, \text{kg}} \right) = 10 \, \text{m/s}^2$.

Mass and weight

Your weight is caused by the gravitational force of attraction between you and the Earth. This force is very slightly weaker at the equator than at the poles, so at the equator you will weigh slightly less than at the poles. However, your mass will be the same no matter where you are.

- The **weight** of an object is the force acting on it due to gravity. Weight is measured in newtons, N.

- The **mass** of an object depends on the quantity of matter in it. Mass is measured in kilograms, kg.

You can measure the weight of an object by using a newton-meter. The weight of an object:

- of mass 1 kg is 10 N
- of mass 5 kg is 50 N.

The gravitational force on a 1 kg object is the **gravitational field strength** at the place where the object is. Gravitational field strength is measured in Newtons per kilogram (N/kg). The Earth's gravitational field strength at its surface is about 9.8 N/kg.

If we know the mass of an object, we can calculate the force due to gravity which acts on it (i.e., its weight) using the equation:

$$\text{weight, } W \quad = \quad \text{mass, } m \quad \times \quad \text{gravitational field strength, } g$$

(newtons, N)   (kilograms, kg)   (newtons per kilogram, N/kg)

Worked example

Calculate the weight in newtons of a person of mass 55 kg.

Solution

Weight = mass × gravitational field strength

= 55 kg × 9.8 N/kg = **540 N**

Terminal velocity

If an object falls in a fluid, the fluid drags on the object because of friction between the fluid and the surface of the moving object. This frictional force increases with speed. At any instant, the resultant force on the object is its weight minus the frictional force on it.

- The acceleration of the object decreases as it falls. This is because the frictional force increases as it speeds up. So the resultant force on it decreases and therefore its acceleration decreases.

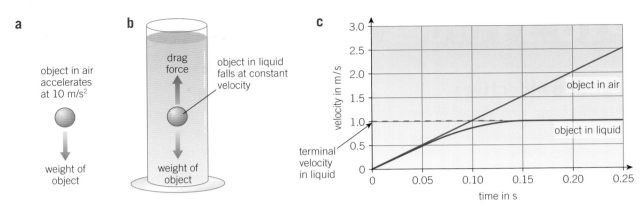

Figure 1 *Falling objects.* **a** *Falling in air,* **b** *falling in a liquid, and* **c** *velocity–time graph for* **a** *and* **b**

- The object reaches a constant velocity when the frictional force on it is equal and opposite to its weight. This velocity is called its **terminal velocity**. The resultant force is then zero, so its acceleration is zero.

When an object moves through the air (instead of water), the frictional force is called air resistance. This is not shown in Figure 1a because the air resistance on the object is much smaller than the frictional force on it when it is falling in the liquid in (Figure 1b). The object in Figure 1a would need to fall much further through the air before it reaches a constant velocity.

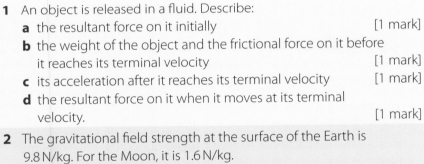

1   An object is released in a fluid. Describe:
   a the resultant force on it initially [1 mark]
   b the weight of the object and the frictional force on it before it reaches its terminal velocity [1 mark]
   c its acceleration after it reaches its terminal velocity [1 mark]
   d the resultant force on it when it moves at its terminal velocity. [1 mark]

2   The gravitational field strength at the surface of the Earth is 9.8 N/kg. For the Moon, it is 1.6 N/kg.
   a Calculate the weight of a person of mass 50 kg on the Earth. [1 mark]
   b Calculate the weight of the same person if she was on the Moon. [1 mark]
   c A lunar vehicle weighs 300 N on the Earth. Calculate its weight on the Moon. [2 marks]

3   A parachutist of mass 70 kg supported by a parachute of mass 20 kg reaches a constant speed.
   a Explain why the parachutist reaches a constant speed. [4 marks]
   b Calculate:
      i the total weight of the parachutist and the parachute [1 mark]
      ii the size and direction of the force of air resistance on the parachute when the parachutist falls at constant speed. [1 mark]

4   **a** Use Figure 1 to determine the acceleration of the object in liquid 0.10 s after it was released. [2 marks]
   b Show that the ratio of the drag force to the weight at 0.1 s is about 0.5. [3 marks]

Study tip

When the upward force acting on an object falling in a fluid balances the downward force, the object continues at a constant speed – it doesn't stop!

Key points

- The weight of an object is the force acting on the object due to gravity. Its mass is the quantity of matter in the object.
- An object acted on only by gravity accelerates at about $10 \, \text{m/s}^2$.
- The terminal velocity of an object is the velocity it eventually reaches when it is falling. The weight of the object is then equal to the frictional force on the object.
- When an object is moving at terminal velocity, the resultant force on it is zero.

P10.3 Forces and braking

Learning objectives

After this topic, you should know:

- the forces that oppose the driving force of a vehicle
- what the stopping distance of a vehicle depends on
- what can increase the stopping distance of a vehicle
- how to estimate the braking force of a vehicle.

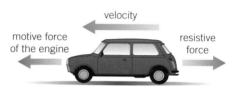

Figure 1 *Constant velocity*

Reaction time challenge

Use a ruler to test your own reaction time. Ask a friend to hold a ruler just above your hand and drop it without warning. You should try to catch it. The distance the ruler drops before you catch it can be used to work out your reaction time. Typical reaction times range from 0.4 s to 0.9 s.

- Try this when concentrating and when distracted by something else, such as a conversation.

Forces on the road

For any car travelling at constant velocity, the resultant force on it is zero. This is because the driving force of its engine is balanced by the resistive forces, which are mostly caused by air resistance.

A car driver uses the accelerator pedal to vary the driving force of the engine.

The braking force needed to stop a vehicle in a given distance depends on:

- the speed of the vehicle when the brakes are first applied
- the mass of the vehicle.

You can see this using the equation 'resultant force = mass × acceleration', which you first met in Topic P10.1, where the braking force is the resultant force.

- The greater the speed, the greater the deceleration needed to stop the vehicle within a given distance. So the braking force required to stop a car travelling at a high speed must be greater than the braking force required to stop a car travelling at a low speed, within the same distance.
- The greater the mass, the greater the braking force needed for a given deceleration.

Stopping distances

Driving tests always ask about **stopping distances**. This is the shortest distance a vehicle can safely stop in, and is in two parts:

1. The **thinking distance** – the distance travelled by the vehicle in the time it takes the driver to react (i.e., during the driver's reaction time). Because the car moves at constant speed during the reaction time, the thinking distance is equal to the speed × the reaction time. This shows that the thinking distance is proportional to the speed.

2. The **braking distance** – the distance travelled by the vehicle during the time the braking force acts.

stopping distance = thinking distance + braking distance

Figure 2 shows the stopping distance for a vehicle on a dry flat road travelling at different speeds. Check for yourself that the stopping distance at 31 m/s (70 miles per hour) is 96 m.

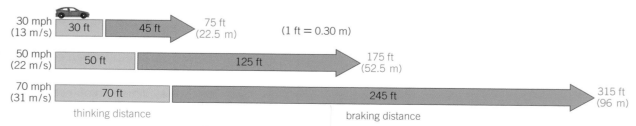

Figure 2 *Stopping distances*

Factors affecting stopping distances

1 Tiredness, alcohol, and drugs affect the brain and increase reaction times. Distractions like using a mobile phone also increase reaction time and cause serious accidents. All these factors increase the thinking distance (because thinking distance = speed × reaction time). So the stopping distance is greater.

2 The faster a vehicle is travelling, the further it travels before it stops. This is because the thinking distance and the braking distance both increase with increased speed.

3 In adverse road conditions, for example on wet or icy roads, drivers have to brake with less force to avoid skidding. So stopping distances are greater in poor weather and road conditions.

4 Poorly maintained vehicles, for example with worn brakes or tyres, take longer to stop because the brakes and tyres are less effective.

Road vehicles and forces

The deceleration of a road vehicle depends on the friction between the road and the car tyres. To avoid skidding on a dry flat road, the deceleration should be no more than about 6 m/s². Check this yourself using the equation $a = \dfrac{v^2 - u^2}{2s}$ and data from Figure 2.

Vehicle masses range from about 1000 kg for a car to about 38 000 kg for a heavy truck. So braking forces vary from about 6 kN for a car to about 250 kN for a heavy truck. The same range of forces are needed to accelerate a road vehicle.

1 For each of the following factors, identify which distance of a vehicle (thinking distance or breaking distance) is affected:
 a the road surface [1 mark]
 b the tiredness of the driver [1 mark]
 c poorly maintained brakes. [1 mark]

2 a Use the chart in Figure 2 to calculate, in metres, the effect of an increase from 13 m/s (30 mph) to 22 m/s (50 mph) on the following:
  i the thinking distance [1 mark] ii the braking distance [1 mark]
  iii the stopping distance. [1 mark]
 b A driver has a reaction time of 0.8 s. Calculate the change in her thinking distance if she travels at 15 m/s instead of 30 m/s. [2 marks]

3 a When the speed of a car is doubled:
  i explain why the thinking distance of the driver is doubled, assuming that the driver's reaction time is unchanged [2 marks]
  ii explain why the braking distance is more than doubled. [2 marks]
 b A student thinks that braking distance is proportional to the square of the speed. Use the chart in Figure 2 to decide whether or not this is a valid claim. [3 marks]

4 A car of mass 1500 kg moving at 31 m/s decelerates and stops in a distance of 75 m.
 a Ⓗ Calculate the deceleration of the vehicle. [2 marks]
 b Estimate the braking force on the car. [2 marks]

Deceleration

The deceleration a of a vehicle can be calculated using the equation $v^2 = u^2 + 2as$, where s is the distance travelled, u is the initial speed, and v is the final speed. You first met this equation in Topic P9.4. Rearranging the equation with $v = 0$ for a vehicle that stops gives

$$s = -\frac{u^2}{2a} \text{ or } a = -\frac{u^2}{2s}.$$

- $s = -\dfrac{u^2}{2a}$ shows that s is proportional to u^2 for constant deceleration.

- $a = -\dfrac{u^2}{2s}$ can be used to calculate the deceleration when the values for u and s are given.

Braking force

To work out the braking force, use the equation $F = ma$ with given values or estimates for m and a.

Synoptic links

You learnt about friction in Topic P8.3.

Key points

- Friction and air resistance oppose the driving force of a vehicle.
- The stopping distance of a vehicle depends on the thinking distance and the braking distance.
- High speed, poor weather conditions, and poor vehicle maintenance all increase the braking distance. Poor reaction time (due to tiredness, alcohol, drugs, or using a mobile phone) and high speed both increase the thinking distance.
- $F = ma$ gives the braking force of a vehicle.

P10.4 Momentum

Learning objectives

After this topic, you should know:

- how to calculate momentum
- the unit of momentum
- what momentum means for a closed system
- that two objects that push each other apart move away with equal and opposite momentum.

Figure 1 *A contact sport*

Worked example

Calculate the momentum of a sprinter of mass 50 kg running at a velocity of 10 m/s.

Solution

Momentum = mass × velocity
= 50 kg × 10 m/s
= **500 kg m/s**

Synoptic links

Momentum is a vector quantity because it has size and direction. See Topic P8.1.

Momentum is important to anyone who plays a contact sport. In a game of rugby, a player with a lot of momentum is very difficult to stop.

momentum of a moving object, *p* (kg m/s) = **mass, *m*** (kg) × **velocity, *v*** (m/s)

Momentum has both a size and a direction so it is a vector quantity.

Investigating collisions

When two objects collide, the momentum of each object changes. Figure 2 shows how to use a computer and a motion sensor to investigate a collision between two trolleys.

Trolley A is given a push so that it collides with a stationary trolley B. The two trolleys stick together after the collision. The computer gives the velocity of A before the collision and the velocity of both trolleys afterwards.

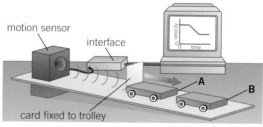

Figure 2 *Investigating collisions*

1   For two trolleys of the same mass, the velocity of trolley A is halved by the impact. The combined mass after the collision is twice the moving mass before the collision. So the momentum (= mass × velocity) after the collision is the same as the momentum before the collision.

2   For a single trolley pushed into a double trolley, the velocity of A is reduced to one-third. The combined mass after the collision is three times the initial mass. So in this test as well, the momentum after the collision is the same as the momentum before the collision.

In both tests, the total momentum is unchanged (i.e., is conserved) by the collision. This is an example of the **conservation of momentum**. It applies to any system of objects as long as the system is a closed system, which means that no resultant force acts on it.

In general, the law of conservation of momentum says that:

In a closed system, the total momentum before an event is equal to the total momentum after the event.

You can use this law to predict what happens whenever objects collide or push each other apart in an 'explosion'. Momentum is conserved in any collision or explosion as long as no external forces act on the objects.

If you are a skateboarder, you will know that your skateboard can shoot away from you when you jump off it. Its momentum is in the opposite direction to your own momentum. What can you say about the momentum of objects when they fly apart from each other?

Investigating a controlled explosion

Figure 3 shows a controlled explosion using trolleys. When the trigger rod is tapped, a bolt springs out, and the trolleys recoil (spring back) from each other.

You can also test what happens if one of the trolleys is a 'double trolley', as shown in Figure 4.

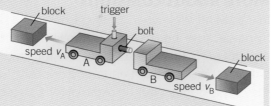

Figure 3 *Investigating explosions*

Using trial and error, you can place blocks on the runway so that the trolleys reach them at the same time. This lets you compare the speeds of the trolleys. Some results are shown in Figure 4.

Safety: Protect yourself and the bench from falling objects.

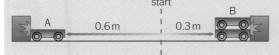

Figure 4 *Using different masses*

- Two single trolleys travel equal distances in the same time. This shows that they recoil at equal speeds.

- A double trolley travels only half the distance that a single trolley does. Its speed is half that of the single trolley.

In each test:

1 the mass of the trolley × the speed of the trolley is the same, and

2 they recoil in opposite directions.

So momentum has size and direction. The results show that the trolleys recoil with equal and opposite momentum.

1 a Define momentum and give its unit. [1 mark]
  b Calculate the momentum of a 40 kg person running at 6 m/s. [1 mark]

2 a Calculate the momentum of an 80 kg rugby player running at a velocity of 5 m/s. [1 mark]
  b An 800 kg car moves with the same momentum as the rugby player in **a**. Calculate the velocity of the car. [1 mark]
  c Calculate the velocity of a 0.40 kg ball that has the same momentum as the rugby player in **a**. [1 mark]

3 A 60 kg skater and an 80 kg skater standing in the middle of an ice rink push each other away (Figure 5).
  Describe:
  a the force they exert on each other when they push apart
  b the momentum each skater has just after they separate
  c each of their velocities just after they separate
  d their total momentum just after they separate. [5 marks]

4 In Question **3**, the 60 kg skater moves away at 2.0 m/s. Calculate:
  a her momentum [1 mark]
  b the velocity of the other skater. [3 marks]

Figure 5

Key points

- The momentum of a moving object is $p = m\,v$
- The unit of momentum is kg m/s.
- A closed system is a system in which the total momentum before an event is the same as the total momentum after the event. This is called conservation of momentum.

P10.5 Using conservation of momentum

Learning objectives

After this topic, you should know:

- how momentum can be described as having direction as well as size
- why two objects that push each other apart move away at different speeds
- what happens to the total momentum of two objects when they collide.

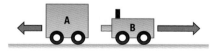

Figure 1 *Momentum in a bowling alley*

Figure 2 *Worked example*

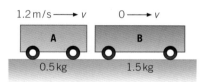

Figure 3

More about collisions

In a bowling alley, when a ball crashes into a skittle the skittle is sent flying because it gains momentum in the crash. The ball loses an equal amount of momentum, because the total momentum is conserved. The ball and the skittle exert equal and opposite forces on each other when they are in contact but no external forces act on them.

Conservation of momentum applies to any collision where no external forces act. The worked examples below show how we can use the principle of conservation of momentum to work out the velocity of an object after a collision with another object. The same method can be applied to any collision where two objects collide in a straight line.

> **Worked example**
>
> A 0.5 kg trolley **A** is pushed at a velocity of 1.2 m/s into a stationary trolley **B** of mass 1.5 kg as shown in Figure 2. The two trolleys stick to each other after the impact. Calculate:
>
> **a** the momentum of the 0.5 kg trolley before the collision
>
> **b** the velocity of the two trolleys straight after the impact.
>
> **Solution**
>
> **a** Momentum = mass × velocity = 0.5 kg × 1.2 m/s = **0.6 kg m/s**
>
> **b** The momentum after the impact = the momentum before the impact = 0.6 kg m/s
>
> (1.5 kg + 0.5 kg) × velocity after the impact = 0.6 kg m/s
>
> the velocity after the impact = $\dfrac{0.6 \text{ kg m/s}}{2 \text{ kg}}$ = **0.3 m/s**

Applying conservation of momentum

In the 'controlled explosion' examples in topic P10.4 and in Figure 3:

- momentum of A after the explosion = (mass of A × velocity of A)
- momentum of B after the explosion = (mass of B × velocity of B)
- total momentum before the explosion = 0 (because both trolleys were at rest).

Using conservation of momentum gives:

(mass of A × velocity of A) + (mass of B × velocity of B) = 0

So:

(mass of A × velocity of A) = −(mass of B × velocity of B)

The minus sign after the equals sign tells you that the momentum of B is in the opposite direction to the momentum of A. The equation tells you that A and B move apart with equal and opposite amounts of momentum. So the total momentum after the explosion is the same as the total momentum before the explosion.

Momentum in action

When a shell is fired from an artillery gun, the gun barrel recoils backwards. The recoil of the gun barrel is slowed down by a spring. This lessens the backwards motion of the gun.

Worked example

An artillery gun of mass 2000 kg fires a shell of mass 20 kg at a velocity of 120 m/s. Calculate the recoil velocity of the gun.

Solution

Applying the conservation of momentum gives:

mass of gun × recoil velocity of gun = −(mass of shell × velocity of shell)

If you let V represent the recoil velocity of the gun:

$2000 \, kg \times V = -(20 \, kg \times 120 \, m/s)$

$V = -\dfrac{2400 \, kg \, m/s}{2000 \, kg} = \mathbf{-1.2 \, m/s}$

Worked example

A 4000 kg truck moving at a velocity of 12 m/s crashes into the back of a 1000 kg car moving at a velocity of 2 m/s in the same direction. The two vehicles move together immediately after the impact. Calculate their velocity.

Solution

Let v represent the velocity of the vehicles after the impact.

momentum of truck before impact = 48 000 kg m/s

momentum of car before impact = 2000 m/s

momentum of truck after impact = 4000 kg × v

momentum of car after impact = 1000 kg × v

$4000 \, v + 1000 \, v = 48\,000 + 2000$

= 50 000

$5000 \, v = 50\,000$.

So $v = \mathbf{10 \, m/s}$

1 A 1000 kg rail wagon moving at a velocity of 5.0 m/s on a level track collides with a stationary 1500 kg wagon. The two wagons move together after the collision.

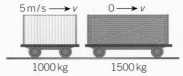

5 m/s →v 0 →v

1000 kg   1500 kg

   a Calculate the momentum of the 1000 kg wagon before the collision. [1 mark]
   b Calculate the velocity of the wagons after the collision. [3 marks]

2 An object is attached to a trolley of mass 0.80 kg, which is then pushed into an identical stationary trolley at a speed of 1.1 m/s. The two trolleys couple together and move at a velocity of 0.70 m/s after the collision. Calculate the mass of the object. [6 marks]

3 A 600 kg cannon recoils at a speed of 0.5 m/s when a 12 kg cannon ball is fired from it.
   a Calculate the velocity of the cannon ball when it leaves the cannon. [3 marks]
   b Suggest how the recoil velocity of the cannon would have been different if a 4 kg cannon ball was used instead, and was fired at the same speed as the 12 kg ball. [1 mark]

4 Look at Figure 3 and Topic P10.4. Calculate how far from the start you would place the right-hand block if A is a double trolley and B is a triple trolley. Assume that the left-hand block is 0.60 m from the start. [3 marks]

Key points

- Momentum is defined as mass × velocity, and has both size and direction.
- When two objects push each other apart, they move with different speeds if they have unequal masses, and with equal and opposite momentum, so their total momentum is zero.
- Use the equation $m_A \, v_A + m_B \, v_B = 0$ when two objects, A and B, recoil from each other.

P10.6 Impact forces

Learning objectives

After this topic, you should know:

- what affects the force of impact when two vehicles collide
- how the impact force depends on the impact time
- what can be said about the impact forces and the total momentum when two vehicles collide
- why an impact force depends on the impact time.

Figure 1 *A crash test. Car makers test the design of a crumple zone by driving a remote control car into a brick wall*

Crumple zones at the front end and rear end of a car are designed to lessen the force of an impact. The force changes the momentum of the car.

- In a front-end impact, the momentum of the car is reduced.
- In a rear-end impact (where a vehicle is struck from behind by another vehicle), the momentum of the car is increased.

In both cases, the effect of a crumple zone is to increase the impact time and so lessen the impact force.

Impact time

Why does making the impact time longer reduce the impact force?

Suppose a moving trolley hits another object and then stops.

- The impact force on the trolley acts for a given time (the impact time) and then makes it stop.
- A soft pad on the front of the trolley would act as a cushion to increase the impact time.
- The momentum of the trolley would be reduced over a longer time, and so the change of momentum per second would be less.
- So because the impact force is equal to the change of momentum per second, the impact force would be less.

The longer the impact time, the more the impact force is reduced.

If you know the impact time, you can calculate the impact force as follows:

- From Topic P9.2, you know that

$$\text{acceleration} = \frac{\text{(final velocity – initial velocity)}}{\text{time taken}} = \frac{\text{change of velocity}}{\text{time taken}}$$

- From Topic P10.1, you know that

$$\text{force} = \text{mass} \times \text{acceleration}$$

Because mass × change of velocity = change of momentum, then:

$$\textbf{force, } F \text{ (N)} = \frac{\textbf{mass, } m \text{ (kg)} \times \textbf{change of velocity, } \Delta v \text{ (m/s)}}{\textbf{time taken, } t \text{ (s)}}$$

This equation shows how much the impact force can be reduced by increasing the impact time. Car safety features such as crumple zones and side bars increase the impact time and so reduce the impact force on the car.

Worked example

A bullet of mass 0.004 kg moving at a velocity of 90 m/s is stopped by a bulletproof vest in 0.0003 s.

Calculate **a** the deceleration of the bullet, **b** the change of momentum, and **c** the impact force.

Solution

a Initial velocity of bullet = 90 m/s

Final velocity of bullet = 0

Change of velocity = final velocity − initial velocity = 0 − 90 m/s

= −90 m/s

(the minus sign tells you that the change of velocity is a decrease)

$$\text{Deceleration} = \frac{\text{change of velocity}}{\text{impact time}} = \frac{-90\,\text{m/s}}{0.0003\,\text{s}} = \mathbf{-300\,000\,\text{m/s}^2}$$

b Change of momentum = mass × change of velocity

= 0.004 kg × (0 − 90 m/s) = **−0.36 kg m/s**

c $$\text{Force} = \frac{\text{change of momentum}}{\text{time taken}} = \frac{-0.36\,\text{kg m/s}}{0.0003\,\text{s}} = \mathbf{-1200\,N}$$

(the minus sign tells you that the force decelerates the bullet)

Two-vehicle collisions

When two vehicles collide, they exert equal and opposite impact forces on each other at the same time. So the change of momentum of one vehicle is equal and opposite to the change of momentum of the other vehicle. The total momentum of the two vehicles is the same after the impact as their total momentum before the impact, so momentum is conserved – assuming that no external forces are acting on them.

1 **a** In a crash test, when a passenger wears a seat belt, explain why it reduces the impact force on him. [2 marks]

   b A ball of mass 0.12 kg moving at a velocity of 18 m/s is caught by a person in 0.0003 s. Calculate the impact force. [3 marks]

2 **a** An 800 kg car travelling at 30 m/s is stopped safely when the brakes are applied. Calculate the braking force that would be needed to stop it in:

    i 6.0 s   **ii** 30 s. [4 marks]

   b If the vehicle in **a** had been stopped in a collision lasting less than 1 s, explain by referring to momentum why the force on it would have been much greater. [2 marks]

3 A 2000 kg van moving at a velocity of 12 m/s collides with a stationary truck of mass 10 000 kg. Immediately after the impact, the two vehicles move together.

   a Show that the velocity of the van and the truck immediately after the impact was 2 m/s. [3 marks]

   b The impact lasted for 0.3 s. Calculate:

    i the deceleration of the van [2 marks]

    ii the change of momentum of the van [2 marks]

    iii the force of the impact on the van. [2 marks]

4 Playgrounds have cushioned surfaces to reduce the risk of injury in a fall. Explain in terms of force and momentum why a cushioned surface is safer than a hard surface. [2 marks]

Force and time

The equation $F = \dfrac{m\Delta v}{t}$ tells you that for a given change of momentum, the force F is inversely proportional to time t. For example, if t is doubled, F is halved.

Synoptic links

See Maths skills for more about inverse proportion.

Key points

- When vehicles collide, the force of the impact depends on mass, change of velocity, and the length of the impact time.
- The longer the impact time, the more the impact force is reduced.
- When two vehicles collide:
  - they exert equal and opposite forces on each other
  - their total momentum is unchanged.
- Impact force = change of momentum ÷ impact time, so the shorter the impact time, the greater the impact force.

P10.7 Safety first

Learning objectives

After this topic, you should know:

- why cycle helmets and cushioned surfaces reduce impact forces
- why seat belts and air bags reduce the force on people in car accidents
- how side impact bars and crumple zones work
- how to work out if a car in a collision was speeding.

Figure 1 *An air bag in action*

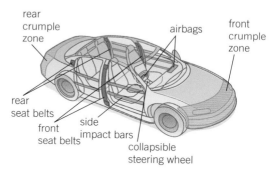

rear crumple zone
airbags
front crumple zone
rear seat belts
front seat belts
side impact bars
collapsible steering wheel

Figure 2 *Car safety features*

Figure 3 *A child car seat*

When you go on a cycle ride or travel in a car, you want to feel safe if you crash. Even falling over in a playground can cause serious injury. Different safety features have been designed to keep you safe in such situations.

Safety helmets

Cyclists and horse riders need to wear suitable safety helmets in case they fall off when they are riding. A rider who falls without a helmet could suffer serious head damage, as well as broken bones and bruises. The rider's head could hit the ground at a speed of more than 10 m/s. The helmet is designed to increase the impact time. With a longer impact time, the rate of change of momentum per second is reduced. So the helmet reduces the force of the impact.

Car safety

A seat belt stops its wearer from continuing forwards when the car suddenly stops. Someone without a seat belt would hit the windscreen in a 'short sharp' impact and suffer major injury.

- The time taken to stop someone in a car is longer if they are wearing a seat belt than if they are not. So the decelerating force is reduced by wearing a seat belt.
- The seat belt acts across the chest, so it spreads the force out. Without the seat belt, the force would act on the person's head when it hit the windscreen.

Air bags in a car are designed to protect the driver and the front passenger. In a car crash, an inflated air bag spreads the force of an impact across the upper part of a person's body. It also increases the impact time. So the effect of the force is lessened compared with just using a seat belt. Crumple zones, side impact bars, and collapsible steering wheels also increase the impact times to lessen the force of an impact.

Child car seats are essential when babies or children travel in a car, otherwise the driver can be fined. The law on child car seats applies to children up to twelve years old or up to 1.35 metres in height. Different types of child car seat must be used for babies up to nine months old, infants up to about four years old, and children over four years.

- Baby seats must face backwards.
- Children under four years old should usually be in a child car seat fitted to a back seat.

The law was brought in to reduce deaths and serious injuries of children in cars. Before the law was passed, children were more likely to be seriously injured in car collisions. Many of these accidents happened during the school run. Drivers are responsible for making sure every child in their car is seated safely in the correct type of seat.

Road safety

Braking too harshly causes skidding, which can result in a collision. The stopping distance is the average smallest distance for a vehicle to make an emergency stop:

stopping distance = thinking distance + braking distance.

Table 1 shows these distances for speeds of 10 m/s, 20 m/s, and 30 m/s.

The table shows that:

- thinking distance is proportional to speed
- braking distance is proportional to the square of the speed.

Table 1 *Stopping distances*

Speed (m/s)	10	20	30
Thinking distance (m)	7	14	21
Braking distance (m)	8	32	74
Stopping distance (m)	15	56	96

Playground safety

Playgrounds are fitted with cushioned surfaces underneath swings and slides in case children fall off. When a child falls on such a soft surface, the duration of the impact is longer than it would be if the surface had been hard like concrete. So the child's momentum is reduced to zero over a longer time by the cushioning effect of the softer surface. The change of momentum per second is therefore reduced, so the impact force is reduced compared with an impact of the same initial momentum on a hard surface.

Gymnasium crash mats contain cushioning material so they have the same effect as a cushioned playground surface when someone falls on a crash mat.

1 Explain why a cyclist should always wear a safety helmet when riding a bicycle. [3 marks]

2 a Explain why rear-facing car seats are safer than front-facing seats for babies. [2 marks]

b Explain why an inflated air bag in front of a car user reduces the force on a user in a head-on crash. 🚫 [3 marks]

3 Explain why a seat belt reduces the force on a car user when the car suddenly stops. [2 marks]

4 A car collided into a lorry that was crossing a busy road. The speed limit on the road was 60 miles per hour (27 m/s) (Figure 4). The following measurements were made by police officers at the scene of the crash:
The car and lorry ended up 6 m from the point of impact.
The car's mass was 750 kg, and the lorry's mass was 2150 kg.
The speed of a vehicle for a braking distance of 6 m is 9 m/s.

a Use this speed to calculate the momentum of the car and the lorry immediately after the impact. [1 mark]

b Use conservation of momentum to calculate the velocity of the car immediately before the collision. [3 marks]

c Work out whether or not the car was travelling over the 60 mph speed limit before the crash. [1 mark]

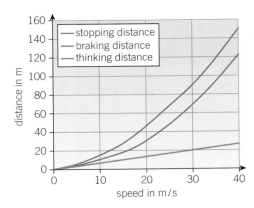

Figure 4 *Stopping distances*

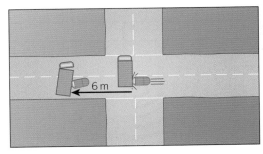

Figure 5 *Car collision*

Key points

- Cycle helmets and cushioned surfaces (e.g., in playgrounds) reduce impact forces by increasing the impact time.
- Seat belts and air bags spread the force across the chest and increase the impact time.
- Side impact bars and crumple zones give way in an impact, and so increase the impact time.
- Conservation of momentum can be used to find the speed of a car before an impact.

P10.8 Forces and elasticity

Learning objectives

After this topic, you should know:

- what is meant when an object is called elastic
- how to measure the extension of an object when it is stretched
- how the extension of a spring changes with the force applied to it
- what is meant by the limit of proportionality of a spring.

Squash players know that hitting a squash ball changes the ball's shape briefly. The shape of an object can be changed by stretching, bending, twisting, or compressing it. A squash ball is **elastic** because it goes back to its original shape. The ball is said to have been elastically deformed.

A rubber band is also elastic because it returns to its original length after it is stretched and then released. Rubber is an example of an elastic material. An object such as a polythene bag does not return to its original shape after being deformed and so is said to have been inelastically deformed.

An object is elastic if it returns to its original shape when the forces deforming it are removed.

Stretch tests

You can investigate how easily a material or a spring stretches by hanging weights from it (Figure 1).

- The spring to be tested is clamped at its upper end. An empty weight hanger is attached to the spring to keep it straight.
- The length of the spring is measured using a metre ruler. This is its original length.
- The weight hung from the spring is increased by adding weights one at a time. The spring stretches each time more weight is hung from it.
- The length of the spring is measured each time a weight is added. The spring should be measured from the same points each time to ensure accurate results. The total weight added and the total length of the spring are recorded in a table.

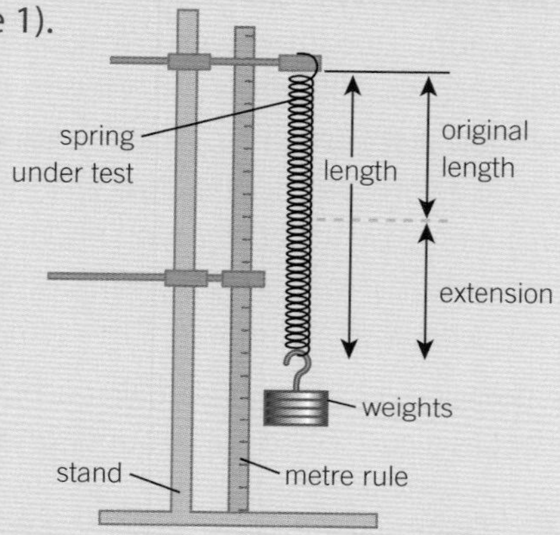

Safety: Clamp the stand to the bench and take care with falling weights. Wear eye protection.

Figure 1 *Investigating stretching*

Table 1 *Weight versus length measurements for a rubber strip*

Weight in N	Length in mm	Extension in mm
0.0	120	0
1.0	152	32
2.0	190	70
3.0	250	
4.0		

The increase of length from the original is called the **extension**. This is calculated each time a weight is added and recorded, as shown in Table 1.

$$\text{extension of the strip of material or spring at any stage} = \text{length at the stage} - \text{original length}$$

The measurements can be plotted on a graph of weight on the vertical *y*-axis against extension on the *x*-axis. Figure 2 shows the results for strips of different materials and a steel spring plotted on the same axes.

The steel spring gives a straight line through the origin. This shows that the weight hung on the steel spring is **directly proportional** to the extension of the spring. For example, doubling the weight from 2.0 N to 4.0 N doubles the extension of the steel spring.

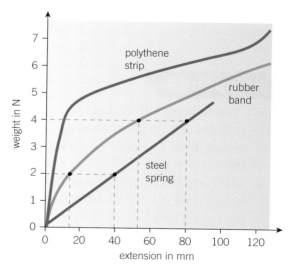

Figure 2 *Extension versus hung weight for different materials*

Hooke's law

In the tests above, the extension of a steel spring is directly proportional to the force applied to it. You can use the graph to predict what the extension would be for any given force. But if the force is too big, the spring stretches more than predicted. This is because the spring has been stretched beyond its **limit of proportionality**.

The extension of a spring is directly proportional to the force applied, as long as its limit of proportionality is not exceeded.

The above statement is known as Hooke's law. If the extension of any stretched object or material is directly proportional to the stretching force, the object or material is said to obey Hooke's law.

1   The lines on the graph in Figure 2 show that rubber and polythene have a low limit of proportionality. Beyond this limit, they do not obey Hooke's law. The relationship between their extension and the force applied is non-linear. A steel spring has a much higher limit of proportionality. Its relationship to the force applied stays linear for much longer.

2   Hooke's law can be written as an equation:

force applied, F =   spring constant, k   × extension, e
    (newtons, N)   (newtons per metre, N/m)   (metres, m)

The spring constant is equal to the force per unit extension needed to extend the spring, assuming that its limit of proportionality is not reached. The stiffer a spring is, the greater its spring constant.

3   Hooke's law also applies to an object when it is compressed. In this situation, the change of length is a compression, *not* an extension.

1   What is meant by:
   a the limit of proportionality of a spring   [1 mark]
   b the spring constant of a spring   [1 mark]
   c the extension of a stretched spring.   [1 mark]
2   **a** Describe what happens to a strip of polythene if it is stretched beyond its limit of proportionality.   [1 mark]
   b Compare the result of stretching then releasing a rubber band with the result of stretching a strip of polythene.   [1 mark]
3   **a** Look at Figure 2. When the weight is 4.0 N, calculate the extension of:
      i the spring   [1 mark]
      ii the rubber band   [1 mark]
      iii the polythene strip.   [1 mark]
   b i Calculate the extension of the spring when the weight is 3.0 N.   [1 mark]
      ii Calculate the spring constant of the spring.   [2 marks]
4   **a** Write Hooke's law.   [1 mark]
   b A spring has a spring constant of 25 N/m.
      i Calculate how much force is needed to make the spring extend by 0.10 m.   [2 marks]
      ii Calculate the extension of the spring when it hangs vertically from a fixed point and supports a 5.0 N weight at its lower end. [2 marks]

Using maths

You can write the word equation for Hooke's law using symbols as

$F = k \times e$

where

F = force in newtons, N

k = the spring constant in newtons per metre, N/m

e = extension in metres, m.
To determine the spring constant of a spring from a set of force and extension measurements, plot the force F on the y-axis (and the extension e on the x-axis). The gradient of the straight line you obtain is the spring constant k.

Synoptic link

You learnt about elastic potential energy stores, including the equation to calculate them, in Topic P1.5.

Key points

- An object is called elastic if it returns to its original shape after removing the force deforming it.
- The extension is the difference between the length of the object and its original length.
- The extension of a spring is directly proportional to the force applied to it, as long as the limit of proportionality is not exceeded. This relationship is linear.
- Beyond the limit of proportionality, the extension of a spring is no longer proportional to the applied force. This relationship becomes non-linear.

P10 Force and motion

Summary questions

1 **a** Give the reason why the stopping distance of a car is increased if:
   i the road is wet instead of dry [1 mark]
   ii the driver is tired instead of alert. [1 mark]
   b A driver travelling at 18 m/s takes 0.7 s to react when a dog walks into the road 40 m ahead. The braking distance for the car at this speed is 24 m.
   i Calculate the distance travelled by the car in the time it takes the driver to react. [2 marks]
   ii Calculate how far in front of the dog the car stops. [2 marks]
   iii The total mass of the car and its contents is 1200 kg. Calculate the car's deceleration when the brakes are applied, and so calculate the braking force. **H** [3 marks]

2 A space vehicle of mass 200 kg rests on its four wheels on a flat area of the lunar surface. The gravitational field strength at the surface of the Moon is 1.6 N/kg.
   a Calculate the weight of the space vehicle on the lunar surface. [2 marks]
   b Calculate the force that each wheel exerts on the lunar surface. [1 mark]

3 **a** A racing cyclist accelerates at 5.0 m/s^2 when she starts from rest. The total mass of the cyclist and her bicycle is 45 kg. Calculate:
   i the resultant force that produces this acceleration [2 marks]
   ii the total weight of the cyclist and the bicycle. [2 marks]
   b Explain why she can reach a higher speed by crouching than by staying upright. [4 marks]

4 In a Hooke's law test on a spring, the following results were obtained.

Weight in newtons	Length in millimetres	Extension in millimetres
0.0	245	0
1.0	285	40
2.0	324	
3.0	366	
4.0	405	
5.0	446	
6.0	484	

   a Copy and complete the third column of the table. [1 mark]
   b Plot a graph of the extension on the vertical axis against the weight on the horizontal axis. [3 marks]

c If a weight of 7.0 N is suspended on the spring, work out what the extension of the spring would be. [2 marks]
d **i** Calculate the spring constant of the spring. [2 marks]
   ii An object suspended on the spring gives an extension of 140 mm. Calculate the weight of the object. [2 marks]

5 **H** A car of mass 1500 kg is moving at a speed of 30 m/s on a horizontal road when the driver applies the brakes and the car stops 12 s later.

   a **i** Calculate the initial momentum of the car before the brakes are applied. [1 mark]
   ii Calculate the braking force. [2 marks]
   b Describe how the momentum of the car changes when the brakes are applied. [1 mark]
   c Discuss the effect on the motion of the car if the brakes had been applied with much greater force. [2 marks]

6 **H** A 2000 kg truck moving at a velocity of 18.0 m/s on a level road collides with a stationary vehicle of mass 1200 kg. The velocity of the truck is reduced to 10.0 m/s as a result of the collision.

   a Calculate the momentum of the truck:
   i before the collision [1 mark]
   ii after the collision. [1 mark]
   b For the 1200 kg vehicle after the collision, calculate:
   i the momentum [1 mark]
   ii the velocity. [2 marks]

7 **H** When a stationary football of mass 0.44 kg was kicked, its velocity increased to 19 m/s as a result of the impact.
   a Calculate the gain of momentum of the football due to the impact. [2 marks]
   b The impact lasted 0.0384 s. Show that the impact force was 220 N. [2 marks]

Practice questions

01 A student uses a reaction time tester linked to digital timer. The student started the test by pressing a button. After a random amount of time time has elapsed an LED came on and the student pressed the button again. The digital timer recorded the time between the LED coming on and the student pressing the button.

01.1 Suggest one reason why this reaction time test is better than using the dropping ruler method. [1 mark]

01.2 Name the independent variable in the test. [1 mark]

01.3 The stopping distance of a car is thinking distance + braking distance.
Which two factors increase the thinking distance? Choose the correct answers.

tiredness	alcohol	brakes	tyres

[2 marks]

01.4 The acceleration of a car on a straight road is 3.28 m/s². The mass of the car is 1200 kg. Calculate the resultant force on the car. Give your answer correct to 2 significant figures. [2 marks]

01.5 Complete the sentence using the correct word from the box.

acceleration	speed	deceleration

The greater the braking force on a car the greater the _____ of the car. [1 mark]

01.6 Describe the dangers of a very large deceleration on the motion of a car. [2 marks]

02 A student investigated how the pulling force on a trolley affected the acceleration of the trolley between two points. The apparatus he used is shown in **Figure 1**.

Figure 1

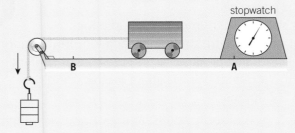

The trolley was held at position **A** and weights were suspended by a hook. The trolley was then released and the time taken to travel between **A** and **B** was recorded. The test was repeated using weights with different values.

02.1 List the factors that should remain constant (the control variables) during the investigation. [3 marks]

02.2 The results of the investigation were recorded in **Table 1**.

Table 1

Force in N	100	200	300	400	500
Acceleration in m/s²	0.25	0.50	0.65	1.00	1.25

Plot a graph of the results. [3 marks]

02.3 Identify which one of the results is anomalous and suggest a reason for this anomalous result. [2 marks]

02.4 The student predicted that if the force was doubled the acceleration would also be doubled. Describe how the graph can be used to confirm his prediction. [2 marks]

02.5 A stopwatch was used to measure the time taken for the trolley to travel between points A and B. Another student suggested that light gates attached to a data logger would improve the investigation. Give two advantages of using light gates and a data logger to measure the acceleration. [2 marks]

03 A baby bouncer consists of a harness seat for a toddler, attached to a spring. The idea is for the baby to hang in the seat with his feet just touching the floor, so that a good push up will get the baby bouncing.

Figure 2

03.1 The spring is tested in the laboratory. When a force of 10 N is suspended from the spring, the extension of the spring is 0.02 m. Calculate the spring constant of the spring and give the unit. [3 marks]

03.2 The mass of the baby is 9 kg. Calculate the weight of the baby.
gravitational field strength = 9.8 N/kg. [2 marks]

03.3 The baby is placed unsuspended in the harness. His feet are 0.68 m from the lower end of the spring. Calculate the height above the floor the lower end of the stretched spring needs be if his feet are just to touch the floor. Use the information from question **03.2**. [3 marks]

03.4 The maximum permissible mass of the baby on the bouncer is 15 kg. Explain why it is important to test the spring beyond this maximum figure. [2 marks]

11 Force and pressure
11.1 Pressure and surfaces

Learning objectives

After this topic, you should know:

- what is meant by pressure
- what the unit of pressure is
- how to use the pressure equation
- know why the area of contact is important in pressure applications.

Figure 1 *Caterpillar tracks*

What is pressure?

If you have ever stood barefoot on a sharp object, you will have found out about pressure in a very painful way. Because all of your weight acts on the very tip of the object, there is a huge amount of pressure on your foot at the area of contact.

Pressure is caused when objects exert forces on each other, or when a fluid (which can be either a liquid or a gas) exerts a force on an object in contact with the fluid. The pressure caused by any force depends on the area of contact on which the force acts, as well as on the size of the force.

Caterpillar tracks fitted to vehicles are essential on sandy or muddy ground or on snow-covered ground. The reason is that the contact area of the tracks on the ground is much bigger than it would be if the vehicle had wheels instead. So the tracks reduce the pressure of the vehicle on the ground because its weight is spread over a much bigger contact area.

Pressure is the force per unit area. The unit of pressure is the pascal (Pa), which is equal to one newton per square metre (N/m²).

For a force **F** acting evenly at right angles to a surface of area **A** (i.e., 'normal' to the surface), you can find out the pressure **p** on the surface using the equation:

$$\textbf{pressure, } \textbf{\textit{p}} \text{ (pascals, Pa)} = \frac{\textbf{force, } \textbf{\textit{F}} \text{ (newtons, N)}}{\textbf{area, } \textbf{\textit{A}} \text{ (metres squared, m}^2\text{)}}$$

Rearranging equations

You can rearrange the pressure equation $p = \dfrac{F}{A}$ to give $F = pA$ or $A = \dfrac{F}{p}$.

Worked example

A caterpillar vehicle of weight 12 000 N is fitted with tracks that have an area of 3.0 m² in contact with the ground. Calculate the pressure of the vehicle on the ground.

Solution

$$\text{pressure} = \frac{\text{force}}{\text{area}} = \frac{12\,000\,\text{N}}{3.0\,\text{m}^2} = \textbf{4000 Pa}$$

Prefixes

Because you regularly use big numbers in physics calculations, powers of ten are used a lot, especially in this topic, where 1 kPa = 10³ Pa = 1000 Pa, and 1 MPa = 10⁶ Pa = 1 000 000 Pa. See Maths skills for Physics for more information.

Measure your foot pressure

Draw around your shoes on centimetre squared paper. Count the number of centimetre squares in each footprint (ignoring any square that is less than half-filled) to find the area of contact in square centimetres. Convert this to square metres using the conversion 1 m² = 10 000 cm².

Use suitable scales (e.g., bathroom scales) to measure your weight. If the scales read mass in kilograms, find your weight in newtons using g = 9.8 N/kg.

- Work out your pressure using the equation $\text{pressure} = \dfrac{\text{weight}}{\text{area}}$

In hospital

A sharp knife cuts more easily than a blunt knife. Surgical knives used in operating theatres need to be very sharp. A sharp knife has a much smaller area of contact than a blunt knife does. So the pressure (= the force ÷ the contact area) of a sharp knife is much bigger than the pressure of a blunt knife for the same force. So when the same force is applied with a sharp knife, it has a much bigger cutting effect than it would have with a blunt knife.

A pressure test

The arrangement shown in Figure 2 can be used to measure the pressure inside an inflatable bag.

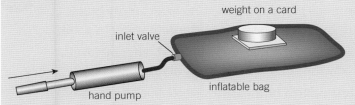

weight on a card

inlet valve

inflatable bag

hand pump

Figure 2 *A pressure test*

With the bag deflated, place a weight on a card on the bag. Use the hand pump to inflate the bag until the card and weight is raised. The pressure of the air in the bag is then equal to the pressure of the weight on the card.

- Measure the area of the card, and then use the equation
  Pressure (Pa) = Weight (N) ÷ area of the card (m²) to calculate the pressure of the air in the bag.

1 Explain each of the following:
  a When you do a handstand, the pressure on your hands is greater than the pressure on your feet when you stand upright. [2 marks]
  b Snowshoes like the one shown in Figure 3 are useful for walking across soft snow. [2 marks]

2 A rectangular concrete paving slab of weight 1200 N has sides of length 0.60 m and 0.40 m and a thickness of 0.05 m. Calculate the pressure of the paving slab on the ground when it is:
  a laid flat on a bed of sand [2 marks]
  b standing upright on its short side. [2 marks]

3 A pressure vessel is fitted with a valve that opens if the difference in the pressure in the vessel and the outside pressure is greater than 45 kPa. The valve contains a spring that exerts a force on a disc in the valve to keep the valve closed. The pressure acts on a disc area of 0.0002 m². Calculate the force needed to open the valve. [1 mark]

4 The four tyres of a car of weight 9400 N are inflated to a pressure of 180 kPa. Calculate the contact area of each tyre on the ground. [3 marks]

Figure 3 *Spreading the weight*

Key points

- Pressure is the force normal to a surface ÷ area of the surface.
- Pressure $p = \dfrac{F}{A}$.
- The unit of pressure is the pascal (Pa), which is equal to $1\,N/m^2$.
- The force F or area A can be calculated by rearranging the pressure equation $p = \dfrac{F}{A}$ to give $F = pA$ or $A = \dfrac{F}{p}$.

P11.2 Pressure in a liquid at rest

Learning objectives

After this topic, you should know:

- how the pressure in a liquid increases with liquid depth
- why the pressure along a horizontal line in a liquid is constant
- what the pressure in a liquid depends on
- how to calculate the pressure caused by a liquid column.

Figure 1 *Using a snorkel tube*

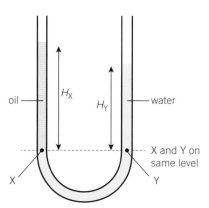

Figure 3 *Comparing densities*

An underwater swimmer using a snorkel tube can breathe safely provided that the top of the snorkel tube stays above water. However, a deep-sea diver would not be able to breathe through a very long snorkel tube, because the pressure of the water increases as it gets deeper. At depths of more than a few metres, the diver's chest muscles would not be strong enough to expand her chest muscles against the water pressure on her body.

Pressure tests
Make several small holes down the side of an empty plastic bottle (Figure 2a) and then fill the bottle with water in a sink.

- Observe the water jets from the holes at different levels. Explain why the strength of the jets differs.

Repeat the test using a bottle with several small holes round the bottle at the same level, like that shown in Figure 2b.

- Describe and explain what you see this time.

The pressure of a liquid increases with depth. Figure 2a shows water jets from the holes down the side of an open plastic bottle filled with water. The further the hole is below the level of water in the bottle, the greater the force with which the jet leaves the bottle. This is because the water pressure at each hole is greater the further the hole is below the water level.

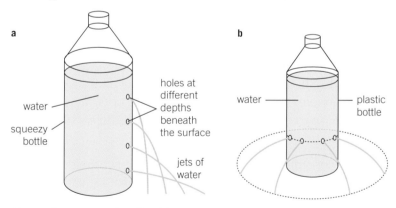

Figure 2 *Pressure in a liquid at rest*
a *Pressure increases with depth* **b** *Same pressure at the same depth*

The pressure along a horizontal line in a liquid is constant. You can see this for yourself by making several holes around the bottle at the same depth, as shown in Figure 2b. The jets from these holes are at the same pressure.

The pressure in a liquid depends on the density of the liquid. Suppose water is poured into one side of a U-shaped tube, and then oil is carefully poured into the other side, as shown in Figure 3. When the liquid settles, the oil level is higher than the water level on the other side. This is because oil is less dense than water, so a greater depth of oil is needed to create the same pressure.

The pressure of a liquid column

The pressure p at the bottom of a column of liquid depends on the height of the column and the density of the liquid. For a column of liquid of density ρ and height h, the pressure p caused by the liquid at the base of the column is given by the equation:

$$p = h \times \rho \times g$$

where p represents pressure (pascals, Pa), h represents height (metres, m), ρ represents density (kilograms per cubic metre, m³), and g represents gravitational field strength (newtons per kilogram, N/kg).

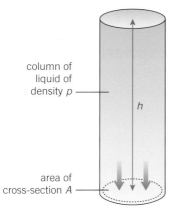

column of liquid of density ρ

h

area of cross-section A

Figure 4 *Calculating liquid pressure*

Worked example

Calculate the pressure due to sea water of density 1050 kg/m³ at a depth in the sea of 200 m.

$g = 9.8$ N/kg

Solution

pressure $p = h\rho g$

$\qquad = 200\,\text{m} \times 1050\,\text{kg/m}^3 \times 9.8\,\text{N/kg}$

$\qquad = 2.1 \times 10^6\,\text{Pa}$

Go further!

Proof of $p = h \rho g$

Consider the column of liquid in the container shown in Figure 4. The pressure caused by the liquid column on the bottom of the container is due to the weight of the liquid.

For a column of height h and area of cross-section A:

volume of liquid in the container $= hA$

the mass of liquid =
volume of liquid × density of liquid
$= hA\rho$, where ρ is the density of the liquid.

The weight of the liquid = mass × g
$\qquad\qquad\qquad\qquad = hA\rho\, g$,

where the gravitational field strength g
$\qquad\qquad\qquad\qquad = 9.8$ N/kg.

The pressure p at the base of the liquid

$$= \frac{\text{weight}}{\text{area of cross-section}} = \frac{hA\rho\, g}{A}$$

Therefore, cancelling A gives $p = h \rho g$

You *don't* need to know how to prove this in your course. But it might give you a deeper understanding of pressure in a liquid.

$g = 9.8$ N/kg

1 **a** Explain why the wall of a dam needs to be thicker at the base than at the top. [2 marks]

 b A water tank at the top of a tall building supplies water to taps in the building. Explain why the pressure of the water from a tap on the ground floor is greater than the pressure from a tap on a higher floor. [2 marks]

2 A rainwater gutter collects rainwater that runs off a sloped roof. The water in the gutter runs into a vertical downpipe joined to one end of the gutter. Explain why the downpipe end of the gutter needs to be no higher than the other end. [1 mark]

3 A sink plug has an area of 0.0006 m². It is used to block the outlet of a sink filled with water to a depth of 0.090 m. Calculate:

 a the pressure on the plug due to the water. [2 marks]

 b the force needed to remove the plug from the outlet (density of water = 1000 kg/m³). [1 mark]

4 Explain how you would use the arrangement in Figure 3 to find out the density of sea water compared with tap water. ⬤ [3 marks]

Study tip

Make sure you know the difference between the symbol for pressure (p), and the symbol for density (ρ, the greek letter rho).

Key points

- The pressure in a liquid increases with increasing liquid depth.
- A liquid flows until the pressure along the same horizontal level is constant.
- The greater the density of a liquid, the greater the pressure in the liquid.
- The pressure p due to the column of height h of liquid of density ρ is given by the equation $p = h \rho g$.

P11.3 Atmospheric pressure

Learning objectives

After this topic, you should know:

- why the atmosphere exerts a pressure
- how and why atmospheric pressure changes with altitude
- how the density of the atmosphere changes with altitude
- how to calculate the force on a flat object due to a pressure difference.

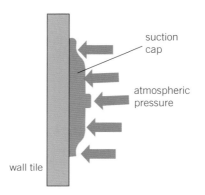

Figure 1 *The atmospheric pressure acting on a rubber suction cap*

suction cap

atmospheric pressure

wall tile

Figure 2 *The summit of Mount Everest. The biggest dangers posed by climbing Mount Everest are actually caused by the high altitude rather than the difficulty of the climb. The majority of deaths of climbers on Mount Everest occur in the 'Death Zone' – where the atmospheric pressure is too low to support life – above 8000 metres*

Atmospheric pressure

The Earth's atmosphere is a layer of air round the Earth. It exerts a pressure of about 100 kPa at sea level. This means that the force on each square metre of surface at sea level is about 100 000 N.

Atmospheric pressure is due to air molecules colliding with surfaces. Each impact exerts a tiny force on a surface, but the number of molecules that collide with the surface each second is very large. The density of the atmosphere decreases with increasing height above the ground (i.e., increasing altitude). This is because the weight of air pressing down on a horizontal surface at any altitude decreases with altitude. So there are fewer molecules per cubic metre and therefore fewer impacts per second at a higher altitude. So atmospheric pressure decreases with increasing altitude. In effect, the weight of air above any given altitude exerts pressure on the air below. At increased altitude, there is less weight of air above, so atmospheric pressure decreases with increasing altitude.

The Earth's atmosphere extends more than 100 km into space. The average (mean) pressure of the atmosphere at sea level is about 100 kPa. Atmospheric pressure changes slightly from day to day, changing with the local weather conditions. Fine clear weather is usually associated with high pressure.

Using atmospheric pressure

A rubber suction cap pressed onto a wall tile stays on the tile and does not fall of. This is because atmospheric pressure acts on the outside of the cap but not on the inside between the cap and the wall. The action of pressing the cap on the wall squeezes any air trapped between the cap and the wall out. The force due to atmospheric pressure acts on the outside surface of the cap only and keeps the cap on the wall (Figure 1).

A drinking straw only works when the air in the straw is sucked out. Without any air in the straw, atmospheric pressure acting on the liquid surface outside the straw pushes liquid up the straw.

Atmospheric pressure at high altitude

Figure 3 shows how atmospheric pressure changes with height above sea level – the altitude. The top of Mount Everest, the highest mountain on the Earth, is about 8848 m above sea level. The atmospheric pressure at the top of Mount Everest is about 30 kPa compared with about 100 kPa at sea level.

The concentration of oxygen in the atmosphere decreases with increasing altitude. This is why commercial jets that fly at high altitudes have air-filled pressurised cabins and emergency oxygen masks in case the cabins lose air at high altitude. For safety reasons, the cabin pressure at high altitude is adjusted to be less than atmospheric pressure at sea level (which is what most people are used to). However, to avoid health problems due to low pressure and for passenger comfort, cabin pressures are not allowed to fall below about 70 kPa. For example, at an altitude of 12 km, the cabin pressure is adjusted to about 70 kPa, whilst the atmospheric pressure at that altitude is only 20 kPa.

A simple model of the Earth's atmosphere
The atmosphere is a thin layer of air around the Earth. Assuming that the density of air in the Earth's atmosphere is the same at different altitudes, the amount (and therefore the weight) of air above any level would decrease with increasing altitude so the pressure would decrease with increasing altitude.

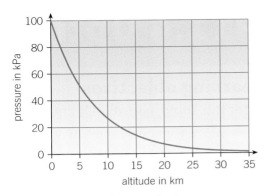

Figure 3 *Graph of atmospheric pressure against altitude*

 The pressure p at sea level would then be given by the equation:

pressure, p	=	**height, h**	×	**density, ρ**	×	**gravity, g**
(pascals, Pa)		(metres, m)		(kilograms per cubic metre, kg/m³)		(newtons per kilogram, N/kg)

Remember that h is the overall height of the model atmosphere, and ρ is the density of the air.

Worked example

$g = 9.8$ N/kg

The atmospheric pressure at an altitude of 10 km is about 25 kPa, and at 30 km is less than 2 kPa. Estimate the average density of the atmosphere between these two altitudes.

Solution
The equation $p = h\rho g$ can be rearranged to $\rho = p/hg$, and can be applied to a model atmosphere of height h, where ρ is the average density of the air in the atmosphere, and p is the pressure difference between the two altitudes.

$$h = 30\,km - 10\,km = 20\,km = 20\,000\,m$$

$$p = 25\,kPa - 2\,kPa = 23\,kPa$$

Therefore:

$$\rho = 23\,000\,Pa \div (20\,000\,m \times 9.8\,N/kg)$$

$$= 0.12\,kg/m^3$$

Worked example
$g = 9.8$ N/kg

Each window of a passenger jet has an area of 0.040 m². When the jet is at a certain altitude, the outside pressure of the atmosphere is 40 kPa and the cabin pressure is 75 kPa.

Calculate the force on a window due to this pressure difference.

Solution
The force on a window due to this pressure difference
= pressure difference × window area
= (75 kPa − 40 kPa) × 0.040 m²
= 35 000 Pa × 0.040 m² = **1400 N**

For the following questions, use the data: $g = 9.8$ N/kg and atmospheric pressure at sea level = 100 kPa to 2 significant figures.

1 Explain how atmospheric pressure acts when you drink through a straw.   [2 marks]

2 Explain why atmospheric pressure decreases with height above sea level.   [4 marks]

3 A suction cap on a wall tile has a surface area of 1.2×10^{-3} m² in contact with the tile. Estimate the force needed to pull the suction cap off the tile.   [2 marks]

4 **H** Use Figure 3 to estimate the mean density of the air in the atmosphere between sea level and an altitude of 30 km.   [4 marks]

Key points
- Air molecules collide with surfaces and create pressure on them.
- Atmospheric pressure decreases with higher altitude because there is less air above a given altitude than there is at a lower altitude.
- The density of the atmosphere decreases with increasing altitude.
- The force on a flat object due to a pressure difference = the pressure difference × the area of the flat surface.

P11.4 Upthrust and flotation

Learning objectives

After this topic, you should know:

- what is meant by an upthrust on an object in a fluid
- what causes an upthrust
- what the pressure in a fluid depends on
- how to explain whether an object in a fluid floats or sinks.

When you go swimming, have you noticed that you feel lighter in the water? People with mobility problems often find it much easier to move in water than in air. Water exerts an upward force on the body. This force is called an **upthrust**.

Investigating upthrust

Use a newton-meter to weigh a metal object in air.

- Repeat the test by weighing the same object when it is completely in the water.

You should find that the newton-meter reading is less when the object is in water. This is because when the metal object is in the water, it experiences an upthrust. The difference between the two newton-meter readings is equal to the upthrust on the object.

Repeat the test with the same object only partly immersed in the water. You should find that the newton-meter reading is in between the two earlier readings. This is because the upthrust is less when the object is only partly immersed in the water.

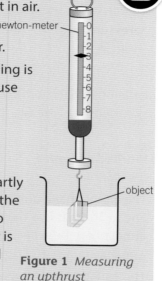

Figure 1 *Measuring an upthrust*

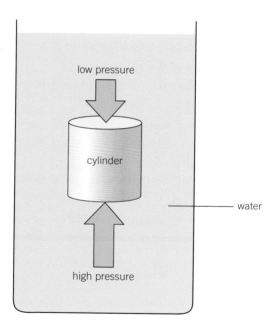

Figure 2 *Explaining upthrust*

Figure 3 *A loaded ship*

Explaining upthrust

The water level in a water container rises when an object is lowered into the water. This is because the object *displaces* some of the water.

- The more the object is lowered into the water, the bigger the volume of water displaced, and the bigger the upthrust.
- When the object is fully immersed, the volume of water displaced is equal to the volume of the object.

Figure 2 shows a cylinder fully immersed in water. Because pressure increases with depth, the pressure of the water at the bottom of the cylinder is greater than the pressure on the top of the cylinder. So the upward force of the water on the bottom of the cylinder is greater than the downward force of the water on the top of the cylinder. The upthrust is the resultant of these two forces.

Float or sink?

A ship being loaded with cargo will float lower and lower in the water as the load is increased. The ship displaces more water when the load increases, so the upthrust increases. At any instant, the upthrust on the boat is equal to the weight of the ship and its cargo. If the ship is loaded too much, it sinks because it has displaced as much water as it possibly can, and because the upthrust can not support the total weight.

An object floats when its weight is equal to the upthrust.

An object sinks when its weight is greater than the upthrust.

Density tests

Objects made of materials such as cork or wood float in water, but metal objects sink in water. Objects that float have a density less than the density of water. Objects that sink have a density greater than the density of water. By observing if an object floats or sinks, you can tell if its density is less than or greater than the density of water.

To understand why density matters in float tests, remember that the pressure in a liquid increases with increase of density as well as with increase of depth. Using the equation $p = h \rho g$, it can be shown that the upthrust is equal to the weight of liquid displaced.

- An object that is more dense than the liquid sinks because its weight is greater than the weight of the liquid displaced. So the weight of the object is greater than the upthrust on the object when it is fully immersed.

- An object that is less dense than the liquid floats because its weight is less than the weight of the liquid it displaces. So the weight of the object is less than the upthrust on the object when it is fully immersed.

1 a Explain why it is difficult to hold an inflated plastic ball under water. 🖉 [4 marks]

   b Explain why cork is a suitable material for filling a life belt. 🖉 [3 marks]

2 When an object is weighed using a newton-meter, the reading on the newton-meter is 5.2 N when the object is in air, and 4.7 N when the object is immersed in water.

   a Explain why the reading on the newton-meter is less when the object is in water. [2 marks]

   b Describe how the newton-meter reading changes as the object is lowered into water. [2 marks]

3 a Ice floats on water. Explain what this tells you about the density of ice compared with the density of water. [1 mark]

   b Three blocks A, B, and C are released in a bowl of a water.
Block A sinks to the bottom
Block B floats with its top half above the water
Block C floats with a small proportion above the water.

     i Which object has the greatest density? Give a reason for your answer. [2 marks]

     ii Which object has the lowest density? Give a reason for your answer. [2 marks]

4 Figure 4 shows a weighted test tube floating vertically in water.

   a Make a reasoned prediction about how the length of tube above the water depends on the total weight of the tube. [2 marks]

   b Design an experiment to test your prediction, using the test tube and any other apparatus necessary. 🖉 [6 marks]

Go further!

In Figure 2, the pressure difference between the top and the bottom of the cylinder = $L\rho g$, where L is the length of the cylinder, ρ is the density of the liquid, and $g = 9.8$ N/kg.

So the difference between the force on the bottom and the top of the cylinder = the pressure difference × the cylinder's cross-sectional area A
= $(L\rho g) \times A$
= $LA\rho g = V\rho g$, where $V = L \times A$ gives the cylinder's volume. Therefore, the upthrust = $V\rho g$.

Because mass m = volume V × density ρ, the upthrust = $V\rho g = mg$, where m is the mass of liquid displaced by the cylinder. So the upthrust is equal to the weight of liquid displaced by the cylinder.

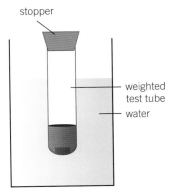

stopper

weighted test tube

water

Figure 4

Key points

- The upthrust on an object in a fluid:
  - is an upward force on the object due to the fluid
  - is caused by the pressure of the fluid.
- The pressure at a point in a fluid depends on the density of the fluid and the depth of the fluid at that point.
- An object sinks if its weight is greater than the upthrust on it when it is fully immersed.

Summary questions

Assume $g = 9.8$ N/kg unless otherwise stated in a question.

1 A sink plug is used to block the outlet of a sink filled with water.

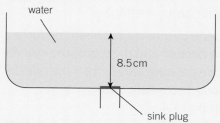
water

8.5 cm

sink plug

Figure 1

a The sink plug is 8.5 cm below the water surface. Calculate the pressure due to the water on the sink plug. [2 marks]

b The sink plug has an area of 6.0×10^{-4} m². Calculate the force needed to remove the sink plug (density of water = 1000 kg/m³). [2 marks]

2 Blood pressure is usually measured in millimetres of mercury, not pascals. This unit is used because the first blood pressure gauges contained a tube of mercury. density of mercury = 13 600 kg/m³

a The blood pressure of a healthy person is 120 mm of mercury. Calculate this pressure in pascals. [2 marks]

b The precision of the reading of a mercury gauge is 2 mm. Discuss whether the gauge is more precise than an electronic gauge that gives readings with an accuracy of 50 Pa. [3 marks]

3 The density of air at sea level is 1.3 kg/m³. Atmospheric pressure at sea level is 101 kPa.
$g = 9.8$ N/kg

a Show that the overall height of this model atmosphere would be no more than 7800 m. [1 mark]

b What conclusions can you draw about the validity of this model atmosphere? [3 marks]

4 Ⓗ A barge floats because the upthrust on it is equal and opposite to its weight.

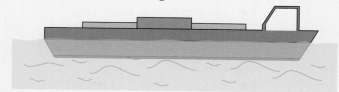

Figure 2

a Explain why the barge is acted on by an upthrust. [2 marks]

b When the barge is loaded, the hull is 2.5 m below the water level.
i Calculate the pressure of the water on the hull. [2 marks]
ii The barge has a flat hull of area 80 m². Calculate the upthrust on the barge, and so determine the total weight of the barge and its load (density of water = 1000 kg/m³). [2 marks]

c When the barge is unloaded, explain why it floats higher in the water than when it is loaded. [4 marks]

5 Ⓗ Salt water and tap water have different densities. Describe how you would use a weighted test tube to investigate how the density of salt water varies with the amount of salt dissolved in the water. In your account, you should give a reasoned prediction about the investigation and explain: ✏
- what you would measure, what measuring equipment you would use, and how you would use the measurements
- how you would present and use your measurements to test your prediction. [6 marks]

6 A submersible craft is an underwater vehicle that is used by the people inside it to explore or investigate objects on the floor of the sea.

a A submersible craft is at a depth of 250 m below the surface of the sea.
i Calculate the pressure due to the sea water on the craft at this depth. The density of sea water is 1100 kg/m³. [2 marks]
In this type of submersible craft, the air is maintained at a pressure of 100 kPa. An observation window in the craft has an area of 0.043 m².
The pressure on the outside of the window is the sum of the pressure due to the sea water and the atmospheric pressure on the surface of the sea, which is about 100 kPa.
ii Estimate the force on the window due to the pressure difference between the inside and outside of the window. [4 marks]

Practice questions

01 The pressure of water changes as the depth of the water changes.

01.1 **Figure 1** shows different designs for the walls of a deep dam.

Figure 1

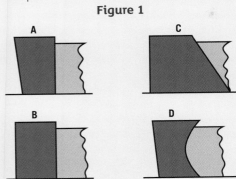

Choose the most suitable design and give a reason for your choice. [2 marks]

01.2 A student investigated how water in a can flowed from spouts at different heights. Copy the diagram and draw lines to show how the water flowed out of the spouts.

Figure 2

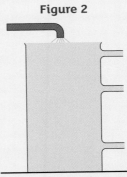

[3 marks]

01.3 Why was it important to keep the can filled with water in the investigation. [2 marks]

01.4 New scuba divers are only allowed to dive to a depth of about 18 m. Calculate the water pressure at a depth of 18 m. The density of sea water is 1025 kg/m³. Use g = 9.8 N/kg. [2 marks]

02 A hand cart with two wheels is being used to transport some equipment across wet sand.

02.1 Explain why it is important for the hand cart to have wide wheels. [2 marks]

02.2 The weight of the handcart and equipment is 85 kg. Each wheel has an area of 0.015 m² in contact with the sand. Calculate the pressure of the handcart and equipment on the wet sand. Give your answer correct to 3 significant figures. [3 marks]

03.1 A pycnometer is a container used for determining the density of a liquid or powder, having a specific volume and often provided with a thermometer to indicate the temperature of the contained substance. Describe how you would use the pycnometer to find the density of several samples of oil. Your answer should include any precautions you would take to ensure valid results. [4 marks]

03.2 **Table 1** gives the density of four liquids at a temperature of 25 °C.

Table 1

Liquid	Density in kg/m³
castor oil	956
methanol	786
glycerol	1260
sea water	997

The four liquids are poured into a large measuring cylinder. After a time the liquids separate into individual layers. Draw a diagram of the measuring cylinder and label the liquids. [4 marks]

03.3 A plimsoll line is drawn on merchant ships to show the ship is not overloaded with cargo. Explain how a ship is able to float when loaded with heavy cargo. [3 marks]

04.1 Passenger airlines fly at about 12 000 m. At this height the cabin is pressurised to a pressure equal to the atmospheric pressure at 2000 m. Use **Figure 3** to determine the cabin pressure. [1 mark]

Figure 3

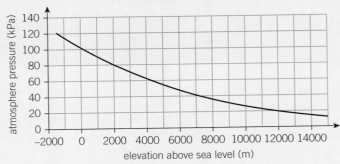

04.2 A design goal for many newer aircraft is to lower the cabin altitude, which can be beneficial for passenger comfort. Suggest a reason why. [2 marks]

04.3 A passenger drank from a plastic bottle of water whilst flying in the aircraft. He screwed the top back on and returned the bottle to his rucksack when the bottle was empty. The passenger found the plastic bottle had crushed on landing. Explain why. [3 marks]

4 Waves, electromagnetism, and space

When you speak into a mobile phone, you create sound waves that carry information. These waves are detected by a microphone that produces electrical waves in the phone circuits. Your phone then sends out radio waves that carry the information to your mobile phone network and then to the person you are calling. Medical doctors use radio waves in scanners to obtain 3D images of organs. They also use X-rays and ultrasonic waves to visualise objects inside the body.

In this section, you will learn about waves and their properties, and the many ways they are used. You will also learn about magnetic fields and how we use them to produce electrical waves.

Key questions

- How do we measure waves and how fast do they travel?

- What happens when waves reach a boundary between two substances?

- What are electromagnetic waves and how do they differ from sound waves?

- How do waves carry information?

Making connections

- Astronomers use non-optical telescopes (like radio telescopes) as well as optical telescopes to obtain images of distant objects in space. You will learn more about non-optical telescopes and detectors when you study **P16 Space**.

- Power stations generate alternating currents not direct currents. In this section you will learn how alternating currents are generated and why transformers can only work using alternating current. When you study these topics, look back at **P5 Electricity in the home** to remind yourself about alternating currents.

I already know...

The top of a water wave is called a crest and the bottom is called a trough.

Light travels much faster than sound and can travel through space whereas sound cannot.

The spectrum of white light is continuous from red to orange, yellow to green, and blue to violet.

There are different types of waves, such as sound waves and electromagnetic waves, but they all have common properties such as refraction.

A magnet lines up with the Earth's magnetic field.

An electric motor is used to turn objects. An electric generator produces an electric current when it turns.

Satellites orbit the Earth.

I will learn...

How the wavelength of a wave depends on its speed and its frequency.

How to measure the speed of sound waves in air and in a solid.

How electromagnetic waves carry information and how they are used to form images.

What we mean by refraction of waves when they cross a boundary between different substances.

How the strength of a magnetic field is measured and what a solenoid is.

How an electric motor and an electric generator work.

How satellites stay in their orbits and what we mean by a geostationary satellite.

Required Practicals

Practical		Topic
8	Investigating plane waves in a ripple tank and waves in a solid.	P12.4
9	Investigating the reflection and refraction of light.	P14.2 P14.3
10	Investigating infrared radiation	P13.2

Learning objectives

After this topic, you should know:

- what waves can be used for
- what transverse waves are
- what longitudinal waves are
- which types of waves are transverse and which are longitudinal.

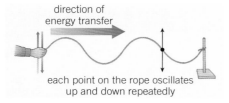

Figure 1 *Waves in water are examples of mechanical waves*

direction of energy transfer

each point on the rope oscillates up and down repeatedly

Figure 2 *Transverse waves*

Waves transfer energy without transferring matter. You can also use waves to transfer information, for example when you use a mobile phone or listen to the radio.

There are different types of waves. These include:

- sound waves, water waves, waves on springs and ropes, and seismic waves produced by earthquakes. These are all examples of **mechanical waves**, which are vibrations that travel through a medium (a substance).

- light waves, radio waves, and microwaves. These are all examples of **electromagnetic waves**, which can all travel through a vacuum at the same speed of 300 000 kilometres per second. No medium is needed.

Observing mechanical waves

Figure 2 shows how you can make waves on a rope by moving one end up and down.

Tie a ribbon to the middle of the rope. Move one end of the rope up and down. You will see that the waves move along the rope but the ribbon doesn't move along the rope – it just moves up and down. This type of wave is known as a **transverse wave**. It is said that the ribbon vibrates or oscillates. This means that it moves repeatedly between two positions. When the ribbon is at the top of a wave, it is said to be at the peak (or crest) of the wave.

Repeat the test with the slinky. You should observe the same effects if you move one end of the slinky up and down.

However, if you push and pull the end of the slinky as shown in Figure 3, you will see a different type of wave, known as a **longitudinal wave**. Notice that there are areas of **compression** (coils squashed together) and areas of **rarefaction** (coils spread further apart) moving along the slinky.

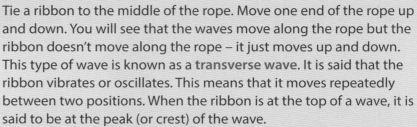

direction of energy transfer

each point on the slinky oscillates backwards and forwards repeatedly   compression

Figure 3 *Making longitudinal waves on a slinky*

- Describe how the ribbon moves when you send longitudinal waves along the slinky.

Safety: Handle the slinky spring carefully.

More wave tests

When waves travel through a substance, the substance itself doesn't travel. You can see this with waves on a rope and by observing:

- waves spreading out on water after a small object is dropped in the water. The waves travel across the surface but the water does not travel away from the object.

- a tuning fork vibrating so that it makes sound waves travel through the air away from the tuning fork. The air itself doesn't travel away from the vibrating object – if it did, a vacuum would be created.

Transverse waves

Imagine sending waves along a rope that has a white spot painted on it. You would see the spot move up and down without moving along the rope. In other words, the spot would oscillate perpendicular (at right angles) to the direction in which the waves are moving. The waves on a rope and the ripples on the surface of water are called transverse waves because the vibrations (called oscillations) move up and down or from side to side. All electromagnetic waves are transverse waves.

The oscillations of a transverse wave are perpendicular to the direction in which the waves transfer energy.

Longitudinal waves

The slinky spring in Figure 3 is useful to demonstrate how sound waves travel. When one end of the slinky is pushed in and out repeatedly, vibrations travel along the spring. These oscillations are parallel to the direction in which the waves transfer energy. Waves that travel in this way are called longitudinal waves.

Sound waves travelling through air are longitudinal waves. When an object vibrates in air, it makes the air around it vibrate as it pushes and pulls on the air. The oscillations (compressions and rarefactions) that travel through the air are sound waves. The oscillations are along the direction in which the wave travels.

The oscillations of a longitudinal wave are parallel to the direction in which the waves transfer energy.

Mechanical waves can be transverse or longitudinal.

1 a What is the difference between a longitudinal wave and a transverse wave? [1 mark]
 b Give *one* example of:
   i a transverse wave [1 mark] ii a longitudinal wave. [1 mark]
 c When a sound wave passes through air, describe what happens to the air particles at a point of compression. [1 mark]

2 A long rope with a knot tied in the middle lies straight along a smooth floor. A student picks up one end of the rope. This sends waves along the rope.
 a State whether the waves on the rope are transverse or longitudinal waves. [1 mark]
 b Describe:
   i the direction of energy transfer along the rope [1 mark]
   ii the movement of the knot. [1 mark]

3 Describe how to use a slinky spring to demonstrate to a friend the difference between longitudinal waves and transverse waves. [2 marks]

4 Describe and explain the motion of a small ball floating on a pond when waves travel across the pond. 🖊 [3 marks]

Synoptic links

You will learn more about electromagnetic waves in Topic P13.1.

Study tip

- Make sure you can know how to describe the difference between transverse waves and longitudinal waves.
- Remember that electromagnetic waves are transverse, and sound waves are longitudinal.

Key points

- Waves can be used to transfer energy and information.
- Transverse waves oscillate perpendicular to the direction of energy transfer of the waves. Ripples on the surface of water are transverse waves. So are all electromagnetic waves.
- Longitudinal waves oscillate parallel to the direction of energy transfer of the waves. Sound waves in air are longitudinal waves.
- Mechanical waves need a medium (a substance) to travel through. They can be transverse or longitudinal waves.

P12.2 The properties of waves

Learning objectives

After this topic, you should know:

- what is meant by the amplitude, frequency, and wavelength of a wave
- how the period of a wave is related to its frequency
- the relationship between the speed, wavelength, and frequency of a wave
- how to use the wave speed equation in calculations.

If you want to find out how much energy or information waves carry, you need to measure them. Figure 1 shows a snapshot of waves on a rope. The crests, or peaks, are at the top of the wave. The troughs are at the bottom. They are equally spaced.

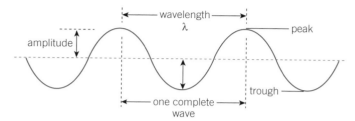

Figure 1 *Waves on a rope*

The **amplitude** of a wave is the maximum displacement of a point on the wave from its undisturbed position. For example, in Figure 1, this is the height of the wave crest (or the depth of the wave trough) from the middle.

The bigger the amplitude of the waves, the more energy the waves carry.

The **wavelength** of a wave is the distance from a point on the wave to the equivalent point on the adjacent wave. For example, in Figure 1, this is the distance from one wave crest to the next wave crest.

Frequency

If you made a video of the waves on the rope in Figure 1, you would see the waves moving steadily across the screen. The number of waves passing a fixed point every second is called the **frequency** of the waves.

The unit of frequency is the hertz (Hz). For the waves on the rope, one wave crest passing each second is equal to a frequency of 1 Hz.

The period of a wave is the time taken for each wave to pass a fixed point. For waves of frequency f, the period T is given by the equation:

$$\textbf{period, } T \text{ (seconds, s)} = \frac{1}{\textbf{frequency, } f \text{ (hertz, Hz)}}$$

Wave speed

Figure 2 shows a ripple tank, which is used to study water waves in controlled conditions. You can make straight waves by moving the long edge of a ruler up and down on the water surface in a ripple tank. Straight waves are called plane waves. The waves all move at the same speed and stay the same distance apart.

The **speed** of the waves is the distance travelled by each wave every second through a medium. Energy is transferred by the waves at this speed.

For waves of constant frequency, the speed of the waves depends on the frequency and the wavelength as follows:

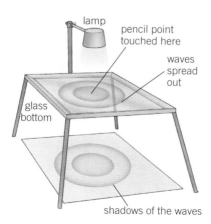

Figure 2 *The ripple tank*

$$\text{wave speed, } v \quad = \quad \text{frequency, } f \quad \times \quad \text{wavelength, } \lambda$$
$$\text{(metres per second, m/s)} \qquad \text{(hertz, Hz)} \qquad \qquad \text{(metres, m)}$$

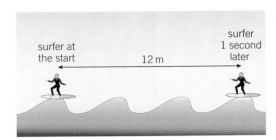

Figure 3 *Surfing*

To understand what the wave speed equation means, look at Figure 3. The surfer is riding on the crest of some unusually fast waves.

Suppose the frequency of the waves is 3 Hz and the wavelength of the waves is 4.0 m.

● At this frequency, three wave crests pass a fixed point once every second (because the frequency is 3 Hz).

● The surfer therefore moves forward a distance of three wavelengths every second, which is $3 \times 4.0\,\text{m} = 12\,\text{m}$.

So the speed of the surfer is 12 m/s.

This speed is equal to the frequency × the wavelength of the waves.

Measuring the speed of sound in air

You need two people for this. You and a friend need to stand on opposite sides of a field at a measured distance apart. You should be as far apart as possible but within sight of each other.

If your friend bangs two cymbals together, you will see them crash together straightway, but you won't hear them straightaway. The crashing sound will be delayed because sound travels much slower than light. Use a stopwatch to time the interval between seeing the impact and hearing the sound. Repeat the test several times to get an average value of the time interval.

Calculate the speed of sound in air using the equation

$$\text{speed} = \frac{\text{distance}}{\text{time taken}}$$

1 State what is meant by the frequency of a wave. [1 mark]

2 Figure 4 shows a wave travelling from left to right along a rope.

Figure 4

  a Copy the figure and mark on your diagram:
    i one wavelength [1 mark]
    ii the amplitude of the waves. [1 mark]
  b Describe the motion of point P on the rope when the wave crest at P moves along by a distance of one wavelength. [2 marks]

3 a A speedboat on a lake sends waves travelling across a lake at a frequency of 2.0 Hz and a wavelength of 3.0 m. Calculate the speed of the waves. [2 marks]
  b If the waves had been produced at a frequency of 1.0 Hz and travelled at the speed calculated in **a**:
    i calculate what their wavelength would be [2 marks]
    ii calculate the distance travelled by a wave crest in 60 s. [2 marks]

4 Sound waves in air travel at a speed of 340 m/s.
  a Calculate how far they travel in air in 5.0 s. [2 marks]
  b Calculate their wavelength if their frequency is 3.0 kHz. [2 marks]

Key points

● For any wave, its amplitude is the maximum displacement of a point on the wave from its undisturbed position, such as the height of the wave crest (or the depth of the wave trough) from the position at rest.

● For any wave, its frequency is the number of waves passing a point per second.

● The period of a wave = $\dfrac{1}{\text{frequency}}$

● For any wave, its wavelength is the distance from a point on the wave to the equivalent point on the next wave (e.g., from one wave trough to the next wave trough).

● The speed of a wave is $v = f \times \lambda$.

P12.3 Reflection and refraction

Learning objectives

After this topic, you should know:

- the patterns of reflection and refraction of plane waves in a ripple tank
- whether plane waves that cross a boundary between two different materials are refracted
- how the behaviour of waves can be used to explain reflection and refraction
- what can happen to a wave when it crosses a boundary between two different materials.

Investigating waves using a ripple tank

Reflection of waves can be investigated using the ripple tank. Each ripple is called a wavefront because it is the front of each wave as it travels across the water surface. Plane (i.e., straight) waves, produced by repeatedly dipping the long edge of a ruler in water, are directed at a metal barrier in the water. These waves are called the incident waves to distinguish them from the reflected waves. The incident waves are reflected by the barrier.

In Figure 1, the incident wavefront is not parallel to the barrier before or after reflection. The reflected wavefront moves away from the barrier at the same angle to the barrier as the incident wavefront.

A reflection test

Use a ruler to create and direct plane waves at a straight barrier (Figure 1). Find out if the reflected waves are always at the same angle to the barrier as the incident waves. You could align a second ruler with the reflected waves and measure the angle of each ruler to the barrier. Repeat the test for different angles.

Safety: Mop up any water spillages.

Refraction of waves is the change of the direction in which they are travelling when they cross a boundary between one medium and another medium. You can see this in a ripple tank when water waves cross a boundary between deep and shallow water. Plane waves directed at a non-zero angle to the boundary change direction as they cross the boundary, as shown in Figure 2.

Wavefronts at a non-zero angle to the barrier

Incident wavefront

Reflected wavefront

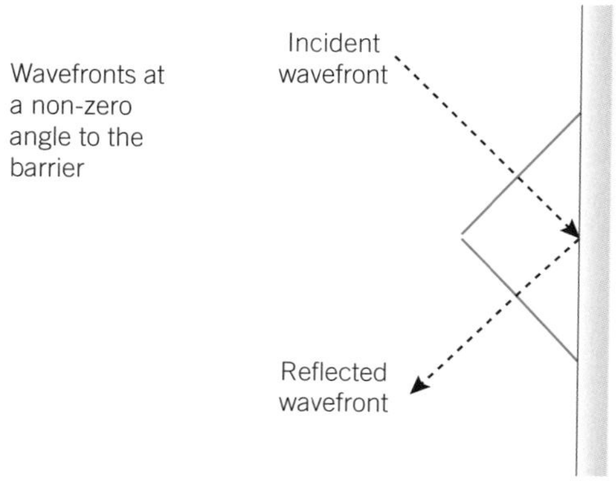

Figure 1 *Reflection of plane waves*

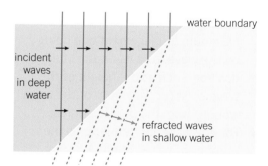

water boundary

incident waves in deep water

refracted waves in shallow water

Figure 2 *Refraction*

Refraction tests

Use a vibrating beam to create plane waves continuously in a ripple tank containing a transparent plastic plate.

Arrange the plate so that the waves cross a boundary between the deep and shallow water. The water over the plate needs to be very shallow.

At a non-zero angle to a boundary. The waves change their speed and direction when they cross the boundary. Find out if plane waves change direction towards or away from the boundary when they cross from deep to shallow water.

Perpendicular to a boundary (at normal incidence). The waves cross the boundary without changing direction. However, their speed changes.

- Find out if the waves travel slower or faster when they cross the boundary.

Safety: Mop up any water spillages.

Explaining refraction

To explain how a wavefront moves forward, imagine that each tiny section creates a wavelet (a little wave) that travels forward (Figure 3). The wavelets move forward together to recreate the wavefront that created them.

Refraction

When plane waves cross a boundary at a non-zero angle to the boundary, each wavefront experiences a change in speed and direction.

In Figure 3, the wavefronts move more slowly after they have crossed the boundary. So the refracted wavefronts are closer together and are at a smaller angle to the boundary than the incident wavefronts.

The refracted waves and the incident waves have the same frequency, but they travel at a different speeds, so they have different wavelengths.

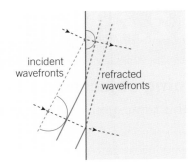

Figure 3 *Explaining refraction*

Materials and waves

When a wave is directed at a substance, some or all of the wave may be reflected at the surface. What happens is dependent on the wavelength of the wave and also on the substance (e.g., its surface). For example, microwaves are reflected by metal surfaces but they can pass through paper.

Of the waves that go into a substance, some or all of them may be absorbed by the substance. This would heat the substance because it would gain energy from the waves. For example, food is heated in microwave ovens because the microwaves are absorbed by the food.

As waves travel through a substance, the amplitude of the waves gradually decreases as the substance absorbs some of the waves' energy.

Waves that are not absorbed by the substance they are travelling through are transmitted by it. For example, light is mostly **transmitted** by ordinary glass, but is almost completely absorbed by darkened glass.

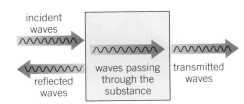

Figure 4 *Waves and substances*

---

1 When plane waves reflect from a straight barrier, describe the angle of each reflected wavefront to the barrier and the angle of each incident wavefront to the barrier. [1 mark]

2 Draw a diagram that shows plane waves passing from deep to shallow water at a non-zero angle to a straight boundary. Draw some refracted wavefronts, indicating their direction. [3 marks]

3 **a** Sea waves rolling up a sandy beach are not reflected. Explain why the sides of a ripple tank are sloped. [1 mark]

b Describe what would happen if the sides of the ripple tank were vertical instead of sloped. [1 mark]

4 Sunglasses have lenses made of dark glass that reduce the amount of daylight entering your eyes. Design a test using a light meter and a lamp to find out if the two lenses in a pair of sunglasses are equally effective. ✓ [4 marks]

Key points

- Plane waves in a ripple tank are reflected from a straight barrier at the same angle to the barrier as the incident waves because their speed and wavelength do not change on reflection.
- Plane waves crossing a boundary between two different materials are refracted unless they cross the boundary at normal incidence.
- Refraction occurs at a boundary between two different materials because the speed and wavelength of the waves change at the boundary.
- At a boundary between two different materials, waves can be transmitted or absorbed.

Learning objectives

After this topic, you should know:

- what sound waves are
- how to investigate waves

Figure 1 *Making sound waves – the buzzing of a bee is caused by the vibrations of their wings*

Sound waves are easy to produce. Your vocal cords vibrate and produce sound waves every time you speak. Any object vibrating in air makes the layers of air near the object vibrate, which make the layers of air next to them vibrate. The vibrating object pushes and pulls repeatedly on the air. This sends out vibrations of air in waves of compressions and rarefactions. When the waves reach your ears, they make your eardrums vibrate in and out so that you hear sound.

The vibrations travelling through the air are sound waves. The waves are longitudinal because the air particles vibrate (or oscillate) along the direction in which the waves transfer energy. The speed of sound waves in air is 330 m/s.

Investigating sound waves

You can use a loudspeaker to produce sound waves. Figure 2 shows how to do this using a signal generator.

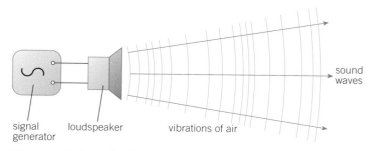

Figure 2 *Using a loudspeaker*

If you observe the loudspeaker closely, you can see it vibrating. It produces sound waves as it pushes the surrounding air backwards and forwards.

If you alter the frequency dial of the signal generator, you can change the frequency of the sound waves.

Sound waves cannot travel through a vacuum. You can test this by listening to an electric bell in a bell jar (Figure 3). As the air is pumped out of the bell jar, the ringing sound fades away.

A sound test

Sound waves reflect from smooth hard surfaces, such as bare walls. If you clap your hands together in a large hall with bare walls, for example, a school gymnasium, you will hear the **echo** a short time later. The time delay is because the sound waves travel to the wall and back before you hear the echo.

If the distance d to the nearest wall is measured and the time delay t is also measured, the speed of sound in air can be calculated using the equation:

$$\text{speed, } s = \frac{\text{distance to the wall and back, } 2d}{\text{time delay, } t}$$

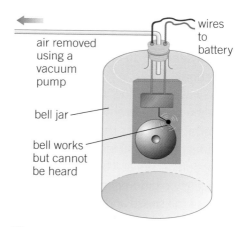

Figure 3 *A sound test – sound waves can't travel through a vacuum*

Investigating waves

Investigate waves on a stretched string using the apparatus shown in Figure 4. The oscillator sends waves along the string. You can adjust the frequency of the oscillator until there is a single loop on the string. Its length is half the length of one wavelength. The vibrating string sends out sound waves at the same frequency into the surrounding air.

- Note the frequency of the oscillator.
- Make suitable measurements to find the length, L, of a single loop and calculate the wavelength of the waves (= 2L)
- Calculate the speed of the waves on the string using the equation: wave speed = frequency × wavelength
- Increase the frequency to obtain more loops on the string. Make more measurements to see if the wave speed is the same.

To measure the speed of the waves in a ripple tank (Figure 2, Topic P12.2), use a ruler to create plane waves that travel towards one end of the ripple tank.

- Use a stopwatch to measure the time it takes for a wave to travel from one end of the ripple tank to the other.
- Measure the distance the waves travel in this time.
- Use the equation speed = distance ÷ time to calculate the speed of the waves.

Observe the effect on the waves of moving the ruler up and down faster. More waves are produced every second and they are closer together.

- Determine whether the speed of the waves has changed.

Safety: Take care not to spill any liquids and, if you do, let your teacher know. You should also take care with hanging weights – clamp the stands to the bench and wear eye protection.

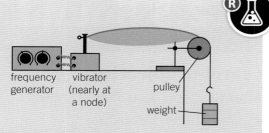

Figure 4 *Investigating waves on a string*

1 A lighting strike was heard by a student 4 seconds after he saw the lightning flash. Calculate the distance of the student to the point where the lightning strike occurred. The speed of sound in air is 330 m/s. [2 marks]

2 A student standing at a fixed distance from a large bare wall clapped her hands repeatedly in time with the echoes she heard. Her friend timed 11 claps in 2.4 s. Estimate the shortest distance from the student to the wall. The speed of sound in air = 340 m/s [2 marks]

3 A stretched string of length 1.24 m was made to oscillate as shown in Figure 4. A pattern of three equal loops was seen on the string when the oscillator vibrated at a frequency of 180 Hz.
   a i Calculate the wavelength of the waves on the string. [2 marks]
   ii Calculate the speed of the waves on the string. [2 marks]
   b Explain why the vibrations of the string caused sound waves in the surrounding air. [3 marks]

Key points

- Sound waves are vibrations that travel through a medium (a substance).
- Sound waves cannot travel through a vacuum (e.g., in outer space).
- To investigate waves, use:
  - a ripple tank for water waves
  - a stretched string for waves in a solid
  - a signal generator and a loudspeaker for sound waves.

P12.5 Sound waves

Learning objectives

After this topic, you should know:

- what affects the loudness of a musical note
- how sound waves are detected by the ear
- why human hearing is limited.

Figure 1 *Making music*

What type of music do you like? Whatever your taste in music is, when you listen to it you usually hear sounds produced by specially designed instruments. Even your voice is produced by a biological organ that has the job of producing sound.

- Musical notes are easy to listen to because they are rhythmic. The sound waves change smoothly, and the wave pattern repeats regularly.
- General noise is made up of sound waves that vary in frequency without any pattern.

Investigating different sounds

Use a microphone connected to an oscilloscope to display the waveforms of different sounds.

Figure 2 *Investigating different sound waves*   **Figure 3** *Tuning-fork waves*

1 Test a tuning fork to see the waveform of a sound of constant frequency. To make the sound louder, hold the base of the tuning fork on the table so that the table top also vibrates.

2 Compare the pure waveform of a tuning fork with the sound you produce when you talk or sing or whistle. You may be able to produce a pure waveform when you whistle or sing but not when you talk.

3 Use a signal generator connected to a loudspeaker to produce sound waves. The waveform on the oscilloscope screen should be a pure waveform.

Your investigations should show you that increasing the loudness of a sound increases the amplitude of the sound waves. So the waves on the screen become taller (Figure 4).

The ear

Your ear can detect an enormous range of sound waves of different intensities as well as a wide range of frequencies, from 20 Hz to about 20 kHz (20 000 Hz). When sound waves make your ear drum vibrate, your ear sends signals to your brain about what you are hearing.

- Sound waves entering a solid are converted to vibrations and travel through the solid as vibrations. The conversion of sound waves to vibrations of solids only works over a limited frequency range. So the frequency range of the human ear is limited.

a louder sound

a quieter sound

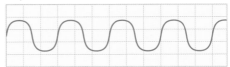

Figure 4 *As the amplitude of a sound wave increases, the loudness of the sound will increase*

Sound waves travel through different substances at different speeds. When sound waves travel from one substance to another, their frequency does not change but their speed may change depending on the two substances, and so their wavelength may change. Therefore, the frequency of the sounds you listen to do not change when the sound waves and vibrations pass through the air and the different parts of your ear.

Figure 5 shows how a normal ear responds to sounds of different frequencies. The graph shows how the intensity of sounds the ear can just detect varies with frequency. It shows that a normal human ear hears best at a frequency of about 3 kHz because the intensity the ear can just detect is least at this frequency. However, the human ear is unable to hear sounds at frequencies of less than 20 Hz and more than about 20 KHz, no matter how intense the sound is.

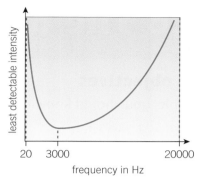

Figure 5 *Frequency response of a normal ear*

Echo sounding

Echo sounding uses pulses of high-frequency sound waves to detect objects in deep water and to measure water depth below a ship (Figure 6). An echo is the reflection of sound waves from a smooth surface. The pulses from the transmitter are reflected at the sea bed directly below the ship and detected by a receiver at the same depth as the transmitter. The time taken, t, by each wave to travel to the sea bed and back is measured. The total distance travelled by the wave = vt, where v is the speed of sound in water. This is twice the depth of the sea bed below the surface. So, the depth of water below the ship = $\frac{1}{2}vt$

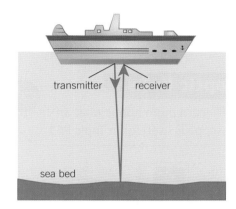

Figure 6 *Ray diagram of an echo from the sea bed to a ship*

1   A tuning fork creates a note of sound when it is struck briefly. Describe how the waveform of the sound from a tuning fork changes as the sound becomes quieter. [1 mark]

2   A microphone and an oscilloscope are used to investigate sound from a loudspeaker connected to a signal generator. Describe the change you would expect to see on the oscilloscope screen if the sound is made quieter. [2 marks]

3   A depth finder was used to send measure the depth of the sea bed. The sound pulses from the depth finder took 0.36 seconds to travel from the surface to the sea bed and back. Calculate the depth of the sea bed. The speed of sound in water is 1350 m/s. [2 marks]

4   a   Explain why the sound from a vibrating tuning fork is much louder if the base of the tuning fork is held on a table. 🚫 [2 marks]

    b   Design a test to find out if sound travels through walls. List the equipment you would use. [3 marks]

Key points

- The pitch of a note increases if the frequency of the sound waves increases.
- The loudness of a note increases if the amplitude of the sound waves increases.
- Sound waves cause the ear drum to vibrate, and the vibrations send signals to the brain.
- The conversion of sound waves to vibrations of solids only works over a limited frequency range, so human hearing is limited.

P12.6 The uses of ultrasound

Learning objectives

After this topic, you should know:

- what ultrasound waves are
- why ultrasound waves can be used to scan the human body
- how ultrasound waves are used to measure distances in medicine and in industry
- why an ultrasound scan is safer than taking an X-ray image.

The human ear can detect sound waves in the frequency range of about 20 Hz to about 20 kHz. Sound waves above the highest frequency that humans can detect are called **ultrasound waves**.

Ultrasound scanners

Ultrasound waves are used for prenatal scans of a baby in the womb. They are also used to get an image of organs in the body such as a kidney, or damaged ligaments and muscles. An ultrasound scanner is made up of an electronic device called a transducer placed on the body surface, a control system, and a display screen. The transducer produces and detects sets (or pulses) of ultrasound waves.

Each ultrasound wave pulse from the transducer:

- is partially reflected from the different tissue boundaries in its path
- returns to the transducer as a sequence of ultrasound waves reflected by the tissue boundaries, arriving back at different times.

The transducer is moved across the surface of part of the body. The ultrasound waves are then detected by the transducer. They are used to build up an image on a screen of the internal tissue boundaries in the body.

The advantages of using ultrasound waves for certain types of medical scanning are that ultrasound waves, unlike X-rays, are:

- reflected at boundaries between different types of tissue (different media), so they can be used to scan organs and other soft tissues in the body
- non-ionising. Non-ionising radiation is radiation that does not have enough energy to remove an electron to ionise an atom or molecule. So it is harmless when used for scanning.

Measuring ultrasound waves

Ultrasound waves can be very useful in industrial imaging. Flaws in metal castings can be detected using ultrasound waves. A flaw might be an internal crack, which creates a boundary inside the metal. The ultrasound waves are partly reflected from this boundary. A transducer at the surface of the block sends ultrasound waves into the block. The reflected waves are detected by the transducer and displayed on an oscilloscope screen or on a computer monitor, as shown in Figure 2.

Synoptic links

You willl learn more about what an oscilloscope is used for in Topic P5.1.

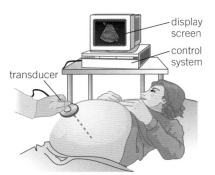

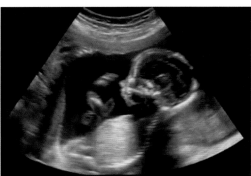

Figure 1 a *An ultrasound scanner system*
b *An ultrasound image of a baby in the womb*

Study tip

Ultrasound is often reflected – it goes there and back – so in calculations be careful about distance.

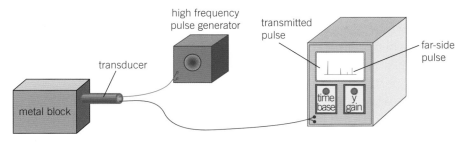

Figure 2 *Detecting flaws in a metal*

The oscilloscope screen in Figure 2 shows:

- a strong ultrasound wave pulse due to partial reflection of the transmitted pulse at the surface, then
- some further ultrasound wave pulses – in this case, two due to partial reflection at internal boundaries, and the last wave pulse due to partial reflection at the far side of the metal object.

The further away a boundary is from the transmitter, the longer a reflected wave takes to return. You can use the oscilloscope to measure the time taken by the wave to travel from the transmitter at the surface to and from the boundary that reflected it. To calculate distance travelled, use the equation:

$$\begin{array}{c}\textbf{distance travelled}\\\textbf{by the wave, } s\\(\text{metres, m})\end{array} = \begin{array}{c}\textbf{speed of ultrasound waves}\\\textbf{in body tissue, } v\\(\text{metres per second, m/s})\end{array} \times \begin{array}{c}\textbf{time taken, } t\\(\text{seconds, s})\end{array}$$

Because the ultrasound waves travel from the surface to the boundary then back to the surface, the depth of the boundary below the surface is half the distance travelled by each wave to and from the boundary. So:

$$\begin{array}{c}\textbf{the depth of the boundary}\\\textbf{below the surface, } s\\(\text{metres, m})\end{array} = \frac{1}{2} \times \begin{array}{c}\textbf{speed of the}\\\textbf{ultrasound waves, } v\\(\text{metres per}\\\text{second, m/s})\end{array} \times \begin{array}{c}\textbf{time taken, } t\\(\text{seconds, s})\end{array}$$

This principle is useful in medicine. Ultrasound waves are used by eye surgeons when they need to know how long an eyeball is.

1 **a** Explain why ultrasound waves are partly reflected by body organs. [2 marks]
   b Explain why an ultrasound scanner is better than an X-ray scanner for scanning a body organ. [3 marks]
   c Give one reason why the amplitude of the ultrasonic waves becomes smaller as the waves travel across the body. [2 marks]

2 Look at the screen in Figure 2. The oscilloscope shows the reflected wave pulses that are detected for each transmitted wave pulse.
   a How many internal boundaries are present? [1 mark]
   b i The oscilloscope beam takes 2.0×10^{-4} s to travel across the screen. The distance across the block is 90 mm. Calculate the speed of ultrasound in the block. [3 marks]
   ii Calculate the distance from the flaw nearest to the transducer to the surface of the block where the transducer is placed. [2 marks]

3 In a test to measure the depth of the sea bed, ultrasound wave pulses took 0.40 s to travel from the surface to the sea bed and back. Given that the speed of sound in sea water is 1350 m/s, calculate the depth of the sea bed below the surface. [2 marks]

4 In an ultrasound scan of an eye, the distance from the front of the eye to the retina of the eye was known to be 24 mm. Figure 3 shows the oscilloscope display for the eye. Pulses A and D show the front of the eye and the retina, respectively. Pulses B and C show the lens surfaces.
   a Calculate the distance from the back of the eye lens to the retina. [2 marks]
   b Discuss the accuracy of the distance you calculated in **a**. [2 marks]

Worked example

In Figure 2 the distance across the block is 90 mm. Estimate the distance from the transducer at the surface to the nearest internal boundary.

Solution

On the screen, the pulse due to the nearest internal boundary is halfway between the transmitted pulse and the far-side pulse.

Therefore, the distance d from the transducer to the nearest internal boundary is about half the distance across the block.

Distance $d = 0.5 \times 90$ mm = **45 mm**

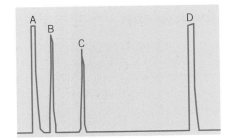

Figure 3 *A scan of the eye*

Key points

- Ultrasound waves are sound waves of frequency above 20 kHz.
- Ultrasound waves are partly reflected at a boundary between two different types of body tissue.
- Ultrasound waves reflected at boundaries are timed, and the timings are used to calculate distances.
- An ultrasound scan is non-ionising, so it is safer than an X-ray.

P12.7 Seismic waves

Learning objectives

After this topic, you should know:

- what seismic waves are
- how seismic waves are produced
- what primary seismic waves and secondary seismic waves are
- what information seismic waves give about the structure of the Earth.

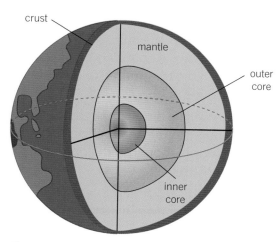

Figure 1 *Inside the Earth*

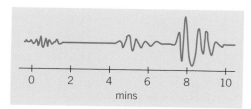

Figure 2 *A seismometer trace*

The study of earthquakes is called seismology. About half a million earthquakes occur every year but only a few cause serious damage. An earthquake happens suddenly when forces inside the Earth increase and become strong enough to break and move layers of rock. The energy transferred makes shock waves called seismic waves that travel through the Earth and across its surface. Studying seismic waves provides information about the structure of the Earth.

Inside the Earth

Earthquakes are generated in the Earth's crust. This is a solid layer of rock about 50 km thick that surrounds a much thicker layer of molten rock called the mantle. The Earth's crust beneath the oceans is much thinner and younger than the crust below the continents. The Earth has a solid inner core and a liquid outer core beneath the mantle (Figure 1).

Recording seismic waves

Earthquakes happen inside the Earth's crust. The point where an earthquake originates from is called its focus. The nearest point on the surface to the focus is called the epicentre of the earthquake. Earthquakes are recorded by detectors on the surface of the Earth called seismometers.

Analysing seismic waves

A trace from a seismometer is shown in Figure 2. It displays three main types of seismic waves:

- Primary waves (P-waves) cause the initial tremors lasting about one minute. These are longitudinal waves that push or pull on material as they move through the Earth.

- Secondary waves (S-waves) cause more tremors a few minutes later. They are transverse waves that travel more slowly than P-waves. They shake the material that they pass through inside the Earth from side to side.

- Long waves (L-waves) arrive last and cause violent movements on the surface up and down as well as backwards and forwards. They travel more slowly than P-waves or S-waves, and they only happen in the Earth's crust.

Learning about the Earth's structure

When an earthquake happens, seismometer readings from different parts of the world are used to find out where its epicentre is.

- P-waves and S-waves bend as they travel through the mantle. This is because their speed changes gradually with depth, and so their direction changes with depth.

- P-waves refract at the boundary between the mantle and the outer core. This is because their speed changes abruptly at the boundary.
- S-waves are transverse waves, and so they can't travel through the liquid outer core.

When an earthquake happens, some seismometers record only long waves. These seismometers are in the shadow zone of the earthquake (Figure 3), which is a zone from about 105° to 142° where no P-waves or S-waves are recorded.

1   The existence of the shadow zone shows that there is a liquid (outer) core under the mantle because:

- P-waves are refracted at the boundary between the mantle and the outer core when the waves enter the core and when they leave the core. Because the second refraction is further around, the waves can't reach the shadow zone.

- S-waves can't travel through the outer core, because they are transverse waves and can't travel through liquid.

2   Weak P-waves detected in the shadow zone show that the core has a solid inner part that refracts P-waves at the boundary between the outer core and the inner core into the shadow zone.

3   The boundary between the crust and the mantle of the Earth was discovered when it was found that the speed of seismic waves changed at a depth of about 50 km below the surface.

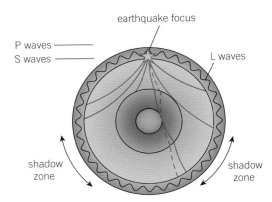

Figure 3 *Refraction of seismic waves*

1   **a** Name the three types of seismic waves in order of increasing speed. [1 mark]
   b Name the type of seismic wave that is:
   i longitudinal only [1 mark]
   ii transverse only. [1 mark]
   c Explain why P-waves and S-waves bend towards the surface as they travel through the Earth. [4 marks]

2   **a** Draw a labelled diagram to show the structure of the Earth. [4 marks]
   b i Explain why S-waves cannot travel through the outer core of the Earth. [1 mark]
   ii Explain why P-waves change their direction at the boundary between the mantle and the outer core. [2 marks]

3   A seismometer at a certain location first detects seismic waves from an earthquake 460 seconds later. The seismic waves travel a distance of 5000 km to reach this location.
   a Calculate the average speed of the primary seismic waves from this earthquake. [1 mark]
   b Secondary waves arrive about three minutes later than the first primary waves. Estimate how much slower S-waves are than P-waves. [1 mark]

4   Describe and explain the evidence obtained from seismic waves for the existence of a liquid outer core at the centre of the Earth. [6 marks]

Key points

- Seismic waves are waves that travel through the Earth.
- Seismic waves are produced in an earthquake and spread out from the epicentre
- Primary seismic waves (P-waves) are longitudinal waves. Secondary seismic waves (S-waves) are transverse waves.
- The Earth has a liquid inner core surrounded by a solid outer core, which is surrounded by the Earth's mantle. The mantle is surrounded by the Earth's crust.

P12 Wave properties

Summary questions

1 a Figure 1 shows transverse waves on a string. Copy the diagram and label distances on it to show what is meant by:
   i the wavelength [1 mark]
   ii the amplitude of the waves. [1 mark]

Figure 1

b Describe the difference between a transverse wave and a longitudinal wave. [1 mark]
c Give **one** example of:
   i a transverse wave [1 mark]
   ii a longitudinal wave. [1 mark]

2 A speedboat on a lake creates waves that make a buoy bob up and down.
   a The buoy bobs up and down three times in one minute. Calculate the frequency of the waves. [2 marks]
   b The waves travel 24 metres in one minute. Calculate the speed of the waves in metres per second. [2 marks]
   c Calculate the wavelength of the waves. [2 marks]

3 a When a wave is refracted at a boundary where its speed is reduced, state what change, if any, happens to:
   i its wavelength [1 mark]
   ii its frequency. [1 mark]
   b When a wave is reflected, what change, if any, happens to:
   i its wavelength? [1 mark]
   ii its frequency? [1 mark]
   iii **H** its speed? [1 mark]

4 a Copy Figure 2a and complete it to show the reflection of a straight wavefront at a straight reflector. [2 marks]
   b Copy and complete Figure 2b to show the refraction of a straight wavefront at a straight boundary as the wavefront moves from deep to shallow water. [1 mark]

Figure 2

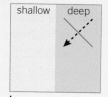

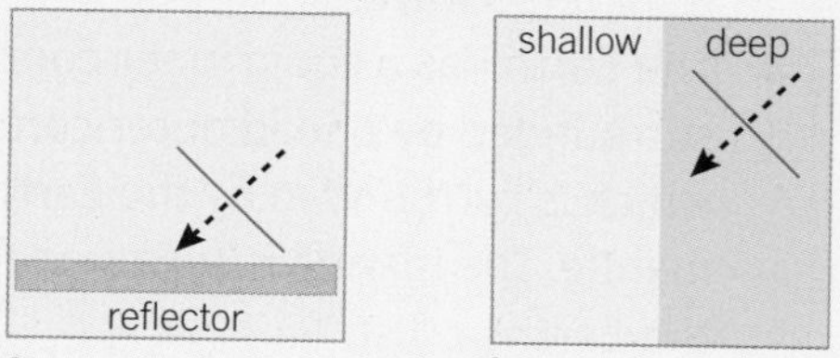

a   b

c State and explain the change in direction of the wave in Figure 2 when it crosses the boundary. [3 marks]

5 H a A loudspeaker is used to produce sound waves.
   i Describe how sound waves are created when an object in air vibrates. [3 marks]
   ii In terms of the amplitude of the sound waves, explain why the sound is fainter further away from the loudspeaker. [1 mark]
   b A microphone is connected to an oscilloscope. Figure 3 shows the display on the screen of the oscilloscope when the microphone detects sound waves from a loudspeaker which is connected to a signal generator.

Figure 3

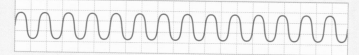

Describe how the waveform displayed on the oscilloscope screen changes if the sound from the loudspeaker is:
   i made louder [1 mark]
   ii reduced in pitch. [1 mark]
   c Describe how you would use the arrangement to measure the upper frequency limit of a person's hearing. **✏** [5 marks]

6 A person is standing a certain distance from a flat side wall of a tall building. She claps her hands and hears an echo.
   a Explain the cause of the echo. [2 marks]
   b **H** She hears the sound 0.30 s after clapping her hands. Calculate how far she is from the nearest point of the wall. The speed of sound in air = 340 m/s. [2 marks]

7 H Ultrasound waves used for medical scanners have a frequency of 2000 kHz.
   a Use the equation 'speed = frequency × wavelength' to calculate the wavelength of these ultrasound waves in human tissue. The speed of ultrasound in human tissue is 1500 m/s. [2 marks]
   b Ultrasound waves of this frequency in human tissue are not absorbed much. Explain why it is important in a medical scanner that they are not absorbed. [1 mark]

Practice questions

01.1 Sound travels in waves. Choose **two** statements that are correct.

> Sound waves are longitudinal.
> Sound waves are transverse.
> Sound wave oscillations are parallel to the direction of energy transfer.
> Sound wave oscillations are perpendicular to the direction of energy transfer.
> Sound waves travel through a vacuum.

[2 marks]

01.2 Label a compression and rarefaction on **Figure 1**.

Figure 1

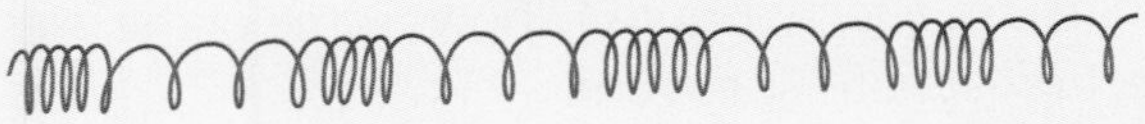

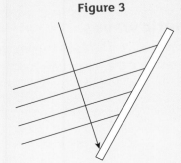

[2 marks]

01.3 The frequency of a sound wave is 440 oscillations every second. Calculate the time period of the tuning fork. [2 marks]

02 A student observes a large stone falling into a pool of water. A ripple of water is formed on the surface (Figure 2).

Figure 2

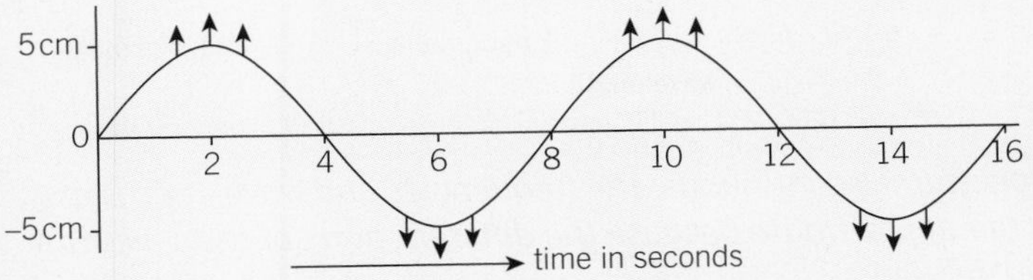

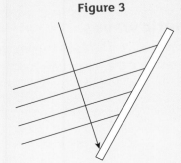

02.1 State whether the wave is a transverse or a longitudinal wave. [1 mark]

02.2 Determine the amplitude of the wave. [1 mark]

02.3 The student observes a plastic boat only moving up and down on the surface of the water. Explain why. [2 marks]

02.4 The wavelength of the wave is 28 cm. Calculate the speed of the wave and give the unit. [3 marks]

02.5 The wave hits a straight wall and is reflected. Copy and complete the ray diagram in **Figure 3**. [2 marks]

Figure 3

03.1 Ⓗ A student decided to investigate sound-proofing materials.
They placed a bell in a container and covered one end in the sound-proofing material. They then placed a sound detector on the other side of the sound-proofing material.
Describe how the student used this equipment to measure the sound detected through three different materials. You should include details of variables that will be changed and kept the same. [4 marks]

03.2 Ⓗ Give the normal range of human hearing. [2 marks]

04 Ⓗ Metals used in the nuclear industry usually have to be free from defects. Ultrasound is used to check for hidden defects in high grade stainless steel. The results of a test are shown in **Figure 4**.

Figure 4

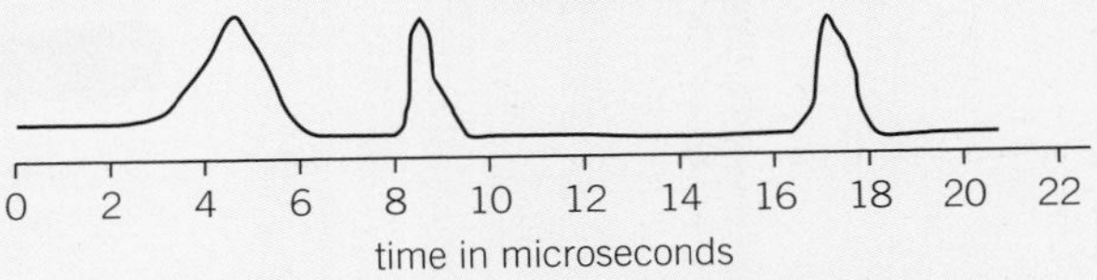

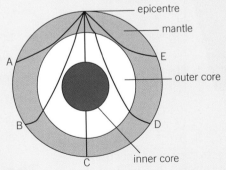

time in microseconds

Use the graph to calculate the depth of the defect below the metal surface.
Speed of sound in stainless steel is 5800 m/s. [2 marks]

05 Ⓗ Earthquakes occur at an epicentre and result in seismic waves, called P-waves and S-waves, being transmitted through the Earth. **Figure 5** shows the structure of the Earth with seismic waves from an epicentre.

Figure 5

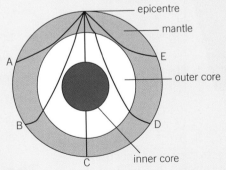

05.1 State at which of the places, A, B, C, D, or E, P-waves be detected. Explain your answer. [2 marks]

05.2 State at which of the places, A, B, C, D, E, will S-waves be detected. Explain your answer. [2 marks]

05.3 The speed of seismic waves in the mantle is often higher than those detected in the crust of the Earth. Suggest **two** reasons why. [2 marks]

Learning objectives

After this topic, you should know:

- the parts of the electromagnetic spectrum
- the range of wavelengths within the electromagnetic spectrum that the human eye can detect
- how energy is transferred by electromagnetic waves
- how to calculate the frequency or wavelength of electromagnetic waves.

Electromagnetic waves are electric and magnetic disturbances that can be used to transfer energy from a source to an absorber. You use waves from different parts of the electromagnetic spectrum in everyday devices and gadgets, including:

- microwave ovens – energy is transferred from a microwave source to the food in the oven, heating it
- radiant heaters – infrared radiation transfers energy from the heater to heat the surroundings.

Electromagnetic waves do not transfer matter. The energy they transfer depends on the wavelength of the waves. This is why waves of different wavelengths have different effects. Figure 1 shows some of the uses of each part of the electromagnetic spectrum.

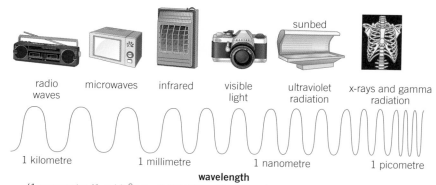

(1 nanometre $(1 \times 10^{-9}\,\text{m})$ = 0.000 001 millimetres, 1 picometre = 0.001 nanometres)

Figure 1 *The spectrum is continuous. The frequencies and wavelengths at the boundaries are approximate because the different parts of the spectrum are not precisely defined.*

Waves from different parts of the electromagnetic spectrum have different wavelengths.

- Long-wave radio waves have wavelengths as long as 10 km ($10^4\,\text{m}$).
- X-rays and gamma rays have wavelengths as short as a millionth of a millionth of a millimetre (= 0.000 000 000 001 mm or $10^{-15}\,\text{m}$).
- Your eyes detect visible light, which is only a limited part of the electromagnetic spectrum (wavelengths of about 350 nm to about 650 nm).

The speed of electromagnetic waves

All electromagnetic waves travel at a speed of $3.0 \times 10^8\,\text{m/s}$ (300 million m/s) through space or in a vacuum. This is the distance the waves travel each second.

You can link the speed of the waves to their frequency and wavelength by using the **wave speed** equation:

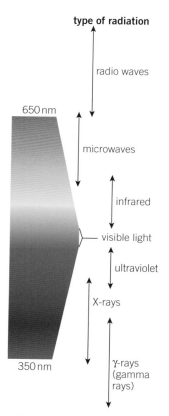

Figure 2 *The electromagnetic spectrum with an expanded view of the visible range*

Synoptic links

You learnt about the wave speed equation in Topic P12.2.

wave speed, v (m/s) = frequency, f (Hz) × wavelength, λ (m)
(metres per second, m/s)   (hertz, Hz)   (metres, m)

You can work out the wavelength λ or the frequency f by rearranging the wave speed equation into:

$$\lambda = \frac{v}{f} \text{ or } f = \frac{v}{\lambda}$$

Worked example

A mobile phone gives out electromagnetic waves of frequency 900 million Hz. Calculate the wavelength of these waves.

The speed of electromagnetic waves in air = 300 million m/s.

Solution

$$\text{Wavelength } \lambda \text{ (m)} = \frac{\text{wave speed } v \text{ (m/s)}}{\text{frequency } f \text{ (Hz)}}$$

$$= \frac{300\,000\,000 \text{ m/s}}{900\,000\,000 \text{ Hz}} = 0.33\text{m}$$

Energy and frequency

The wave speed equation shows you that since electromagnetic waves all have a speed of 300 million m/s, the shorter the wavelength of the waves, the higher their frequency. The energy of the waves increases as the frequency increases. So as the wavelength decreases along the electromagnetic spectrum from radio waves to gamma rays, the energy and frequency of the waves increase.

1   **a** State which is greater: the wavelength of radio waves or the wavelength of visible light waves.   [1 mark]
    b Describe the speed in a vacuum of different electromagnetic waves.   [1 mark]
    c State which is greater: the frequency of X-rays or the frequency of infrared radiation.   [1 mark]
    d Determine where in the electromagnetic spectrum you would find waves of wavelength 10 millimetres.   [1 mark]

2   **a** Put the following parts of the electromagnetic spectrum in order of increasing frequency:
    infrared   radio   X-rays and gamma rays   [1 mark]
    b Determine which parts of the electromagnetic spectrum are missing from the list in **a**.   [1 mark]

3   Electromagnetic waves travel through space at a speed of 300 million metres per second. Calculate:
    a the wavelength of radio waves of frequency 600 million Hz
    b the frequency of microwaves of wavelength 0.30 m   [4 marks]

4   A distant star explodes and emits visible light and gamma rays simultaneously. Explain why the gamma rays and the visible light waves reach the Earth at the same time.   [2 marks]

Learning objectives

After this topic, you should know:

- the nature of white light
- what infrared radiation, microwaves, and radio waves are used for
- what mobile phone radiation is
- why these types of electromagnetic radiation are hazardous.

Absorption and emission of infrared radiation

To compare emission from two different surfaces, measure how fast two cans of hot water cool. The surface of

thermometer to measure water temperature at intervals as it cools

Figure 1 *Testing different surfaces*

one can is light in colour and shiny, and the other has a dark, matt surface (Figure 1).

To compare absorption by two different surfaces, measure how fast the two cans containing cold water heat up when placed in sunlight for the same time. Each can needs to contain the same amount of water.

At the start of each test , the volume and temperature of the water in each can must be the same.

Write a report on your investigation and use your measurements to compare the two surfaces in terms of absorption and emission of infrared radiation.

Safety: Take care with hot water.

Light

Light from ordinary lamps and from the Sun is called **white light**. This is because it has all the colours of the visible spectrum in it. The wavelength increases across the spectrum as you go from violet to red. When you look at a rainbow, you see the colours of the spectrum. You can also see them if you use a glass prism to split a beam of white light.

Photographers need to know how shades and colours of light affect the photographs they take.

1. In a film camera, the light is focused by the camera lens on to a light-sensitive film. The film then needs to be developed to see the image of the objects that were photographed.

2. In a digital camera or a mobile phone camera, the light is focused by the lens on to a sensor. This is made up of thousands of tiny light-sensitive cells called pixels. Each pixel gives a dot of the image. The image can be seen on a small screen at the back of the camera. When a photograph is taken, the image is stored electronically on a memory card.

Infrared radiation

All objects emit infrared radiation.

- The hotter an object is, the more infrared radiation it emits.
- Infrared radiation is absorbed by your skin. It can damage, burn, or kill skin cells because it heats up the cells.

Infrared devices

- Optical fibres in communications systems usually use infrared radiation instead of visible light. This is because infrared radiation is absorbed less than visible light in the glass fibres.

- Remote control handsets for TV and video equipment transmit signals carried by infrared radiation. When you press a button on the handset, it sends out a sequence of infrared pulses. Infrared radiation is used because suitable infrared pulses can easily be produced and detected electronically.

- Infrared scanners are used in medicine to detect infrared radiation emitted from hot spots on the body surface. These hot areas can mean the tissue underneath is unhealthy.

- You can use infrared cameras to see people and animals in the dark.

  Infrared radiation is used to heat up objects quickly:

  - electric heaters that emit infrared radiation warm rooms quickly
  - electric cookers that have halogen hobs heat up food faster than ordinary hobs because halogen hobs are designed to emit much more infrared radiation than ordinary hobs.

Microwaves

Microwaves have a shorter wavelength than radio waves.

- People use microwaves for communications, for example satellite TV, because they can pass through the atmosphere and reach satellites above the Earth. Microwaves can also carry mobile phone signals.

- Microwave ovens heat food faster than ordinary ovens. This is because microwaves can penetrate into food and are absorbed by the water molecules in the food, heating it. The oven itself does not absorb microwaves as it does not contain any water molecules. It therefore does not become hot like the food it is cooking.

Radio waves

Radio wave frequencies range from about 300 000 Hz to 3000 million Hz (where microwave frequencies start). Radio waves are used to carry radio, TV, and mobile phone signals.

You can also use radio waves instead of cables to connect a computer to other devices such as a printer or a computer mouse. For example, Bluetooth-enabled devices can communicate with each other over a range of about 10 metres without the need for cables.

Microwaves and radio waves can be hazardous because they penetrate people's bodies and can heat the internal parts of the body.

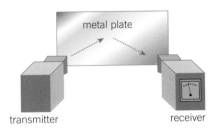

Figure 2 *Detecting microwaves*

1 **a** When you watch a TV programme, name the type of electromagnetic wave that is:
      i detected by the aerial [1 mark]
      ii emitted by the screen. [1 mark]
   b Name the type of electromagnetic wave that is used:
      i to carry signals to and from a satellite [1 mark]
      ii to send signals to a printer from a computer without using a cable. [1 mark]

2 Mobile phones use electromagnetic waves in a wavelength range that includes short-wave radio waves and microwaves.
   a Describe the effect on mobile phone users if remote control handsets operated in this range as well. [1 mark]
   b Explain why the emergency services use radio waves in a wavelength range that no one else is allowed to use. [2 marks]

3 The speed of electromagnetic waves in air is 300 000 km/s. Calculate the wavelength in air of electromagnetic waves of frequency 2400 MHz. [2 marks]

4 Figure 2 shows a microwave receiver being used to detect microwaves from a transmitter. The reading on the receiver meter depends on how much radiation the receiver detects.
   a Describe what the metal plate does to the microwaves. [1 mark]
   b Design a test to find out if microwaves can pass through **i** a metal plate **ii** a thick cardboard sheet. ✏ [4 marks]

P13.3 Communications

Learning objectives

After this topic, you should know:

- why radio waves of different frequencies are used for different purposes
- which waves are used for satellite TV
- how to decide whether or not mobile phones are safe to use
- **H** what carrier waves are
- how optical fibres are used in communications.

Figure 1 *Sending microwave signals to a satellite*

Figure 2 *A mobile phone mast*

Study tip

Remember that in communications, electromagnetic waves carry the information.

Radio communications

When you use a mobile phone, radio waves carry signals between your mobile phone and the nearest mobile phone mast. The waves used to carry any type of signal are called **carrier waves**. They could be radio waves, microwaves, infrared radiation, or visible light. The type of waves used to carry a signal depends on how much information is in the signal and the distance the signal has to travel. For example, microwaves are used to carry signals via satellites to distant countries.

Radio wavelengths

The radio and microwave spectrum is divided into bands of different wavelength ranges. This is because the shorter the wavelength of the waves:

- the more information they can carry
- the shorter their range (due to increasing absorption by the atmosphere)
- the less they spread out.

Microwaves and radio waves of different wavelengths are used for different communications purposes. Examples include:

- Microwaves are used for satellite phone and TV links, and satellite TV broadcasting. This is because microwaves can travel between satellites in space and the ground. Also, they spread out less than radio waves do, so the signal doesn't weaken as much.
- Radio waves of wavelengths less than about 1 metre are used for TV broadcasting from TV masts because they can carry more information than longer radio waves.
- Radio waves of wavelengths from about 1 metre up to about 100 m are used by local radio stations (and for the emergency services) because their range is limited to the area round the transmitter.
- Radio waves of wavelengths greater than 100 m are used by national and international radio stations because they have a much longer range than shorter-wavelength radio waves.

Mobile phones and electromagnetic radiation

A mobile phone sends out a radio signal when you use it. If the phone is very close to your brain, some scientists think the radiation might affect the brain. Because children have thinner skulls than adults, their brains might be more affected by mobile phone radiation. A UK government report published in May 2000 recommended that the use of mobile phones by children should be limited. More research needs to be conducted to find out if mobile phone users are affected.

More about signals and carrier waves

When you speak, you produce sound waves of different frequencies, so you vary (i.e. modulate) the amplitude and the frequency of the sound waves you produce. In a radio station, a microphone produces an alternating current called an audio signal when sound waves reach it. Figure 3 shows how the signal is transmitted and detected.

- An oscillator supplies carrier waves to the transmitter in the form of an alternating current (a current that repeatedly reverses its direction).

- The audio signal is supplied to the transmitter where it's used to modulate the carrier waves.

- The modulated carrier waves from the transmitter are supplied to the transmitter aerial. The varying alternating current supplied to the aerial causes it to emit radio waves that carry the audio signal.

- When the radio waves are absorbed by a receiver aerial, they induce an alternating current in the receiver aerial, which causes oscillations in the receiver. The frequency of the oscillations is the same as the frequency of the radio waves.

- The receiver circuit separates the audio signal from the carrier waves. The audio signal is then supplied to a loudspeaker, which sends out sound waves similar to the sound waves received by the microphone in the radio station.

Optical fibre communications

Optical fibres are very thin glass fibres. They are used to transmit signals carried by light or infrared radiation. The light rays can't escape from the fibre. When they reach the surface of the fibre, they are reflected back into the fibre (Figure 4).

Compared with radio waves and microwaves:

- optical fibres carry much more information as light has a much shorter wavelength than radio waves, and so can carry more pulses of waves

- optical fibres are more secure because the signals stay in the fibre.

1 a Name the types of electromagnetic wave that are used to carry signals along a thin transparent fibre. [1 mark]
   b Explain why signals in an optical fibre are more secure than radio signals. [2 marks]

2 a Explain why children could be more affected by mobile phone radiation than adults. [2 marks]
   b Ⓗ i Explain what is meant by a carrier wave. [2 marks]
        ii Explain why visible light waves can carry more information than radio waves. [2 marks]

3 Explain why microwaves are used for satellite TV and radio waves are used for terrestrial TV. [3 marks]

4 A local radio station broadcasts at a frequency of 105 MHz.
   a Calculate the wavelength of radio waves of this frequency. The speed of electromagnetic waves in air is 300 000 km/s. [2 marks]
   b Explain why national radio stations broadcast at much lower frequencies. [2 marks]

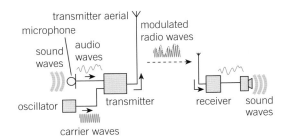

Figure 3 *Using radio waves*

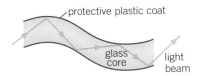

Figure 4 *the reflection of light inside an optical fibre. This is called total internal reflection*

Key points

- Radio waves of different frequencies are used for different purposes because the wavelength (and so the frequency) of waves affects:
  - how far they can travel
  - how much they spread
  - how much information they can carry.
- Microwaves are used for satellite TV signals.
- Further research is needed to evaluate whether or not mobile phones are safe to use.
- Ⓗ Carrier waves are waves that are used to carry information. They do this by varying their amplitude.
- Optical fibres are very thin transparent fibres that are used to transmit communication signals by light and infrared radiation.

P13.4 Ultraviolet waves, X-rays, and gamma rays

Learning objectives

After this topic, you should know:

- the differences between ultraviolet and visible light
- what X-rays and gamma rays are used for
- what ionising radiation is
- why ultraviolet waves, X-rays, and gamma rays are dangerous.

Ultraviolet waves

Ultraviolet (UV) waves lie between violet light and X-rays in the electromagnetic spectrum. Some chemicals emit light as a result of absorbing ultraviolet waves. Posters and ink that glow in ultraviolet light contain these chemicals. Security marker pens containing this kind of ink are used to mark valuable objects. The chemicals absorb ultraviolet waves and then emit visible light.

Ultraviolet waves are harmful to human eyes and can cause blindness. UV wavelengths are smaller than visible light wavelengths. UV waves carry more energy than visible light waves.

Ultraviolet waves are harmful to your skin. For example, too much UV directly from the Sun or from a sunbed can cause sunburn and skin cancer. It can also age the skin prematurely.

- If you stay outdoors in summer, use skin creams to block UV waves and prevent them reaching your skin.
- If you use a sunbed to get a suntan, don't go over the recommended time. You should also wear special goggles to protect your eyes.

Ultraviolet waves

Watch your teacher place different-coloured clothes under an ultraviolet lamp. Observe what happens.

- Describe what white clothes look like under a UV lamp.

Safety: The lamp must point downwards so you can't look directly at the glow from it.

Figure 1 *Using an ultraviolet lamp to detect finger prints*

X-rays and gamma rays

X-rays and gamma rays both travel straight into substances and can pass through them if the substances are not too dense and not too thick. A thick plate made of lead will stop them.

X-rays and gamma rays have similar properties because they both:

- are at the short-wavelength end of the electromagnetic spectrum
- carry much more energy per second than longer-wavelength electromagnetic waves.

They differ from each other because:

- X-rays are produced when electrons or other particles moving at high speeds are stopped – X-ray tubes are used to produce X-rays
- gamma rays are produced by radioactive substances when unstable nuclei release energy
- gamma rays have shorter wavelengths than X-rays, so they can penetrate substances more than X-rays can.

X-rays are often used to detect internal cracks in metal objects. These kinds of application are usually possible because the more dense a substance is, the more X-rays it absorbs from an X-ray beam passing through it. X-rays are also used in medicine to create images of broken limbs. You will learn more about this in Topic P13.5.

Synoptic links

You learnt about radioactive substances in Topic P7.3 and Topic P7.4.

Study tip

Make sure that you know the dangers, as well as the uses, of the different kinds of electromagnetic waves.

Using gamma rays

High-energy gamma rays have several important uses:

Killing harmful bacteria

1 About 20% of the world's food is lost through spoilage, mostly due to bacteria. Bacteria waste products cause food poisoning. Exposing food to gamma rays kills 99% of disease-carrying organisms, including *Salmonella* (found in poultry) and *Clostridium* (which causes botulism).

2 Exposing surgical instruments in sealed plastic wrappers to gamma rays kills any bacteria on the instruments. This helps to stop infection spreading in hospitals.

Killing cancer cells

Doctors and medical physicists use gamma-ray therapy to destroy cancerous tumours. A narrow beam of gamma rays from a radioactive source (cobalt-60) is directed at the tumour. The beam is aimed at it from different directions to kill the tumour but not the surrounding tissue.

Safety matters

X-rays and gamma rays passing through substances can knock electrons out of atoms in the substance. The atoms become charged because they lose electrons. This process is called **ionisation**, and so X-rays and gamma rays are examples of ionising radiation.

If ionisation happens to a living cell, it can damage or kill the cell. For this reason, exposure to too many X-rays or gamma rays is dangerous and can cause cancer. High doses kill living cells, and low doses cause gene mutation and cancerous growth.

People who use equipment or substances that produce any form of ionising radiation (e.g., X-rays or gamma rays) must wear a film badge. If the badge shows that it is over-exposed to ionising radiation, its wearer is not allowed to continue working with the equipment for a period of time.

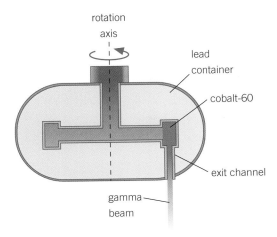

Figure 2 *Gamma treatment – the cobalt-60 source is in a thick lead container. When it is not in use, it is rotated away from the exit channel*

Figure 3 *A film badge tells you how much ionising radiation the wearer has received*

1 a Explain why a crack inside a metal object shows up on an X-ray image. [3 marks]
  b Will gamma rays pass through thin plastic wrappers? [1 mark]
  c Explain why a film badge used for monitoring radiation needs to have a plastic case, not a metal case. [2 marks]

2 a Explain why ultraviolet waves are harmful. [2 marks]
  b i Explain how the Earth's ozone layer helps to protect you from ultraviolet waves from the Sun. [1 mark]
    ii Explain why people outdoors in summer need suncream. [3 marks]

3 a Name the types of electromagnetic radiation that can penetrate thin metal sheets. [1 mark]
  b Name the metal that can be used most effectively to absorb X-rays and gamma rays. [1 mark]

4 a Explain what is meant by ionisation, and describe one way in which ionisation can occur. [3 marks]
  b Name the types of electromagnetic radiation that can:
    i ionise substances they pass through [1 mark]
    ii damage the human eye. [1 mark]

Key points

- Ultraviolet waves have a shorter wavelength than visible light and can harm the skin and the eyes.
- X-rays are used in hospitals to make X-ray images.
- Gamma rays are used to kill harmful bacteria in food, to sterilise surgical equipment, and to kill cancer cells.
- Ionising radiation makes uncharged atoms become charged.
- X-rays and gamma rays damage living tissue when they pass through it.

P13.5 X-rays in medicine

Learning objectives

After this topic, you should know:

- what X-rays are used for in hospitals
- why X-rays are dangerous
- what absorbs X-rays when they pass through the body.

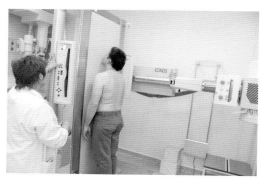

Figure 1 *Taking a chest X-ray*

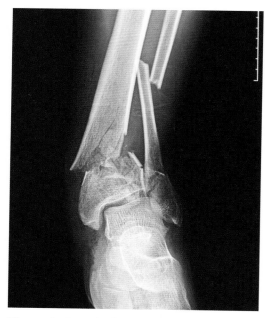

Figure 2 *Spot the break*

Synoptic links

You have learnt about radioactive substances and background radiation in Topics P7.1–P7.6

Have you ever broken one of your bones? If you have, you will have gone to your local hospital for an X-ray photograph. X-rays are electromagnetic waves at the short-wavelength end of the electromagnetic spectrum. They are produced in an X-ray tube when fast-moving electrons hit a target. Their wavelengths are about the same as the diameter of an atom.

To make a radiograph or X-ray photograph, X-rays from an X-ray tube are directed at the patient. A lightproof cassette containing a photographic film or a flat-panel detector is placed on the other side of the patient.

- When the X-ray tube is switched on, X-rays from the tube pass through the part of the patient's body under investigation (Figure 1).

- X-rays pass through soft tissue, but they are absorbed by bones, teeth and metal objects that are not too thin. The parts of the film or the detector that the X-rays reach become darker than the other parts. So the bones appear lighter than the surrounding tissue, which appears dark (Figure 2). The radiograph shows a 'negative image' of the bones. A hole or a cavity in a tooth shows up as a dark area in the bright image of the tooth.

- An organ that consists of soft tissue can be filled with a substance called a **contrast medium** that absorbs X-rays easily. This enables the internal surfaces in the organ to be seen on the radiograph. For example, to obtain a radiograph of the stomach, the patient is given a barium meal before the X-ray machine is used (Figure 3). The barium compound is a good absorber of X-rays.

- Lead plates between the tube and the patient stop X-rays reaching other parts of the body. The X-rays reaching the patient pass through a gap between the plates. Lead is used because it is a good absorber of X-rays.

- A flat-panel detector is a small screen that contains a **charge-coupled device (CCD)**. The sensors in the CCD convert X-rays to light. The light rays then create electronic signals in the sensors that are sent to a computer, which displays a digital X-ray image.

Radiation dose

X-rays, gamma rays, and the radiation from radioactive substances all ionise substances they pass through. There are three types of radiation from radioactive substances. You have already learnt about them. Gamma radiation is one of the three types. The other two types are called alpha and beta radiation.

All the different types of ionising radiation are dangerous. The **radiation dose** received by a person is a measure of the damage done to their body by ionising radiation. The radiation dose depends on:

- the type of radiation used

- how long the body is exposed to it

- The energy per second absorbed by the body from the radiation.

For example, alpha radiation inside the body causes ten times more damage than X-rays, when a person is exposed to them for the same length of time.

Radiation dose is measured in sieverts (Sv) or millisieverts (mSv).

High doses of radiation kill living cells. Low doses can cause gene mutation and cancerous growth. There is no evidence of a safe limit below which living cells would not be damaged.

Everyone is exposed to low levels of ionising radiation from background sources such as cosmic radiation from space and radon gas which seeps through the earth from deep underground. Also, workers who use equipment or substances that produce ionising radiation must wear a film badge that tells them how much ionising radiation they have received.

X-ray therapy

Doctors use X-ray therapy to destroy cancerous tumours in the body. Thick plates between the X-ray tube and the body stop X-rays from reaching healthy body tissues. A gap between the plates allows X-rays through to reach the tumour. X-rays for therapy are shorter in wavelength than X-rays used for imaging.

Higher

The X-rays used for therapy carry much more energy than X-rays used for imaging. Low energy X-rays are suitable for imaging because they are absorbed by bones and teeth but they pass through soft tissue and gaps such as cracks in bones. Low energy X-rays do not carry enough energy to destroy cancerous tumours.

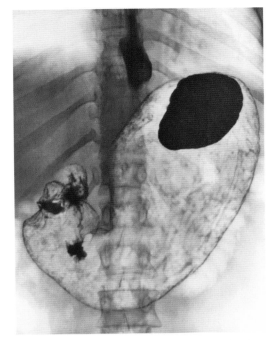

Figure 3 *A coloured X-ray of a stomach ulcer*

Study tip

1 Sv = 1000 mSv.

Study tip

You do not need to remember the unit of radiation dose.

1 **a** Explain what a contrast medium is used for when an X-ray photograph of the stomach is taken. [2 marks]
  b Describe what X-ray therapy can be used for. [1 mark]

2 When an X-ray photograph is taken, explain why it is necessary:
  a to place the patient between the X-ray tube and the film cassette [3 marks]
  b to have the film in a lightproof cassette [1 mark]
  c to shield the parts of the patient not under investigation from X-rays. Explain what would happen to healthy cells if they were not shielded. [4 marks]

3 **a** Name one way in which X-rays used for X-ray therapy differ from X-rays used for X-ray imaging. [1 mark]
  b State why X-rays used for imaging cannot be used for X-ray therapy. [1 mark]

4 The average radiation dose each person receives from ionising radiation is about 2 millisieverts per year. Medical X-rays account for about 13% of this.
  a State what is meant by radiation dose. [1 mark]
  b Estimate the average radiation dose each person receives in one year due to medical X-rays. [1 mark]

Key points

- X-rays are used in hospitals:
  - to make images of your internal body parts
  - to destroy tumours at or near the body surface.
- X-rays are ionising radiation and so can damage living tissue when they pass through it.
- X-rays are absorbed more by bones and teeth than by soft tissues.

P13 Electromagnetic waves

Summary questions

1 **a** Place the five types of electromagnetic wave listed below in order of increasing wavelength.

   A Infrared waves

   B Microwaves

   C Radio waves

   D Gamma rays

   E Ultraviolet waves   [1 mark]

  b Name the type/s of electromagnetic radiation listed in **a** that:

   i can be used to send signals to and from a satellite   [1 mark]

   ii ionise substances when they pass through them   [1 mark]

   iii are used to carry signals in thin transparent fibres.   [1 mark]

2 **a** The radio waves from a local radio station have a wavelength of 2.9 m in air. The speed of electromagnetic waves in air is 300 000 km/s.

   i Write the equation that links frequency, wavelength, and wave speed.   [1 mark]

   ii Calculate the frequency of the radio waves. [2 marks]

  b A certain local radio station transmitter has a range of 30 km. Describe and explain the effect on the range if the power supplied to the transmitter is reduced.   [3 marks]

3 Mobile phones send and receive signals using electromagnetic waves near or in the microwave part of the electromagnetic spectrum. New mobile phones are tested for radiation safety and given an SAR value before being sold. The SAR is a measure of the energy per second absorbed by the head while the phone is in use. For use in the UK, SAR values must be less than 2.0 W/kg. SAR values for two different mobile phones are given below.

Phone **A** 0.2 W/kg

Phone **B** 1.0 W/kg

  a What is the main reason that mobile phones are tested for radiation safety?   [3 marks]

  b Which phone, **A** or **B**, is safer? Give a reason for your answer.   [2 marks]

  c The UK government recommends caution in the use of mobile phones, particularly by children and young people, until scientists and doctors find out more. Explain why children and young people may be more at risk than adults.   [2 marks]

4 The figure shows an X-ray source that is used to direct X-rays at a broken leg. A photographic film in a lightproof wrapper is placed under the leg. When the film is developed, an image of the broken bone is observed.

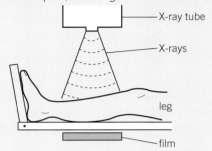

  a **i** Explain why an image of the bone is seen on the film.   [3 marks]

   ii Why is it possible to see the fracture on the image.   [1 mark]

  b When an X-ray photograph of the stomach is taken, the patient is given food containing barium before the photograph is taken.

   i Explain why it is necessary for the patient to be given this food before the photograph is taken.   [3 marks]

   ii The exposure time for a stomach X-ray must be shorter than the X-ray time for a limb. Explain why.   [2 marks]

   iii Low-energy X-rays from the X-ray tube can be absorbed by placing a metal plate between the patient and the X-ray tube. Such X-rays would otherwise be absorbed by the body. Describe the benefit of removing such low-energy X-rays in this way.   [2 marks]

  c An ultrasound scanner is used to observe an unborn baby. Explain why ultrasound is used instead of X-rays to observe an unborn baby.   [2 marks]

5 **a** What is meant by ionisation?   [1 mark]

  b Name the **two** types of electromagnetic radiation that can ionise substances.   [1 mark]

  c Give **two** reasons why ionising radiation is harmful.   [2 marks]

6 **a** Some chemicals can emit light as a result of absorbing ultraviolet waves. Describe and explain how these chemicals could be used as invisible ink. In your explanation, state the type of radiation that is absorbed and the type of radiation that is emitted.   [3 marks]

  b Explain why a beam of infrared radiation cannot be used to carry signals to a detector that is more than a few metres from a transmitter.   [2 marks]

Practice questions

01 **Figure 1** shows the electromagnetic spectrum.

radio waves	microwaves		light	ultraviolet		gamma

01.1 Give the names of the two missing waves. [2 marks]

01.2 Complete the sentences using words from the box below. Each word can be used once, more than once, or not at all.

lower	the same	longer	sound
shorter	energy	higher	faster

Radio waves have a _____ wavelength and _____ frequency than other electromagnetic waves.
[2 marks]

Radio waves have _____ speed in air compared with microwaves. [1 mark]

Radio waves transfer _____ from place to place. [1 mark]

01.3 Electromagnetic waves are used in many applications. Match the correct wave to one use of the wave.

microwave	kills cancer cells
ultraviolet	used in mobile phones
gamma	used in sunbeds

[2 marks]

02 The following headline appeared in a local newspaper. *Councillor Jones says that all mobile phone masts near local schools should be banned. They produce harmful radiation.*

02.1 Satellites are used to send messages around the world using microwaves. Give one reason why microwaves are used but not radio waves. [1 mark]

02.2 A spokesperson from a mobile phone company wrote to the newspaper stating that there are no risks from the masts. Suggest one reason why this statement should be treated with caution. [1 mark]

02.3 A study asked 42 000 people if they used a mobile phone and whether they had cancer. The conclusion was that there was no risk of cancer by using mobile phones. Evaluate the method used to determine whether the conclusion is valid and suggest possible improvements. [4 marks]

02.4 Microwave radiation is classed as non-ionizing radiation. Explain the dangers of ionizing radiation. [2 marks]

03 Gamma radiation and X-rays are used in medicine.

03.1 Surgical instruments are sealed in a plastic bag and then irradiated with gamma rays. Describe how this method keeps the instruments sterile. [2 marks]

03.2 Gamma knife surgery uses many low intensity beams of gamma radiation that come from different directions. All of the beams are focussed onto a tumour in the patient's head. Describe how this procedure targets the tumour but reduces the risk to the patient. [3 marks]

Figure 1

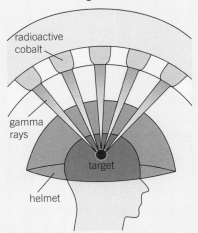

03.3 Describe the properties of X-rays that make them suitable for detecting a broken arm. [2 marks]

04 Sunglasses manufacturers predict that within 10 years, everyone will wear UV protection sunglasses when outside.

04.1 Which one of the following statements is the most likely reason for the prediction made by the manufacturers?

Opticians and other experts will make people aware of the dangers.

The style of sunglasses will be very modern.

The level of sunlight in summer will increase.

[1 mark]

04.2 Exposure to too much UV radiation is known to increase the risk of skin cancer. Suggest what precautions a golfer and snow skier should take when playing their sports. [2 marks]

04.3 Calculate the wave speed of a beam of ultraviolet radiation.
The frequency of the wave is 8×10^{14} Hz and the wavelength is 3.75×10^{-7} m.

Learning objectives

After this topic, you should know:

- what the normal is in a diagram of light rays
- the law of reflection of a light ray at a plane mirror
- how an image is formed by a plane mirror
- what is meant by specular reflection and diffuse reflection.

If you visit a hall of mirrors at a funfair, you will see some strange images of yourself. A tall, thin image or a short, broad image of yourself means you are looking into a mirror that is curved. If you want to see an ordinary image of yourself, look in a plane mirror. This kind of mirror is perfectly flat. You see an exact mirror image of yourself.

Reflection

Light consists of waves. When plane waves reflect from a flat barrier, the reflected waves are at the same angle to the barrier as the incident waves (Figure 1). When each point on the wavefront reaches the barrier, it creates a wavelet moving away from the barrier. This wavelet lines up with the previous reflected wavelets to form a reflected wavefront moving away from the barrier. All parts of a wavefront move at the same speed. This means that the reflected wavefront is at the same angle to the barrier as the incident wavefront.

The reflected waves and the incident waves have the same frequency and they travel at the same speed so they have the same wavelength.

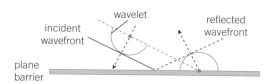

Figure 1 *Explaining reflection*

The law of reflection

Light rays are used to show the direction that light waves are moving in. Figure 2 shows how you can investigate the reflection of a light ray from a ray box by using a plane mirror.

Synoptic link

In Topic P12.3 you learnt about reflection and refraction, which included wavefronts.

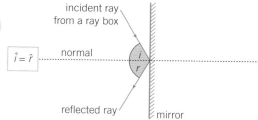

Figure 2 *The law of reflection*

- The line perpendicular to the mirror is called the **normal**.
- The **angle of incidence** is the angle between the incident ray and the normal.
- The **angle of reflection** is the angle between the reflected ray and the normal.

Measurements show that for any light ray reflected by a plane mirror:

the angle of incidence = the angle of reflection.

Image formation by a plane mirror

Figure 3 shows how an image is formed by a plane mirror. This ray diagram shows the path of two light rays from an object that reflect off the mirror. The image and the object in Figure 3 are at equal distances from the mirror.

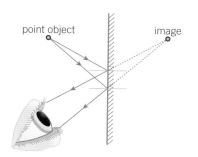

Figure 3 *Image formation by a plane mirror*

Real and virtual images

The image formed by a plane mirror is virtual, upright (the same way up as the object), and laterally inverted (back to front but not upside down). A **virtual image** is formed at a place where light rays appear to come from after they have been reflected (or refracted). It can't be projected onto a screen like the movie images you see at a cinema. An image that can be seen on a screen is described as a real image because it is formed by focusing light rays onto the screen.

Specular and diffuse reflection

A mirror has a smooth surface that reflects light rays without scattering them. This is why you can see a clear image when you look in a mirror. Reflection from a smooth surface is called **specular reflection** because parallel light rays are reflected in a single direction.

Parallel light rays reflected from a rough surface are scattered in different directions. If you polish a rough surface like a dusty table top to make it smooth, you might see a reflection in the surface. Reflection from a rough surface is called **diffuse reflection** because the light is scattered in different directions. Figure 5 shows the difference between specular and diffuse reflection.

1 a In Figure 2, if the angle of reflection of a light ray from a plane mirror is 20°, work out:
   i   the angle of incidence [1 mark]
   ii the angle between the incident ray and the reflected ray. [1 mark]
   b If the mirror is turned so that the angle of incidence is increased to 21°, work out the angle between the incident ray and the reflected ray. [1 mark]

2 An object O is placed in front of a plane mirror, as shown.
   a Complete the path of the two rays shown from O after they have reflected off the mirror. [2 marks]
   b i Use the reflected rays to locate the image of O. [1 mark]
     ii Show that the image and the object are the same distance from the mirror. [1 mark]

3 Two plane mirrors are placed perpendicular to each other.
   a Draw a ray diagram to show the path of a light ray at an angle of incidence of 60° that reflects off both mirrors. [2 marks]
   b i Measure the angle A between the final reflected ray and the incident ray. [1 mark]
     ii Show that angle A is always equal to 180° whatever the angle of incidence is. Remember the angles of a triangle always add up to 180°. [3 marks]
4 a Distinguish between specular and diffuse reflection. [4 marks]
   b Explain the difference between a real image and a virtual image. [2 marks]

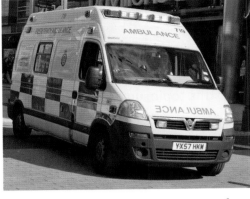

Figure 4 *Ambulances and police cars often carry a mirror image sign on the front. This is so the driver of a vehicle in front can read the sign when they look in their mirror because it gets laterally inverted (back to front but not upside down)*

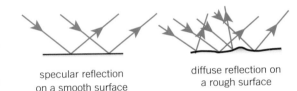

specular reflection on a smooth surface

diffuse reflection on a rough surface

Figure 5 *Reflection at a smooth and at a rough surface*

Key points

- The normal at a point on a mirror is a line drawn perpendicular to the mirror.
- The law of reflection states that the angle of incidence = the angle of reflection.
- For a light ray reflected by a plane mirror:
  - The angle of incidence is the angle between the incident ray and the normal.
  - The angle of reflection is the angle between the reflected ray and the normal.
- Specular reflection is reflection in a single direction without scattering. Diffuse reflection is reflection from a rough surface that scatters the light.

object O

mirror

P14.2 Refraction of light

Learning objectives

After this topic, you should know:

- where refraction of light can happen
- how a light ray refracts when it goes from air into glass or from glass into air.

Synoptic link

In Topic P12.3 you investigated the refraction of water waves in a ripple tank. Water waves travel more slowly in shallow water than in deep water.

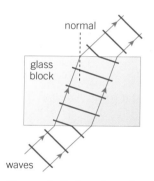

Figure 1 *Refraction of waves.*

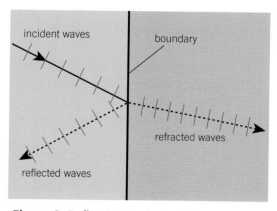

Figure 2 *Reflection and refraction*

When you have your eyes tested, the optician might test different lenses in front of each of your eyes. Each lens changes the direction of light passing through it. This change of direction is called **refraction**.

Refraction is a property of all kinds of waves, including light and sound. Refraction happens to water waves when they cross a boundary between deep and shallow water at a non-zero angle to the boundary. The change of speed at the boundary causes them to change direction.

- Light waves are refracted as shown in Figure 1 when they travel across a boundary between air and a transparent medium or between two transparent media. This is because the speed of light changes at this kind of boundary.

Figure 1 shows light waves (in blue) entering and then leaving a glass block. The direction the light waves are moving in is shown using light rays (in red). The change of direction of each ray relative to the normal (the line at 90° to the boundary) at each boundary is:

- towards the normal when light travels from air into glass
- away from the normal when light travels from glass to air.

Both changes happen because light travels more slowly in glass than in air. Because light travels more slowly in glass than in air, glass is said to be 'optically more dense' than air. In general:

- when light enters a more-dense medium, it is refracted towards the normal
- when light enters a less-dense medium, it is refracted away from the normal.

When waves cross a boundary between two materials, partial reflection can also happen as well as refraction (Figure 2). This is why you might see a faint mirror image of yourself when you look at a window. The waves that cross the boundary lose energy at the boundary and so have a smaller amplitude than that of the incident waves.

Investigating refraction of light

Figure 3 shows how you can use a ray box and a rectangular glass block to investigate the refraction of a light ray when it enters glass. The ray changes direction at the boundary between air and glass (unless it is along the normal).

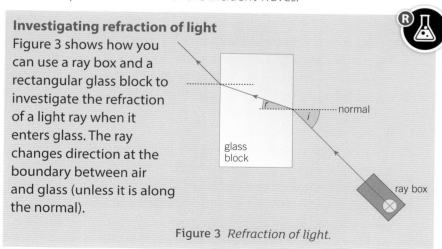

Figure 3 *Refraction of light.*

At the point where the light ray enters the glass, compare the angle of refraction (the angle between the refracted ray and the normal) with the angle of incidence.

You should find that the angle of refraction at the point of entry is always less than the angle of incidence.

Safety: Make sure the glass block does not have any sharp edges.

Refraction rules

Your investigation should show that a light ray:

- changes direction towards the normal when it travels from air into glass. The angle of refraction r is smaller than the angle of incidence i.

- changes direction away from the normal when it travels from glass into air. The angle of refraction r is greater than the angle of incidence i.

1 When a light ray travels from air into glass, its speed changes at the boundary.
  a Determine whether there is an increase or a decrease in the speed of the light waves when they cross the boundary. [1 mark]
  b If the angle of incidence is zero, write the angle of refraction. [1 mark]
  c If the angle of incidence is non-zero, determine whether the angle of refraction is greater than or smaller than the angle of incidence. [1 mark]

2 Copy and complete the path of the light ray through each glass object below:

a 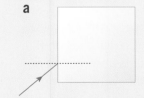 b

[2 marks]   [2 marks]

3 A light ray from the bottom of a swimming pool refracts at the water surface. Its angle of incidence is 40 degrees and its angle of refraction is 75 degrees.
  a Draw a diagram to show the path of this light ray from the bottom of the swimming pool into the air above the pool. [2 marks]
  b Use your diagram to explain why the swimming pool appears shallower than it really is when viewed from above. [2 marks]

4 A white screen is placed in the path of a narrow beam of white light after it has passed through a prism. Describe and explain what is observed on the screen. ✏ [3 marks]

Go further!

Refraction by a prism

Figure 4 shows what happens when a narrow beam of white light passes through a triangular glass prism. The ray comes out of the prism in a different direction to the incident ray and is split into the colours of the spectrum. This happens because the speed of light in glass depends on wavelength, and therefore on the colour of the light. For example, blue light travels slower in glass than red light, so it is refracted more.

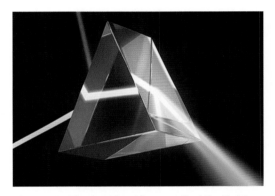

Figure 4 *The incident beam of white light enters the prism at its left hand side.*

Key points

- Refraction is the change in direction of waves when they travel across a boundary from one medium to another.
- When a light ray refracts as it travels from air into glass, the angle of refraction is less than the angle of incidence.
- When a light ray refracts as it travels from glass into air, the angle of refraction is more than the angle of incidence.

P14.3 Light and colour

Learning objectives

After this topic, you should know:

- how the wavelength of light changes across the visible spectrum
- what determines the colour of a surface
- what a translucent object is
- the difference between a translucent object and a transparent object.

Synoptic link

Figure 1 in Topic P13.1 shows how each band of colour in the visible spectrum merges with the next.

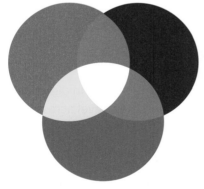

Figure 1 *The spectrum of sunlight.*

Figure 2 *The colours you see when red, green and blue spotlight colours partly overlap on a white screen*

Most people have their own favourite colour, but sometimes colour can be deceptive. Try wearing a blue T-shirt in a room where the lighting is red. Your T shirt will appear black. The colour of your T shirt depends on the colour of the lighting in the room as well as the colour of the material in the T shirt.

Colour and wavelength

Each colour in the visible light spectrum has its own narrow band of wavelength and frequency. Each side of each band merges into the adjacent bands.

The colour of the light coming from a light source depends on the type of light source. For example, the yellow/orange light from a sodium street lamp is different in colour to the light from the Sun. This is because each type of light source emits its own range of wavelengths.

- Stars and lamps such as filament lamps emit light that has a continuous range of wavelengths across the visible spectrum. A filament lamp is an example of a white light source.
- Light sources such as lasers and neon lamps emit a narrow range of wavelengths.

Colour filters work by absorbing certain wavelengths and transmitting other wavelengths. For example, if white light is directed at a red filter, the filter transmits only red light because it absorbs all the parts of the white light spectrum except for red.

Red, green, and blue are called the primary colours of light because they can be mixed to produce any other colour of light (Figure 2).

Surfaces and colour

The colour of the surface of an opaque object depends on chemicals called pigments in the surface materials. Colour also depends on the range of wavelengths in the incident light. Pigments absorb light of specific wavelengths only, and strongly reflect other wavelengths.

- A white surface has no pigments, so it reflects light of any wavelength, either partially or totally. The surface looks white in daylight because daylight is white light and the reflected light includes all the colours of white light (Figure 3).
- The surface of a book that is red in daylight (i.e. in white light) has pigments that absorb all the colours of light except for red. The surface of the book reflects the red component of the incident white light. If you looked at the surface of the book in blue light, it would look black because it would absorb all the incident light.

Surface tests

Investigate the reflection of a narrow beam of light from a ray box by using different surfaces – including a mirror (Figure 2, Topic P14.1), different coloured surfaces (Figure 3), and surfaces of different smoothness.

- In a darkened room, place the ray box on a blank sheet of paper so that you can see the light ray on the paper.
- Use the ray box to direct a light ray as a plane mirror. Repeat at different angles of incidence.
- Then replace the mirror with different coloured surfaces.

Produce a table to record your observations and conclusions.

- Compare the reflection of a narrow beam of light from a shiny surface with the same narrow beam reflected from a rough surface.

Safety: Make sure there are no sharp edges on the objects you are using for your investigation.

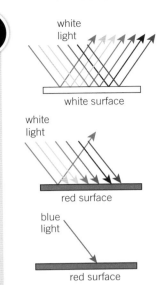

Figure 3 *Surfaces and colour.*

Translucent and transparent objects

- Transparent objects transmit all the incident light that enters the object. No light is absorbed at its surface. The transmitted light travels through the object. This is why you can see clearly through a transparent object. Windows that you can see clear images through are transparent.

- Translucent objects let light pass through them, but the light is scattered or refracted. This happens because the material of the object has lots of internal boundaries that change the direction of the light rays repeatedly. You can see the light that passes through a translucent object, but you can't see images through it. Windows that let light through but don't allow you to see any images through them are made of translucent glass.

- An opaque object is an object that absorbs all the light that reaches it. The light is either reflected, scattered at the surface, or absorbed by the object. No light travels all the way through an opaque object. This is why you can't see through an opaque object.

1 A book observed in daylight has a blue front cover with its title in white. Describe and explain its appearance in red light. [2 marks]

2 In a school play, the leading actress wears a green dress. Describe and explain what colour the dress appears in a scene where
   a the lighting is blue [1 mark]
   b the lighting is blue and the actress also wears a silvery hat. [1 mark]

3 A red filter absorbs all the colours of the white light spectrum except red, which is transmitted by the filter. A blue filter does the same with blue light. They are both positioned so that light passes through one filter then the other filter. If the light directed at the first filter is white light, describe and explain what colour or colours of light are transmitted through both filters. [3 marks]

4 Design an experiment using a light meter and a rectangular glass block to find out how the absorption of light by a glass block depends on the thickness of the block. [6 marks]

Key points

- The wavelength of light increases from red to violet across the visible spectrum.
- The colour of a surface depends on the pigments of the surface materials and the wavelengths of light the pigments absorb.
- A translucent object lets light pass through it but scatters or refracts the light inside it.
- A transparent object lets all the light that enters it pass through it and does not scatter or refract the light inside the object.

P14.4 Lenses

Learning objectives

After this topic, you should know:

- what a convex lens is
- what a concave lens is
- how to calculate magnification.

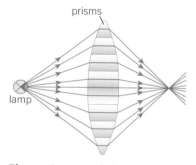

Figure 1 *Investigating lenses.*

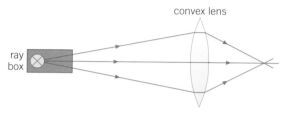

Figure 2 *How a lens works.*

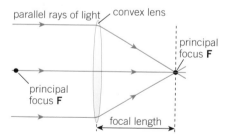

Figure 3 *The focal length of a convex lens.*

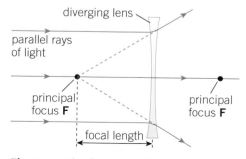

Figure 4 *The focal length of a concave lens.*

Lenses are used in optical devices such as the camera. Although a digital camera is very different from the first cameras made over 160 years ago, they both contain a lens that is used to form an image.

Types of lenses

A lens works by changing the direction of light passing through it. Figure 1 shows the effect of a lens on the light rays from a ray box. The curved shape of the lens surface refracts the rays so they meet at a point. Each section of the lens refracts light as it goes in, and again as it comes out. The overall effect is to make the light rays converge (Figure 2).

Different lens shapes can be tested using this arrangement.

- A **convex lens** (or converging lens) makes parallel rays converge to a focus. The point where parallel rays are focused to is the **principal focus** (or focal point) of the lens (Figure 3). A converging lens is used as a **magnifying glass** and in a camera to form a clear image of a distant object.

- A **concave lens (or diverging lens)** makes parallel rays diverge (spread out). The point where the rays appear to come from is the principal focus of the lens (Figure 4). A diverging lens is used to correct short sight.

- Whether the lens is concave or convex, the distance from the centre of the lens to the principal focus is called the **focal length** of the lens. In ray diagrams, the principal focus is usually shown on each side of the lens.

Investigating the converging lens

Use the arrangement in Figure 5 to investigate the image formed by a convex lens.

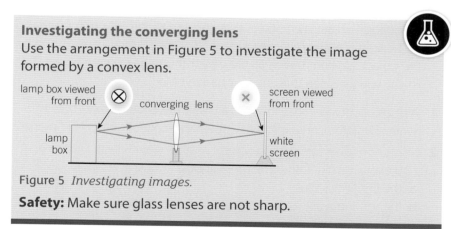

Figure 5 *Investigating images.*

Safety: Make sure glass lenses are not sharp.

The convex lens

1 With the object at different distances beyond the principal focus of the lens, adjust the position of the screen until you see a clear image of the object on it. This is a **real image** because it can be formed on a screen.

- When the object is a long distance away, the image is formed at the principal focus of the lens. This is because the rays from any point on the object are almost parallel to each other when they reach the lens.

- If the object is moved nearer to the lens towards its principal focus, the screen must be moved further from the lens for you to see a clear image. The nearer the object is to the lens, the larger the image.

2 With the object nearer to the lens than the principal focus, a magnified image is formed. The image is a **virtual image** because it is formed where the rays appear to come from. In this situation, the lens acts as a magnifying glass.

Magnification

The **magnification** produced by a lens $= \dfrac{\text{image height}}{\text{object height}}$

If the image is bigger than the object (Figure 6b), the magnification is greater than 1. If the image is smaller than the object (Figure 6a), the magnification is less than 1.

Magnification is a ratio, so it does not have a unit. Image height and object height should both be measured in the *same* unit – *either* millimetres *or* centimetres.

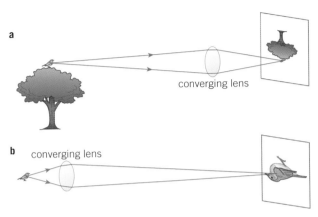

Figure 6 a *The image of a distant object,* **b** *an enlarged image.*

1 **a** Compare the differences between a real image and a virtual image. [2 marks]
  b Determine whether a real image or a virtual image is formed when:
    i a convex lens forms an image of a distant object on a screen [1 mark]
    ii a convex lens is used as a magnifying glass [1 mark]
    iii a concave lens is used to form an image of a distant object. [1 mark]

2 **a** A postage stamp is inspected using a convex lens as a magnifying glass. Describe the image. [1 mark]
  b A convex lens forms a magnified image of a slide on a screen.
    i Describe the image formed by the lens. [1 mark]
    ii The screen is moved away from the lens. Determine the adjustment that must be made to the position of the slide to focus its image on the screen again. [1 mark]
  c i Estimate the magnification of the flower in Figure 7. [1 mark]
    ii Describe how the magnification would change if the lens is moved away from the flower. [2 marks]

3 **a** Describe the image of the bird in Figure 6b and estimate the magnification of the lens. [2 marks]
  b Describe how the image changes if the lens is moved further away from the bird and the card is moved to obtain a new clear image. [2 marks]

4 Design an experiment using the arrangement in Figure 5 to find how the magnification produced by the convex lens depends on the distance from the image to the screen. [6 marks]

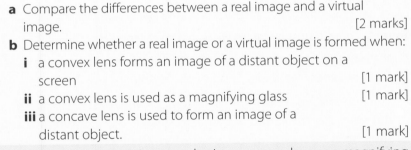

Figure 7 *A magnifying glass.*

Key points

- A convex lens focuses parallel rays to a point called the principal focus.
- A concave lens makes parallel rays spread out as if they had come from a point called the principal focus.
- A real image is formed by a convex lens if the object is further away than the principal focus. A virtual image is formed by a convex lens if the object is nearer than the principal focus.
- Magnification $= \dfrac{\text{image height}}{\text{object height}}$

P14.5 Using lenses

Learning objectives

After this topic, you should know:

- how to find the position and nature of an image formed by a lens
- what type of image is formed by a convex lens when the object is between the lens and its principal focus
- what type of lens is used in a camera and in a magnifying glass
- what type of image is formed in a camera and what type in a magnifying glass.

The position and nature of the image formed by a lens depends on:

- the focal length *f* of the lens
- the distance from the object to the lens.

If you know the focal length and the object's distance, you can find the position and nature of the image by drawing a ray diagram.

Formation of a real image by a convex lens

To form a real image using a convex lens, the object must be beyond the principal focus, F, of the lens. Look at Figure 1. The image is formed on the other side of the lens to the object.

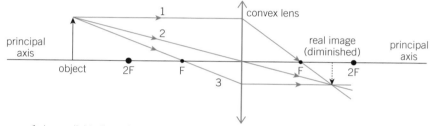

ray 1 is parallel to the axis and is refracted through F
ray 2 passes straight through the centre of the lens
ray 3 passes through F and is refracted parallel to the axis

Figure 1 *Formation of a real image by a convex lens.*

To locate the image and determine its nature, three main rays are used as construction lines from a single point of the object.

- The principal axis of the lens is the straight line that passes along the normal at the centre of each lens surface. The lens is drawn as a straight line with outward arrows to show that it is a convex lens.
- The image is real, inverted, and smaller than the object.

Light acts along the different ray paths:

- *Ray 1* is refracted through F, the principal focus of the lens, because it is parallel to the principal axis of the lens before it passes through the lens.
- *Ray 2* passes through the centre of the lens (its pole) without a change in direction – this is because the lens surfaces at the principal axis are parallel to each other.
- *Ray 3* passes through F, the principal focus of the lens, before it reaches the lens, so it is refracted by the lens parallel to the principal axis.

The camera

In a camera, a convex lens is used to produce a real image of an object on a film (or on an array of pixels for a digital camera). The position of the lens is adjusted to focus the image on the film.

- For a distant object, the distance from the lens to the film must be equal to the focal length of the lens.
- The nearer an object is to the lens, the bigger is the distance from the lens to the film.

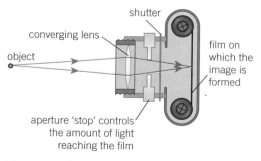

Figure 2 *The camera*

Formation of a virtual image by a convex lens

The object must be between the lens and its principal focus, as shown in Figure 3. The image is formed on the same side of the lens as the object.

The image is virtual, upright, and larger than the object.

The image can be seen only by looking at it through the lens. This is how a magnifying glass works.

Formation of a virtual image by a concave lens

The image formed by a concave lens is always virtual, upright, and smaller than the object. Figure 4 shows why. A concave lens is shown as a line with inward arrows.

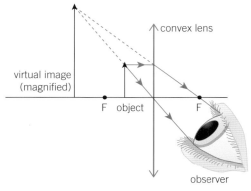

Figure 3 *Formation of a virtual image by a convex lens*

1 **a** Copy and complete the ray diagram in the figure to show how a converging lens forms an image of an object that is smaller than the object, as in a camera. [4 marks]

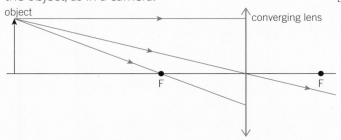

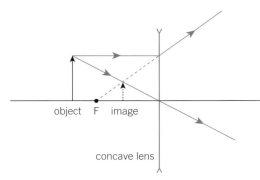

Figure 4 *Image formation by a concave lens*

 b Determine whether the image is:
   i   real or virtual [1 mark]
   ii magnified or diminished [1 mark]
   iii upright or inverted. [1 mark]

2 **a** Draw a ray diagram to show how a convex lens is used as a magnifying glass. [3 marks]
 b Determine whether the image is:
   i real or virtual   [1 mark]   **ii** magnified or diminished   [1 mark]
   iii upright or inverted.   [1 mark]
 c Explain why a concave lens is no use as a magnifying glass. [1 mark]

3 A convex lens produces a magnification of ×2 when it is used to form a real image that is at a distance of 8.0 cm from the object.
 a Draw a scale ray diagram to show the formation of this image. [2 marks]
 b Use your diagram to find the focal length of the lens. [1 mark]
 c Describe how the position and height of the image would change if the object is moved gradually towards the focal point of the lens. [3 marks]

4 **a i** Draw a scale ray diagram to locate the image formed by a convex lens when the object is at 2F [3 marks]
   ii Describe the image formed and determine its magnification. [3 marks]
 b Suggest and discuss an application for the use of the lens in **a**. 🖊 [3 marks]

Key points

- A ray diagram can be drawn to find the position and nature of an image formed by a lens.
- When an object is placed between a convex lens and its principal focus *F*, the image formed is virtual, upright, magnified, and on the same side of the lens as the object.
- A camera contains a convex lens that is used to form a real image of an object.
- A magnifying glass is a convex lens that is used to form a virtual image of an object.

P14 Light

Summary questions

1 a The figure shows an incomplete ray diagram of image formation by a plane mirror.

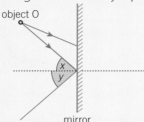

object O

x
y

mirror

  i Describe the angles *x* and *y* in the diagram. [2 marks]

  ii Copy and complete the ray diagram to locate the image. [3 marks]

  iii Compare the distance from the image to the mirror with the distance from the object to the mirror. [1 mark]

b Describe an experiment to test the law of reflection using the ray box and plane mirror. [5 marks]

ray box

plane mirror

2 a The figure shows a light ray directed into a rectangular glass block.

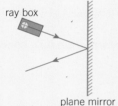

  i Draw the path of the light ray through the block. [2 marks]

  ii Explain why the light ray that emerges from the block is exactly parallel to the incident light ray. [2 marks]

b The figure shows a ray of red light directed into a triangular glass prism.
Copy the drawing and complete the path of the red light ray through the prism. [2 marks]

3 a Explain the different between specular and diffuse reflection. [2 marks]

b i Describe what is meant by translucence. [2 marks]

  ii Explain what causes translucence. [2 marks]

4 The figure shows an incomplete ray diagram of image formation by a lens. The object distance from the lens is 2.5 × the focal length of the lens.

F   F

a i Name the type of lens that is shown in this diagram. [1 mark]

  ii Copy and complete the ray diagram and label the image. [3 marks]

b Describe the image and give an application of the lens used in this way. [2 marks]

5 An object of height 40 mm is placed perpendicular to the principal axis of a convex lens at a distance of 80 mm from the pole of the lens. The focal length of the lens is 50 mm.

a Draw a scale ray diagram to find the distance from the lens to the image. [2 marks]

b Determine whether the image is:

  i real or virtual

  ii upright or inverted. [1 mark]

c Work out the magnification produced by the lens. [1 mark]

6 An object is placed perpendicular to the principal axis of a concave lens.

a Draw a ray diagram to show where the image of the object is formed. [3 marks]

b Determine whether the image is:

  i real or virtual

  ii upright or inverted. [1 mark]

7 a Copy and complete the ray diagram in the figure to show how a convex lens is used as a magnifying glass. [3 marks]

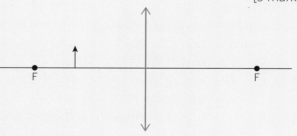

F   F

b Determine whether the image in **a** is:

  i real or virtual

  ii upright or inverted. [1 mark]

Practice questions

01.1 A red and blue rugby shirt hangs on the line in bright sunlight. Explain why the shirt appears red and blue. [3 marks]

01.2 A school uniform list suggests a black skirt and a white polo shirt. Explain why the skirt and shirt appear black and white. [2 marks]

01.3 Objects can be described as transparent, translucent, or opaque. Match each description to the correct example. [3 marks]

transparent	black glass
translucent	clear glass
opaque	frosted glass

02.1 **Figure 1** shows a ray diagram of light striking a plane mirror.

Figure 1

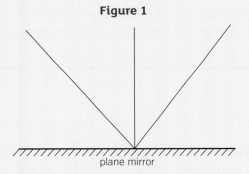

plane mirror

Copy Figure 1 and label the angle of incidence, angle of reflection and the normal. [3 marks]

02.2 Complete the sentences using the correct word from the box.

translucent	specular	diffuse	opaque	random

When light is reflected from a smooth surface it is called _____ reflection. [1 mark]
When light is reflected from a rough surface it is called _____ reflection [1 mark]

03.1 When light is incident on the surface of a transparent material some of the light will be reflected and some will be transmitted into the material.
A student measured the percentage of reflected light from the surface of a material. The results were drawn on a graph in **Figure 2**.

Figure 2

What was the range of the angles of incidence used in the investigation. [1 mark]

03.2 The student repeated the tests and drew two lines on the graph. Suggest one way of improving the investigation. [1 mark]

03.3 The results suggest there has been an error in the investigation. Give the name of the possible error. [1 mark]

03.4 Describe how the percentage of light reflected changes with the angle of incidence. [1 mark]

04 **Figure 3** shows a lens, the position of an object, and the position of the image.

Figure 3

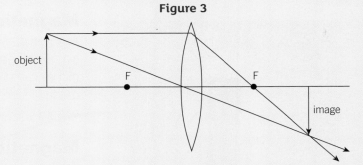

04.1 Give the name of the type of lens shown. [1 mark]

04.2 Give the name of the points, **F**, shown each side of the lens. [1 mark]

04.3 Complete the sentence using the correct words from the box.

inverted	larger	virtual	upright	real	smaller

The image shown is _____ and _____. [2 marks]

Learning objectives

After this topic, you should know:

- the force rule for two magnetic poles near each other
- the pattern of magnetic field lines around a bar magnet
- what induced magnetism is
- why steel, not iron, is used to make permanent magnets.

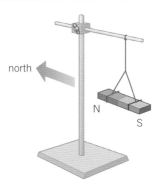

Figure 1 *Checking the poles of a bar magnet*

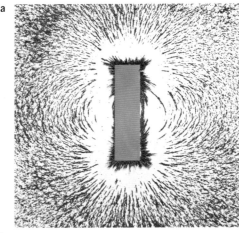

Figure 2 *The magnetic field of a bar magnet* **a** *Using iron filings* **b** *Using a plotting compass*

About magnets

A magnetic compass is a tiny magnetic needle pivoted at its centre. Because of the Earth's magnetic field, one end of the compass always points north, and the other end always points south. The end of a plotting compass or a bar magnet (Figure 1) that points north is the 'north-seeking' pole, usually called its north pole (N-pole), and the other end is the 'south-seeking' pole, its south pole (S-pole).

Investigating bar magnets

1   Suspend a bar magnet as shown in Figure 1 and label the end that points north as its N-pole.

2   Hold the N-pole of a second bar magnet near the suspended bar magnet. You should find it attracts the S-pole of the suspended bar magnet and repels the N-pole.

3   Repeat the above test using the S-pole of the second bar magnet. You should find it attracts the N-pole of the suspended bar magnet and repels its S-pole.

The tests above show the general rule that:

Like poles repel. Unlike poles attract.

Magnetic materials

Any iron or steel object can be magnetised (or demagnetised if it's already magnetised). Only a few other materials (for example cobalt and nickel) can be magnetised and demagnetised. Permanent magnets are made of steel because magnetised steel does not lose its magnetism easily.

Magnetic fields

If a sheet of paper is placed over a bar magnet and iron filings are sprinkled onto the paper, the filings form a pattern of lines. The region around the magnet is called a **magnetic field**. Any other magnetic material placed in this space experiences a force caused by the first magnet.

In Figure 2:

- the iron filings form lines as shown in Figure 2a that end at or near the poles of the magnet. These lines are magnetic field lines, also called lines of force. The lines are more concentrated at the poles than elsewhere. This is because the field is strongest at the poles.

- a plotting compass placed in the magnetic field aligns itself along a magnetic field line, pointing in a direction away from the N-pole of the magnet and towards the magnet's S-pole, as shown in Figure 2b. For this reason, the direction of a line of force is always from the north pole of the magnet to its south pole.

The further the plotting compass is from the magnet, the less effect the magnet has on the plotting compass. This is because the greater the distance from the magnet, the weaker the strength of the magnetic field.

Induced magnetism

An unmagnetised magnetic material can be magnetised by placing it in a magnetic field. The magnetic field is said to induce magnetism in the material. For example, an unmagnetised iron rod placed in line with a bar magnet becomes a magnet with poles at each end. The nearest poles of the rod and the bar magnet always have opposite polarity.

Induced magnetism will cause a force of attraction between any unmagnetised magnetic material placed near one end of a bar magnet. The force is always an attractive force whichever end of the bar magnet is nearest to the material.

1 A bar magnet XY is freely suspended in a horizontal position so that end X points north and end Y points south.
   a Give the magnetic polarity of:
      i end X   ii end Y.   [1 mark]
   b End P of a second bar magnet PQ placed near end X of bar magnet XY repels end X and attracts end Y. Give the magnetic polarity of end P and give a reason for this observation.   [2 marks]

2 The tip of an iron nail is held in turn near each end of a plotting compass needle. Write whether the tip of the nail is a N-pole, a S-pole, or is unmagnetised in each of the following possible observations:
   a i the N-pole of the compass needle is repelled by the tip of the nail and the S-pole is attracted by it
      ii the N-pole of the compass needle is attracted by the tip of the nail and the S-pole is repelled by it
      iii the N-pole of the plotting compass is attracted by the tip of the nail and the S-pole is also attracted by it.   (3 marks)
   b Explain why the tip of the nail in a iii attracts both poles of the plotting compass.   [2 marks]

3 Figure 4 shows a bar magnet XY and a plotting compass near end Y of the bar magnet. The needle of the plotting compass points towards end Y of the bar magnet.

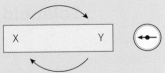

Figure 4

   a Give the magnetic polarity of each end of the bar magnet.   [1 mark]
   b If the bar magnet was rotated gradually about its centre through 180°, describe and explain the effect on the direction of the plotting compass needle.   [3 marks]

4 Design an experiment using a plotting compass and a ruler to find out which of two bar magnets is stronger. Draw a diagram to aid your explanation.   [4 marks]

Plotting a magnetic field

Mark a dot near the north pole of a bar magnet. Place the tail of the compass needle above the dot, and mark a second dot at the tip of the needle. Repeat the procedure with the tail over the new dot each time until the compass reaches the S-pole of the magnet. Draw a line through the dots and mark the direction from the N-pole to the S-pole. Repeat the procedure for further lines.

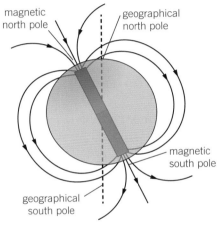

Figure 3 *The Earth's magnetic field. Scientists have plotted the Earth's magnetic field accurately. The pattern is like that of a bar magnet. But the Earth is partly molten inside, and this helps to explain why the magnetic poles move about gradually*

Key points

- Like poles repel, and unlike poles attract.
- The magnetic field lines of a bar magnet curve around from the north pole of the bar magnet to the south pole.
- Induced magnetism is magnetism created in an unmagnetised magnetic material when the material is placed in the magnetic field.
- Steel is used instead of iron to make permanent magnets because steel does not lose its magnetism easily, but iron does.

P15.2 Magnetic fields of electric currents

Learning objectives

After this topic, you should know:

- the pattern of the magnetic field around a straight wire carrying a current and in and around a solenoid
- how the strength and direction of the field varies with position and with the current
- what a uniform magnetic field is
- what an electromagnet is.

Fields around a current carrying wire

Use the arrangement shown in Figure 1 to observe the effect of:

1 **Reversing the current**
You should find that the plotting compass reverses its direction. This shows that the magnetic field lines reverse direction when the direction of the current is reversed.

2 **Moving the plotting compass away from the wire**
You should find that it points more towards 'magnetic north'. The Earth's magnetic field has a bigger effect the further the compass is from the wire. This is because the strength of the magnetic field of the wire decreases as the distance from the wire increases.

3 **Increasing the current**
To vary the current, connect a variable resistor in series with the battery and the wire. As you increase the current you should find the magnetic field becomes stronger everywhere. You can tell this if you increase the current gradually from zero, the plotting compass turns more and more away from the North as the current becomes stronger. This is because the field has a bigger effect on the plotting compass than the Earth's magnetic field.

Safety: Make sure the wire does not get too hot.

The magnetic field near a current-carrying wire

When an electric current passes along a wire, a magnetic field is set up around the wire. Figure 1 shows how you can find the pattern of the magnetic field around a long straight wire by using a plotting compass. The lines of force caused by a straight current-carrying wire are a series of concentric circles. These circles are centred on the wire in a plane that is perpendicular to the wire.

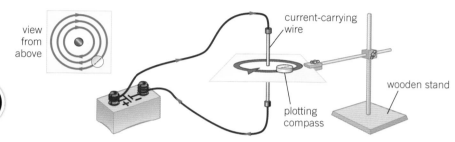

Figure 1 *The magnetic field near a long straight wire. To eliminate magnetism caused by nearby iron objects, use a wooden stand (or any other non-ferrous object) to support the cardboard sheet so that it's horizontal*

You can use the corkscrew rule shown in Figure 2 to remember the direction of the magnetic field for each direction of the current. Reversing the direction of the current reverses the direction of the magnetic field.

Figure 2 *The corkscrew rule*

Solenoids

A solenoid is a long coil of insulated wire (Figure 3). Solenoids are used in lots of devices where a strong magnetic field needs to be produced. The magnetic field is produced in and around the solenoid when a current is passed through the wire. The magnetic field:

- increases in strength if the current is increased
- reverses its direction if the current is reversed.

Inside the solenoid

The magnetic field is much stronger than if the wire was straight. The field lines are parallel to the axis of the solenoid, and they are all in the same direction (i.e., uniform). The magnetic field inside a solenoid is strong and uniform.

Outside the solenoid

The magnetic field lines bend around from one end of the solenoid to the other end of the solenoid. The magnetic field outside is like the field of a bar magnet, except that each field line is a complete loop because it passes through the inside of the solenoid.

Figure 3 shows how you can find the polarity of each end of the solenoid from the direction of the current.

Electromagnets

An **electromagnet** is a solenoid in which the insulated wire is wrapped around an iron bar (the core). When a current is passed along a wire, a magnetic field is created around the wire. Because of this, the magnetic field of the wire magnetises the iron bar. When the current is switched off, the iron bar loses most of its magnetism.

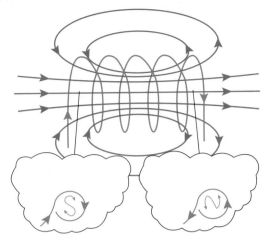

Figure 3 *The magnetic field of a solenoid. Looking at each end, the S-pole is the end where the current is clockwise, and the N-pole is the end at which the current is anticlockwise.*

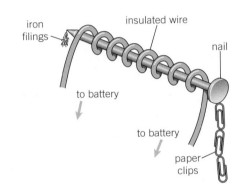

Figure 4 *A simple electromagnet*

1 **a** Sketch the pattern of the magnetic field lines near a vertical wire carrying current upwards [2 marks]
   b Show on your sketch a plotting compass near the wire and indicate the direction which it points when there is a current in the wire. [1 mark]
2 **a** Describe how an insulated wire and an iron bar can be used to produce a strong magnetic field when a current is passed through the wire. [2 marks]
   b Explain why iron, and not steel, is used for the core of an electromagnet. [2 marks]
3 A plotting compass is placed near a straight wire carrying a current as shown in Figure 1.
   a Describe the effect on the plotting compass of reversing the current in the wire. [1 mark]
   b Describe and explain how the direction in which the plotting compass points would change if it was gradually moved away from the wire. [3 marks]
4 **a** Sketch the pattern of the magnetic field lines in and around a solenoid when there is a current in the solenoid. [3 marks]
   b i Show on your sketch a plotting compass at one end of the solenoid and indicate the direction that the plotting compass points. [2 marks]
   ii Describe and explain the effect on the plotting compass of reducing the current in the solenoid. [3 marks]

Key points

- The magnetic field lines *around a wire* are circles centred on the wire in a plane perpendicular to the wire.
- The magnetic field lines *in a solenoid* are parallel to its axis and are all in the same direction. A uniform magnetic field is one in which the magnetic field lines are parallel.
- Increasing the current makes the magnetic field stronger. Reversing the direction of the current reverses the magnetic field lines.
- An electromagnet is a solenoid that has an iron core. It consists of an insulated wire wrapped around an iron bar.

P15.3 Electromagnets in devices

Learning objectives

After this topic, you should know:

- what electromagnets can be used for
- how devices that use electromagnets work.

Electromagnets are used in lots of devices. Four examples are described below.

1 The scrapyard crane.
Scrap vehicles are lifted in a scrap yard using powerful electromagnets attached to cranes. The steel frame of a vehicle sticks to the electromagnet when current passes through the coil of the electromagnet. When the current is switched off, the vehicle frame falls off the electromagnet.

2 A circuit breaker.
A circuit breaker is a switch in series with an electromagnet. The switch is held closed by a spring. When the current is too large, the switch is pulled open by the electromagnet and it stays open until it is reset manually (Figure 2).

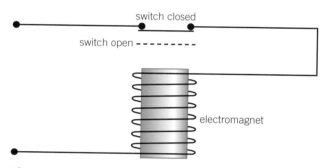

Figure 2 *A circuit breaker*

3 The electric bell.
When an electric bell is connected to a battery, the iron armature is pulled on to the electromagnet. This opens the make-and-break switch, and the electromagnet is switched off. Because of this, the armature springs back and the make-and-break switch closes again, so the whole cycle repeats itself (Figure 3).

Investigating the strength of an electromagnet

Use an electromagnet to attract and hold a flat iron bar. Suspend a known weight from the bar and measure the smallest current needed to hold the load (the bar and the weight).

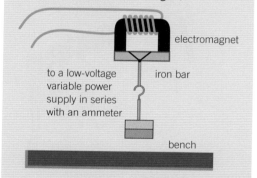

Figure 1 *Testing an electromagnet*

- Increase the weight and repeat the measurement for different known weights.
- Repeat the measurements with more turns on the electromagnet.
- Plot your measurements on a graph of current against weight and use your graph to draw conclusions about the factors that affect the strength of the electromagnet.

Safety: Make sure wires and batteries do not get too hot.

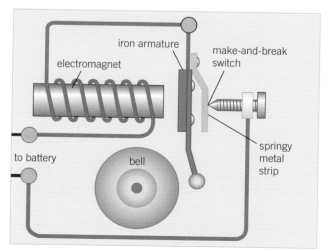

Figure 3 *An electric bell*

4   The relay.
    The relay is used to switch an electrical machine on or off. A small
    current through the coil of the electromagnet magnetises the iron
    core, which then pulls the armature onto the electromagnet. This
    closes the switch gap and switches the machine on (Figure 4). In this
    way, a small current (in the coil) is used to switch on a machine with
    a much bigger current.

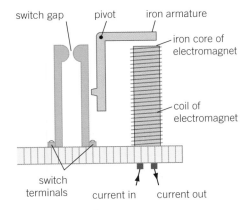

Figure 4 *A relay*

1   A circuit breaker contains an electromagnet in series with a switch.
    a State the purpose of a circuit breaker.   [2 marks]
    b Describe how a circuit breaker works.   [2 marks]

2   List the statements **A–E** below in correct order to explain how
    the circuit breaker in Figure 2 works. Statement **E** is third in the
    correct order.
    A The current is cut off.
    B The iron core of the electromagnet is magnetised.
    C Too much current passes through the coil.
    D The circuit breaker switch is opened.
    E The switch is attracted to the core of the electromagnet. [2 marks]

3   The construction of a buzzer is like that of the electric bell, except
    that the buzzer does not have a striker or a bell.
    a Explain why the armature of the buzzer vibrates when the buzzer
       is connected to a battery.   [4 marks]
    b Explain why the buzzer vibrates at a higher frequency than the
       electric bell.   [3 marks]

4   In an experiment using an electromagnet, different weights were
    attached to an iron plate held by the electromagnet. The
    measurements in the table show the least current needed to
    hold the plate for each weight.

Weight in newtons	0.0	2.0	4.0	6.0	8.0	10.0
Current in amperes	0.10	0.35	0.60	0.80	1.35	2.10

    a Plot a graph of the measurements.   [4 marks]
    b Write your conclusions based on your graph. 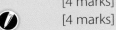   [4 marks]

Key points

- Electromagnets are used in
  scrapyard cranes, circuit breakers,
  electric bells, and relays.
- An electromagnet works in a circuit
  breaker or electric bell or a relay by
  attracting an iron armature which
  opens a switch.

P15.4 The motor effect

Learning objectives

After this topic, you should know:

- how to change the size and reverse the direction of the force on a current-carrying wire in a magnetic field
- how a simple electric motor works
- what is meant by magnetic flux density
- how to calculate the force on a current-carrying wire.

You use electric motors lots of times every day. Using a hairdryer, an electric shaver, a refrigerator pump, and a computer hard drive are just a few examples. All these electrical appliances contain an electric motor. The electric motor works because a force can act on a wire (or any other conductor) in a magnetic field when a current is passed through the wire. This is called the **motor effect**.

- The size of the force can be increased by:
  - increasing the current
  - using a stronger magnet.
- The size of the force depends on the angle between the wire and the magnetic field lines. The force is:
  - greatest when the wire is perpendicular to the magnetic field
  - zero when the wire is parallel to the magnetic field lines.
- The direction of the force is always at right angles to the wire and the field lines. Also, the direction of the force is reversed if either the direction of the current or the magnetic field is reversed. Figure 2 shows **Fleming's left-hand rule**, which tells you how these directions are related to each other.

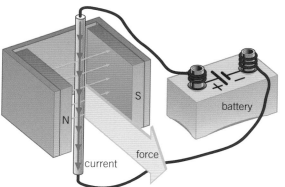

Figure 1 *An experimental setup you could use to investigate the motor effect*

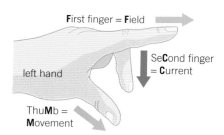

Figure 2 *Fleming's left-hand rule. Hold your fingers at right angles to each other. You can use this rule to work out the direction of the force (i.e. the movement) on the wire.*

Magnetic flux density

The magnetic flux density of a magnetic field is the measure of the strength of the magnetic field. The symbol B is used for magnetic flux density and the unit is the tesla (T).

In Figure 1, the size of the force F on the conductor depends on:

- the current I in the conductor
- the length l of the conductor
- the magnetic flux density B of the magnetic field.

You can write the following equation for the force:

force, F = **magnetic flux density, B × current, I × length, l**
(newtons, N)   (tesla, T)   (amperes, A)   (metres, m)

Worked example

In Figure 1, the magnetic flux density is 0.032 T, the length of the conductor in the field is 40 mm, and the current through the conductor is 2.2 A. Calculate the force on the conductor.

Solution

The length in metres of the conductor is 0.040 m.

Using $F = B I l$ gives:

$$F = 0.032\,\text{T} \times 2.2\,\text{A} \times 0.040\,\text{m}$$
$$= \mathbf{0.028\,N}$$

The electric motor

An electric motor is designed to use the motor effect. You can control the speed of an electric motor by changing the current. Also, you can reverse the direction the motor turns in by reversing the current.

The simple motor in Figure 3 has a rectangular coil of insulated wire (the armature coil) that is forced to rotate. The coil is connected to the battery by two metal or graphite brushes. The brushes press onto a metal split-ring commutator fixed to the coil. Graphite is a form of carbon that conducts electricity and is very slippery. Graphite causes very little friction when it is in contact with the rotating commutator.

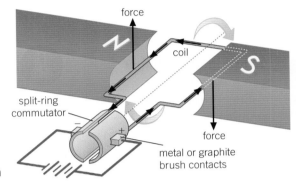

Figure 3 The electric motor

When a current is passed through the coil, the coil spins because:

● a force acts on each side of the coil due to the motor effect

● the force on one side is in the opposite direction to the force on the other side.

The split-ring commutator reverses the current around the coil every half-turn of the coil. Because the sides swap over each half-turn, the coil is pushed in the same direction every half-turn.

1 Explain why the coil of a simple electric motor rotates continuously when the motor is connected to a battery. [3 marks]

2 a Explain why a simple electric motor connected to a battery reverses if the battery connections are reversed. [2 marks]

  b Determine whether or not an electric motor would run faster if the coil was wound on:

  i a plastic block [1 mark]

  ii an iron block, instead of a wooden block. [1 mark]

3 A force is exerted on a straight wire when a current is passed through it and it is at right angles to the lines of a magnetic field. Describe how the force changes if the wire is turned through 90° until it is parallel to the field lines. [3 marks]

4 A straight wire is placed in the magnetic field of a U-shaped magnet as shown in Figure 1. The length of the wire in the field is 35 mm. When a current of 1.8 A is in the wire, a force of 0.024 N acts on the wire due to the field. Calculate the magnetic flux density of the field. [2 marks]

Key points

● In the motor effect, the force is:
  - increased if the current or the strength of the magnetic field or the length of the conductor is increased
  - reversed if the direction of the current or the magnetic field is reversed

● An electric motor has a coil that turns when a current is passed through it.

● Magnetic flux density is a measure of the strength of a magnetic field.

● To calculate the force on a current-carrying conductor at right angles to the lines of a magnetic field, use the equation $F = B I l$.

P15.5 The generator effect

Learning objectives

After this topic, you should know:

- what the generator effect is
- how a potential difference can be induced in a wire
- what affects the size of the induced potential difference
- how to deduce the direction of an induced current.

Investigating a simple generator

Connect some insulated wire to an ammeter (Figure 1). Move the wire between the poles of a U-shaped magnet and observe the ammeter. You should discover that the ammeter pointer deflects as a current is generated when the wire crosses through the magnetic field. This is because a potential difference is induced in the wire when it crosses through the lines of the magnetic field.

Carry out tests to see what difference is made by:

1 holding the wire stationary in the magnetic field

2 moving the magnet instead of the wire

3 moving the wire faster across the magnetic field

4 reversing the direction of motion of the wire.

In your tests above, you should find that:

- no current is generated when the wire is stationary
- a current is also generated when the magnet instead of the wire is moved
- a bigger current is generated when the wire moves faster
- the current is reversed when the direction of motion is reversed.

A hospital has its own electricity generator that is always on standby in case of a power cut. Patients' lives would be put at risk if the mains electricity supply failed and there was no standby generator.

A generator contains coils of wire that spin in a magnetic field. A potential difference, or voltage, is created, or induced, across the ends of the wire when it crosses through the magnetic field lines. This process is called **electromagnetic induction**, which happens when any conductor crosses through magnetic field lines.

If the conductor is part of a complete circuit, the induced potential difference makes an electric current pass around the circuit. This effect is called the **generator effect**.

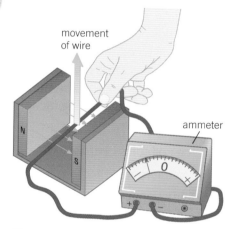

Figure 1 *The generator effect*

A generator test

Figure 2 shows a coil of insulated wire connected to a centre-reading ammeter. When one end of a bar magnet is pushed into the coil, the ammeter pointer deflects. This is because:

- the movement of the bar magnet causes an induced potential difference in the coil

- the induced potential difference causes a current, because the coil is part of a complete circuit.

If a stronger magnet is used, both the induced potential difference and the current will be bigger.

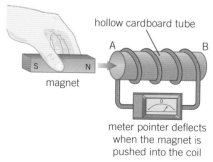

Figure 2 *Testing the generator effect*

In Figure 2, if the bar magnet is then withdrawn from the coil, the ammeter pointer deflects in the opposite direction. This is because the induced potential difference acts in the opposite direction, so the induced current is in the opposite direction.

Table 1 shows the results of testing each direction of motion of the magnet, with the magnet each way around. The table gives the direction of current round end A of the coil.

Table 1

Magnetic pole entering or leaving the coil	Pushed in or pulled out	Direction of current	Induced polarity of A	Magnet and coil
north pole	in	anticlockwise	north pole	repel
north pole	out	clockwise	south pole	attract
south pole	in	clockwise	south pole	repel
south pole	out	anticlockwise	north pole	attract

Moving the coil instead of the magnet has the same effect as described in Table 1.

The induced current generates a magnetic field in and around the coil, but only when the magnet is moving. As Table 1 shows, this induced magnetic field always opposes the original change. So, work has to be done by the person moving the magnet. The electricity generated is the result of the work done by the person moving the magnet.

You can use the solenoid rule to work out from the direction of the induced current whether end A of the coil is like the north pole or the south pole of a bar magnet (Figure 3).

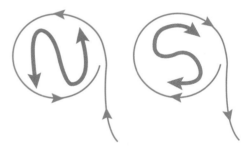

Figure 3 *The solenoid rule*

Key points

- The generator effect is the effect of inducing a potential difference using a magnetic field.
- When a conductor crosses through the lines of a magnetic field, a potential difference is induced across the ends of the conductor.
- The faster a conductor crosses through the lines of a magnetic field, the bigger is the induced potential difference. When a direct-current electromagnet is used, it needs to be switched on or off to induce a potential difference.
- The direction of an induced current always opposes the original change that caused it.

1 When a wire is moved between the poles of a U-shaped magnet as shown in Figure 1, explain why a current passes through the ammeter. [2 marks]

2 A coil of wire is connected to a centre-reading ammeter. A bar magnet is inserted into the coil, making the ammeter pointer flick briefly. Write what you would observe if:
   a the magnet was then held at rest in the coil [1 mark]
   b the coil had more turns of wire wrapped round the tube [1 mark]
   c the magnet was withdrawn rapidly from the coil. [1 mark]

3 Look at Table 1. Explain why this shows that the induced current always opposes the change that causes it. [3 marks]

4 In Figure 2, the tube is turned and clamped so that it is vertical with A at the top end. The bar magnet is held vertically over the top end of the tube. When the bar magnet is released, it drops down the tube and passes through the coil. Describe and explain the effect on the meter. [4 marks]

P15.6 The alternating-current generator

Learning objectives

After this topic, you should know:

- how a simple alternator (alternating-current generator) is constructed and operated
- how the induced potential difference of an a.c. generator varies with time
- how a simple dynamo (direct-current generator) is constructed and operated.

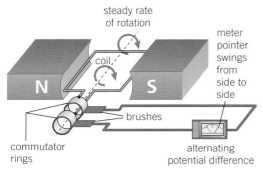

Figure 1 *The a.c. generator*

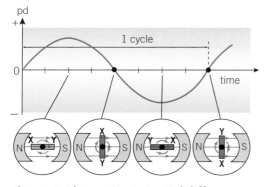

Figure 2 *Alternating potential difference*

Synoptic links

For more information about displaying alternating potential differences on an oscilloscope, look back at Topic P5.1.

The alternator

A simple **alternator** is an alternating-current generator. It is made up of a rectangular coil that is forced to spin in a uniform magnetic field (Figure 1). The coil is connected to a centre-reading meter by metal brushes that press on two metal slip rings. The slip rings and brushes provide a continuous connection between the coil and the meter.

When the coil turns steadily in one direction, the meter pointer deflects first one way, then the opposite way, and then back again. This carries on as long as the coil keeps turning in the same direction. The current in the circuit repeatedly changes its direction through the meter because the induced potential difference in the coil repeatedly changes its direction. So the induced potential difference and the current alternate because they repeatedly change direction.

Figure 2 shows how the induced potential difference varies as the coil rotates. Both the positive and the negative maximum values are called the peak value.

- The size of the induced potential difference is greatest when the plane of the coil is parallel to the direction of the magnetic field. At this position, the sides of the coil (labelled X and Y in Figure 2) cross directly through the magnetic field lines. So the induced potential difference is at its peak value.

- The size of the induced potential difference is zero when the plane of the coil is perpendicular to the magnetic field lines. At this position, the sides of the coil move parallel to the field lines and do not cross through them. So, the induced potential difference is zero.

The faster the coil rotates:

- the bigger the frequency (i.e., the number of cycles per second) of the alternating current. This is because each full cycle of the alternating potential difference takes the same time as one full rotation of the coil.

- the bigger the peak value of the alternating current. This is because the sides of the coil move faster and so they cross through the magnetic field lines at a faster rate.

You can also increase the peak value by using a magnet with a stronger magnetic field and by using a coil with a bigger area and with more turns of wire on it.

The alternating potential difference can be displayed on an oscilloscope screen. If the generator is rotated faster, the screen display will show more waves on the screen and the waves will be taller.

The direct-current dynamo

A **dynamo** is a direct-current generator. A simple dynamo is the same as an alternator except that the dynamo has a split-ring commutator instead of two separate slip rings, as shown in Figure 3. As the coil spins, the split-ring commutator reconnects the coil the opposite way around in the circuit every half-turn. This happens each time the coil is perpendicular to the magnetic field lines. Because of this, the induced potential difference does not reverse its direction as it does in the alternator. The induced potential difference varies from zero to a maximum value twice each cycle, and never changes polarity.

Moving coil sound devices

The moving coil *microphone* generates an alternating potential difference as sound waves make the coil vibrate. The coil is attached to a small diaphragm and is between the poles of a cylindrical magnet (Figure 2). The pressure variations of the sound waves on the diaphragm make it vibrate so the coil vibrates in the magnetic field. The alternating potential difference induced in the coil has the same frequency as the sound waves.

The moving coil *loudspeaker* creates sound waves when an alternating potential difference is applied to its coil. The coil is in a magnetic field as shown in Figure 3. The current in the coil causes a force on the coil due to the **motor effect**. Because the current alternates, this force repeatedly reverses direction and makes the coil and the diaphragm vibrate. This creates sound waves of the same frequency as the alternating potential difference.

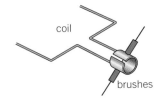

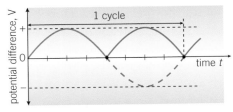

Figure 3 *The d.c. generator*

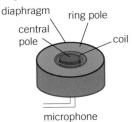

Figure 4 *Inside a microphone*

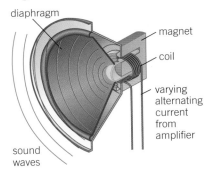

Figure 5 *A loudspeaker*

1 An alternating-current generator has a coil that spins between the poles of a U-shaped magnet. 🖊
   a Explain why an alternating p.d. is induced in the coil. [3 marks]
   b Describe how the alternating p.d. would differ if the coil is made to spin faster. [2 marks]

2 Figure 2 shows how the alternating potential difference produced by an a.c. generator changes with time.
   a Describe how the graph would differ if the coil was rotated more slowly. [2 marks]
   b Give reasons for your answer to **a**. [2 marks]

3 a Describe the function of the split-ring commutator in a simple d.c. generator. [2 marks]
   b Draw a graph to show how potential difference varies with time for a simple d.c. generator. [2 marks]

4 Explain why the a.c. generator in Figure 1 produces its peak potential difference when the coil is parallel to the magnetic field lines. [2 marks]

Key points

- A simple a.c. generator is made up of a coil that spins in a uniform magnetic field.
- The waveform, displayed on an oscilloscope, of the a.c. generator's induced potential difference is at:
  - its peak value when the sides of the coil cross directly through the magnetic field lines
  - its zero value when the sides of the coil move parallel to the field lines.
- A simple d.c. generator has a split-ring commutator instead of two slip rings.

P15.7 Transformers

Synoptic link

The use of transformers in the National Grid was covered in Topic P5.1.

You probably use transformers every day when you use mains electricity to recharge the battery in a low-voltage device such as a mobile phone or a laptop. Mains electricity in your home is 230 V. To recharge a low-voltage battery using mains electricity, a **transformer** has to be used to change the size of the alternating potential difference from 230V to the battery potential difference. You have seen previously that:

- step-up transformers are used to *increase* the size of an alternating potential difference.

- step-down transformers are used to *decrease* the size of an alternating potential difference.

Step-up transformers are used to step p.d. up from about 25 kV at power stations to a much higher p.d. (typically 132 000 V) on the National Grid. Step-down transformers are used to supply electricity from the National Grid to consumers.

How a transformer works

A transformer has two coils of insulated wire, both wound around the same iron core (Figure 1). Iron is used for the core because iron is easily magnetised and demagnetised. The coils are called the primary coil and the secondary coil. The primary coil is connected to an a.c. supply. When the alternating current passes through the primary coil, an alternating potential difference is induced in the secondary coil.

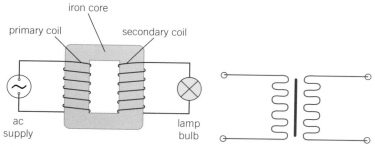

Figure 1 *Transformer action: **a** in a circuit **b** circuit symbol*

This happens because:

- alternating current passing through the primary coil generates an alternating magnetic field in the iron core

- the lines of the alternating magnetic field in the iron core pass through the secondary coil

- the magnetic field in the secondary coil induces an alternating potential difference between the terminals of the secondary coil.

If a bulb is connected across the secondary coil to form a complete circuit, the induced potential difference causes an alternating current in the secondary circuit, so the bulb lights up. This means that energy is transferred from the primary coil to the secondary coil. This happens even though the coils are *not* electrically connected in the same circuit.

Study tip

Make sure you can explain how a transformer works.

Make a model transformer

Wrap a coil of insulated wire around the iron core of a model transformer as the primary coil. Connect the coil to a 1 V a.c. supply. Then connect a second length of insulated wire to a 1.5 V torch bulb. When you wrap enough turns of the second wire around the iron core, the bulb should light up.

● Test whether or not cores made from different materials affect the transformer.

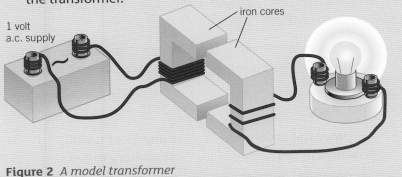

Figure 2 *A model transformer*

Practical transformers

Transformers only work with alternating current. With a direct current, there is no changing magnetic field, so the secondary potential difference is zero.

In the type of transformer described above, the core of the transformer guides the field lines in a loop through the coils. But the field has to be changing to induce a potential difference in the secondary coil.

Figure 3 shows a practical transformer. The primary and secondary coils are both wound around the same part of the iron core.

1 a Describe the purpose of a transformer. [2 marks]
  b A laptop computer can operate with a 14 V battery or with a mains transformer.
    i Suggest the benefit of having a dual power supply. [1 mark]
    ii Determine whether the transformer steps up or steps down the potential difference applied to it. [1 mark]

2 A step-down transformer contains a 200-turn coil and a 4000-turn coil wound on the same iron core.
  a Determine which coil is the primary coil. [1 mark]
  b Permanent magnets are made from steel, not iron. Explain why the transformer would not work as effectively if the core was made of steel instead of iron. [2 marks]

3 a Explain why a transformer does not work with direct current. [2 marks]
  b Explain why it is important that the coil wires of a transformer are insulated. [1 mark]

4 a Describe the construction of a transformer and what a transformer is used for. ✏ [3 marks]
  b Explain how a transformer works. [2 marks]

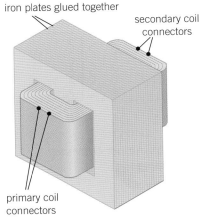

Figure 3 *A practical transformer*

227

P15.8 Transformers in action

Learning objectives

After this topic, you should know:

- how the ratio of the primary potential difference to the secondary potential difference depends on the number of turns on each coil
- how the number of turns on the secondary coil relates to the number of turns on the primary coil for a step-down transformer and for a step-up transformer
- what you can say about a transformer that is 100% efficient
- why less power is wasted by using high potential difference to transfer power through the grid system.

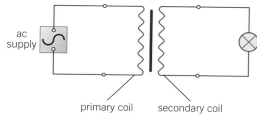

Figure 1 *Transformer efficiency*

Study tip

Practice rearranging and using the equation: $V_p \times I_p = V_s \times I_s$.

Transformer efficiency

To find I_s:

1. divide both sides by V_s to give $\dfrac{V_p \times I_p}{V_s} = \dfrac{V_s \times I_s}{V_s}$, then

2. cancel V_s on the right-hand side of the equation (because it is on the top and the bottom line of the right-hand side, and any number divided by itself equals 1) to give $\dfrac{V_p \times I_p}{V_s} = I_s$, then:

3. swap the two sides of the equation over so that I_s is on the left-hand side to give $I_s = \dfrac{V_p \times I_p}{V_s}$

The transformer equation

A transformer used to recharge the battery in a mobile phone or laptop is very different in size and design to a transformer used in the National Grid. A step-up transformer has more turns in its secondary coil than in its primary coil. A step-down transformer has fewer turns in its secondary coil than in its primary coil.

The secondary potential difference of a transformer depends on the primary potential difference and the number of turns on each coil.

You can use the following equation to calculate any one of these factors if you know the other factors:

$$\frac{\text{potential difference across primary coil, } V_p}{\text{potential difference across secondary coil, } V_s} = \frac{\text{number of turns on primary coil, } n_p}{\text{number of turns on secondary coil, } n_s}$$

- For a step-up transformer, the number of secondary turns n_s is greater than the number of primary turns n_p. So, V_s is greater than V_p.
- For a step-down transformer, the number of secondary turns n_s is less than the number of primary turns n_p. So, V_s is less than V_p.

Transformer efficiency

Transformers are almost 100% efficient. When a device is connected to the secondary coil (Figure 1), almost all the electrical power supplied to the transformer is delivered to the device.

- Power supplied to the transformer (i.e., input power) = primary current I_p × primary potential difference V_p
- Power delivered by the transformer (i.e., output power) = secondary current I_s × secondary potential difference V_s

So, if 100% efficiency is assumed:

$$\begin{array}{c}\text{primary potential} \\ \text{difference, } V_p\end{array} \times \begin{array}{c}\text{primary} \\ \text{current, } I_p\end{array} = \begin{array}{c}\text{secondary potential} \\ \text{difference, } V_s\end{array} \times \begin{array}{c}\text{secondary} \\ \text{current, } I_s\end{array}$$

Worked example

A step-down transformer is used to step down an alternating potential difference of 230 V to 12 V to supply electricity to a 12 V, 2.5 A lamp. Calculate the current in the primary coil of the transformer when the device is switched on. Assume that the transformer is 100% efficient.

Solution

Make I_p the subject of the equation $V_p \times I_p = V_s \times I_s$, and substitute in the known values.

$$I_p = \frac{V_s \times I_s}{V_p} = \frac{12 \times 2.5}{230} = 0.13\,\text{A}$$

Power and the grid potential difference

The electrical power supplied to any appliance depends on the appliance's current and its potential difference. To supply a specific amount of power, the current can be lowered if the potential difference is raised. This is what a step-up transformer does at a power station in the grid system.

The heating effect in a resistor or a wire is proportional to the square of the current. By increasing the grid potential difference, the current through the grid cables is reduced. So the heating effect is smaller and less power is wasted. For example, if the potential difference is made four times bigger by using a step-up transformer:

● the current needed to transfer the same amount of power through the grid cables is four times smaller

● so the heating effect in the grid cables is 16 times smaller (because the heating effect is proportional to the square of the current)

● so the power wasted due to the heating effect of the current in the grid cables is 16 times less than if the potential difference had not been stepped up.

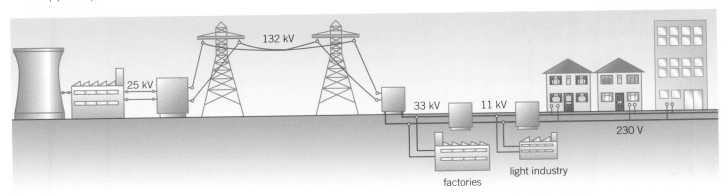

Figure 2 *The grid system*

1 A transformer with 60 turns in the secondary coil is used to step a potential difference of 120 V down to 6 V. Calculate the number of turns on the primary coil. [2 marks]

2 A transformer with a secondary coil of 100 turns is to be used to step a potential difference down from 240 V to 12 V.
a Calculate the number of turns on the primary coil. [2 marks]
b A 12 V, 36 W bulb is connected to the secondary coil. Assume the transformer is 100% efficient. Calculate the current in:
i the bulb   **ii** the primary coil. [2 marks]

3 A 6 V a.c. power supply for an electronic keyboard consists of a transformer that steps the mains potential difference from 230 V to 6.0 V.
a The primary coil has 1150 turns. Calculate the number of turns on the secondary coil. [2 marks]
b The transformer is designed to deliver 1.0 A at 6.0 V. Calculate the primary current needed to deliver this amount of power. Assume the transformer is 100% efficient. [2 marks]

Key points

● The transformer equation is:

$$\frac{\text{primary potential difference } V_p}{\text{secondary potential difference } V_s} = \frac{n_p}{n_s}$$

where n_p = number of primary turns, and n_s = number of secondary turns.

● For a step-down transformer, n_s is less than n_p.
For a step-up transformer, n_s is greater than n_p.

● For a 100% efficient transformer:
$$V_p \times I_p = V_s \times I_s$$
where I_p = primary current, and I_s = secondary current.

● A high grid potential difference reduces the current that is needed, so it reduces power loss and makes the system more efficient.

P15 Electromagnetism

Summary questions

1 Two identical bar magnets are placed end-to-end on a sheet of paper on a table with a gap between them with unlike poles facing each other.

  a Draw the arrangement and the pattern of the magnetic field lines in the gap. [1 mark]

  b A plotting compass is placed in the gap at equal distance from the two magnets at a short distance from the midpoint of the gap. On your drawing, show the plotting compass in this position and show the direction in which it points. [1 mark]

2 Figure 1 shows an electric bell. Explain why the bell rings continuously when a battery is connected to its terminals.

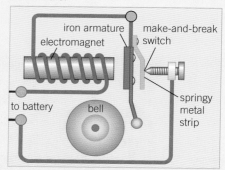

Figure 1 *An electric bell* [5 marks]

3 a 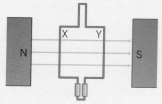 Figure 2 shows a rectangular coil of wire in a magnetic field viewed from above. When a direct current passes clockwise around the coil, a downward force acts on side **X** of the coil.

Figure 2

  i Write the direction of the force on side **Y** of the coil. [1 mark]

  ii Describe the force on each side of the coil parallel to the magnetic field lines. [1 mark]

  iii Describe the effect of the forces on the coil. [1 mark]

  b A vertical wire is in a horizontal magnetic field of magnetic flux density 55 mT. The length of the wire in the field is 45 mm. Calculate the force on the wire when a current of 6.5 A is in the wire. [2 marks]

4 a i Ⓗ The arrangement shown in the figure for Question **3** could be used to generate a direct current if the coil is made to spin and the split-ring commutator is connected to a lamp. Draw a graph to show how the current would vary with time if the coil is turned steadily. [2 marks]

  ii Suggest how you would modify the arrangement shown in the figure for Question **3** to generate an alternating current. [1 mark]

  b Explain why a transformer does not work on direct current. [2 marks]

5 a Cables at a potential difference of 100 000 V are used to transfer 1 000 000 W of electrical power in a grid system.

  i Calculate the current in the cable. [2 marks]

  ii If the potential difference had been 10 000 V, calculate how much current would be needed to transfer the same amount of power. [1 mark]

  b Ⓗ Explain why power is transmitted through the National Grid at a high potential difference rather than a low potential difference. ✐ [3 marks]

6 Ⓗ A transformer has 50 turns in its primary coil and 500 turns in its secondary coil. It is to be used to light a 120 V, 60 W bulb connected to the secondary coil. Assume the transformer is 100% efficient.

  a Calculate the primary potential difference. [2 marks]

  b Calculate the current in the bulb. [2 marks]

  c Calculate the current in the primary coil. [3 marks]

7 Ⓗ A transformer has 3000 turns on its primary coil. An alternating potential difference of 230 V is to be connected to the primary coil, and a 12 V bulb is to be connected to the secondary coil.

  a Calculate the number of turns the secondary coil should have. [2 marks]

  b Calculate what the current through the primary coil would be if the current through the lamp is to be 3.0 A. Assume the transformer is 100% efficient. [3 marks]

Practice questions

01.1 **Figure 1** shows five situations A, B, C, D, and E where bars of different materials are placed next to each other. Some of the bars are magnets and some are non-magnets.

Figure 1

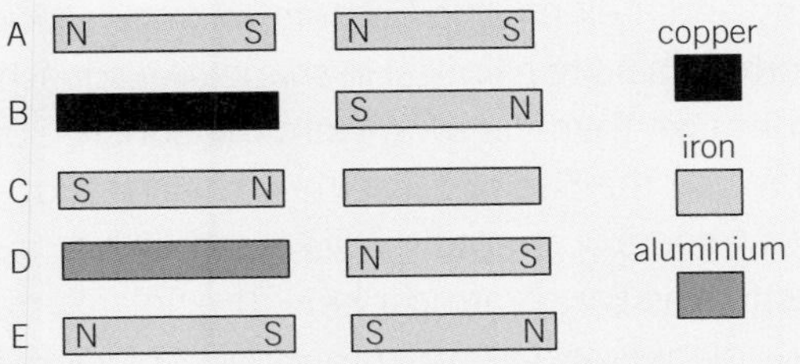

A [N   S] [N   S]   copper ▮

B [▬▬▬▬] [S   N]

C [S   N] [▬▬▬▬]   iron ▯

D [▬▬▬▬] [N   S]

E [N   S] [S   N]   aluminium ▯

State for each situation whether the two bars are repelled, attracted or nothing happens.   [5 marks]

01.2 Complete the sentences using the correct words from the box.

| increases | an induced | loses |
| a permanent | charges | changes |

An induced magnet _____ its magnetism when moved away from a permanent magnet. A soft iron bar becomes _____ magnet when placed next to another magnetic material.
[2 marks]

01.3 A small bar magnet is suspended by thin string so that it is free to spin. The south pole of the magnet is marked. When the magnet comes to rest, describe the direction the ends of the magnet will point to and state why.   [2 marks]

02 A loudspeaker is made by winding coils of insulated copper wire around a paper tube, fitted loosely over the south pole of a circular magnet.

Figure 2

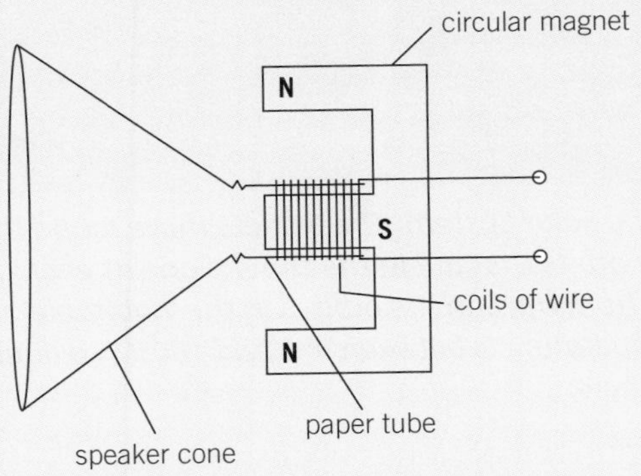

circular magnet
N
S
coils of wire
paper tube
speaker cone

02.1 Give one reason why the copper wire is covered in insulation.
[1 mark]

02.2 Ⓗ Explain how the speaker produces a sound signal when a changing alternating current flows into the coil of wire.   [4 marks]

02.3 Ⓗ Calculate the magnitude of the force acting on the copper wire due to the magnetic field. The total length of the coiled wire is 20 m and a current of 45 mA flows through the wire. The magnetic flux density of the magnet is 3.5×10^{-2} T.   [3 marks]

03 A teacher demonstrated how changing the current in a solenoid affects the magnetic force around the solenoid. He used the equipment in **Figure 3**.

Figure 3

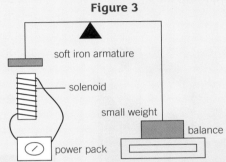

soft iron armature
solenoid
small weight
balance
power pack

03.1 Describe how when a current flows through the solenoid the small weight is raised off the balance.
[3 marks]

03.2 The teacher changed the current through the solenoid and recorded the reading on the balance. Draw a graph of the results.   [3 marks]

Current in A	0	2	4	6	8	10
Weight on balance in N	5	2.5	1.2	0.8	0.5	0.0

03.3 What was the weight of the small weight on the balance.   [1 mark]

03.4 Name three factors that were kept constant.
[3 marks]

03.5 Name the independent variable.   [1 mark]

03.6 Name the dependent variable.   [1 mark]

03.7 Suggest a relationship between the current through the solenoid and the magnetic field produced.   [2 marks]

04 Ⓗ The coil of an electric motor is held between the poles of a magnet and is connected to a switch and a 9 volt battery. When the switch is closed, the left-hand side of the coil experiences an upward force that makes the coil rotate in a clockwise direction.

04.1 Give two ways of making the coil rotate in an anti-clockwise direction   [2 marks]

04.2 Give two ways of increasing the force on the coil in the electric motor.   [2 marks]

Learning objectives

After this topic, you should know:

- how the Solar System formed
- what is meant by a protostar
- how energy is released inside the Sun
- why the Sun is stable.

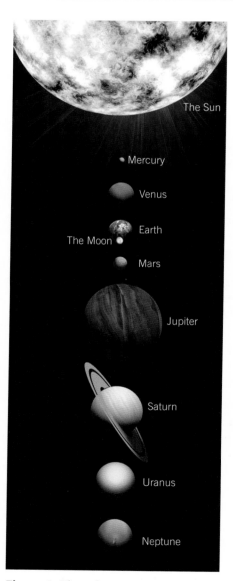

Figure 1 *The solar system*

How the Solar System formed

When you look at the night sky you sometimes see unexpected objects such as comets or meteors. The Solar System contains lots of objects as well as the Sun, the planets, and their moons. Comets are frozen rocks that move around the Sun in orbits that are elliptical in shape (like squashed circles). These elliptical orbits take them far away from the Sun. You only see them when they return near the Sun because then they heat up so much that they emit light. Meteors or shooting stars are small bits of rocks that burn up when they enter the Earth's atmosphere. The Solar System also includes minor (dwarf) planets and asteroids hundreds of kilometres in size orbiting the Sun, mostly between the orbits of Mars and Jupiter.

The birth of a star

The Sun formed billions of years ago from clouds of dust and gas pulled together by gravitational attraction (Figure 2).

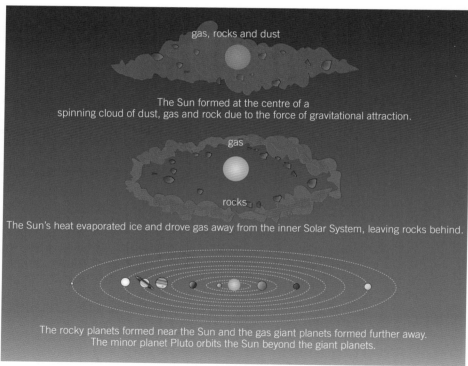

gas, rocks and dust

The Sun formed at the centre of a
spinning cloud of dust, gas and rock due to the force of gravitational attraction.

gas

rocks

The Sun's heat evaporated ice and drove gas away from the inner Solar System, leaving rocks behind.

The rocky planets formed near the Sun and the gas giant planets formed further away.
The minor planet Pluto orbits the Sun beyond the giant planets.

Figure 2 *The formation of the Solar System. The planets move round the Sun in circular or almost-circular orbits that are in the same plane as each other. The Earth is the third planet from the Sun. Its orbit is in the 'habitable' zone round the Sun where the temperature is between 0°C and 100°C, so liquid water can exist on a planet there.*

All stars including the Sun form out of clouds of dust and gas.

- The particles in the clouds are pulled together by their own gravitational attraction so the particles speed up. The clouds merge together and become more and more concentrated to form a **protostar**, which is a star-to-be.

- As a protostar becomes denser, its particles speed up more and collide more, so its temperature increases and it gets hotter. The process transfers energy from the protostar's gravitational potential energy store to its thermal energy store. If the protostar becomes hot enough, the nuclei of hydrogen atoms fuse together, forming helium nuclei. Energy is released in this fusion, so the protostar gets hotter and brighter and starts to shine. A star is born!
- Objects can form that are too small to become stars. These kinds of objects can be attracted by a protostar to become planets orbiting the star.

Shining stars

Stars such as the Sun radiate energy because of hydrogen fusion in the core. They are called **main sequence** stars because this is the main stage in the life of a star. Such stars can maintain their energy output for millions of years until there are no more hydrogen nuclei left to fuse together.

- Energy released in the core (the central part of the star) keeps the core hot, so the process of fusion continues. Radiation such as gamma radiation flows out steadily from the core in all directions.
- The star is stable because the forces within it are balanced. The force of gravity acts inwards trying to make the star contract. This is balanced by the outward force of the radiation from nuclear fusion in its core trying to make the star expand. These forces stay in equilibrium until most of the hydrogen nuclei in the core have been fused together to form helium nuclei.

Planet Earth

The heaviest known natural element is uranium. It has a half-life of 4500 million years. The presence of uranium on the Earth is evidence that the Solar System must have formed from the remnants of a supernova.

Elements such as plutonium are heavier than uranium. Scientists can make these elements by bombarding heavy elements such as uranium with high-speed neutrons. They would have been present in the debris that formed the Solar System. Elements heavier than uranium formed then have long since decayed.

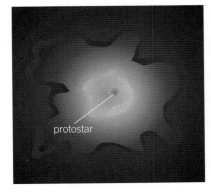

Figure 3 *The birth of a star*

1 a Explain why a comet cannot be seen when it is far away from the Sun. [3 marks]
   b Write one difference and one similarity between a comet and an asteroid. [2 marks]

2 a Write which planet in the Solar System is
      i the largest   ii nearest the Sun. [2 marks]
   b Explain why the Earth is likely to be the only planet in the Solar System where liquid water is always present on the surface. [2 marks]

3 The Earth's orbit is almost circular. State and explain two ways in which conditions on the Earth would differ if the Earth's orbit was like the orbit of a comet. [4 marks]

4 a Describe how the Sun formed from dust and gas clouds in space. ✏ [5 marks]
   b Explain what a main sequence star is and why the Sun is a main sequence star. [3 marks]

Key points

- The Solar System formed from gas and dust clouds that gradually became more and more concentrated because of gravitational attraction.
- A protostar is a concentration of gas and dust that becomes hot enough to cause nuclear fusion.
- Energy is released inside a star because of hydrogen nuclei fusing together to form helium nuclei.
- The Sun is stable because gravitational forces acting inwards balance the forces of nuclear fusion energy in the core acting outwards.

P16.2 The life history of a star

Learning objectives

After this topic, you should know:

- why stars eventually become unstable
- the stages in the life of a star
- what will eventually happen to the Sun
- what a supernova is.

Figure 2 *The Crab Nebula is the remnants of a supernova explosion that was observed in the 11th century. In 1987, a star in the southern hemisphere exploded and became the biggest supernova to be seen for four centuries. Astronomers realised that it was Sandaluk II, a star in the Andromeda galaxy millions of light years from Earth. If a star near the Sun exploded, the Earth would probably be blasted out of its orbit. You would see the explosion before the shock wave hit Earth*

When a star runs out of hydrogen nuclei to fuse together in its core, it reaches the end of its main-sequence stage. Its core collapses, and its outer layers swell out.

Stars about the same size as the Sun (or smaller) swell out, cool down, and turn red.

- These stars are now **red giants**. At this stage, helium and other light elements in the core fuse to form heavier elements.

- When there are no more light elements in the core, fusion stops, and no more radiation is released. Because of its own gravity, the star collapses in on itself. As it collapses, it heats up and turns from red to yellow to white. It becomes a **white dwarf**. This is a hot, dense, white star much smaller in diameter than it was before. Stars such as the Sun then fade out, go cold, and become **black dwarfs**.

Stars much bigger than the Sun end their lives after the main-sequence stage much more dramatically.

- These stars swell out to become **red supergiants**. They then collapse.

- In the collapse, the matter surrounding the star's core compresses the core more and more. Then the compression suddenly reverses in a cataclysmic explosion called a **supernova**. This event can outshine an entire galaxy for several weeks.

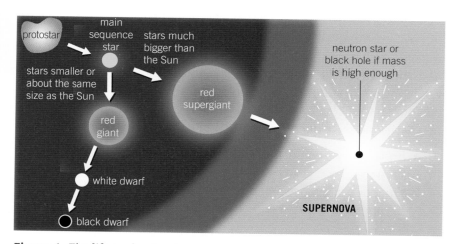

Figure 1 *The life cycle of a star*

The birthplace of the elements

Light elements are formed from fusion in stars. Stars such as the Sun fuse hydrogen nuclei (i.e., protons) into helium and similar small nuclei, including carbon nuclei. When the star becomes a red giant, it fuses helium and the other small nuclei into larger nuclei.

Nuclei larger than iron nuclei cannot be formed by this process because too much energy is needed.

Heavy elements are formed when a massive star collapses then explodes as a supernova. The enormous force of the collapse fuses small nuclei into nuclei bigger than iron nuclei. The explosion scatters the elements throughout the universe.

The debris from a supernova contains all the known elements, from the lightest to the heaviest. Eventually, new stars form as gravity pulls the debris together. Planets form from debris surrounding a new star. Because of this, these planets will be made up of all the known elements too.

The future of the Sun

The Sun is about 5000 million years old and will probably continue to shine for another 5000 million years.

The Sun will turn into a red giant bigger than the orbit of Mercury. By then, the human race will probably have long passed into history.

What's left after a supernova?

The explosion compresses the core of the star into a **neutron star**. This is an extremely dense object made up only of neutrons. If the star is massive enough, it becomes a **black hole** instead of a neutron star. The gravitational field of a black hole is so strong that nothing can escape from it. Not even light, or any other form of electromagnetic radiation, can escape a black hole.

1 **a** The list below shows some of the stages in the life of a star such as the Sun. Put the stages in the correct sequence.
   A white dwarf **B** protostar **C** red giant **D** main sequence. [1 mark]
   b i Name the stage in the above list that the Sun is at now. [1 mark]
   ii Describe what will happen to the Sun after it has gone through the above stages. [1 mark]

2 **a i** Write the force that makes a red supergiant collapse. [1 mark]
   ii Write the force that prevents a main sequence star from collapsing. [1 mark]
   b Explain why a white dwarf eventually becomes a black dwarf. [2 marks]

3 **a** Match each statement below with an element or elements in the list.
   helium   hydrogen   iron   uranium
   i Helium nuclei are formed when nuclei of this element are fused. [1 mark]
   ii This element is formed in a supernova explosion. [1 mark]
   iii Stars form nuclei of these two elements (and others not listed) by fusing smaller nuclei. [1 mark]
   b Describe two differences between a red giant star and a neutron star. [2 marks]

4 **a** Explain why all the uranium in the Earth has not decayed by now. [2 marks]
   b Plutonium-239 has a half-life of 24 000 years. It is formed in a nuclear reactor from uranium-238. Explain why plutonium-239 is not found naturally. [2 marks]

Figure 3 *M87 is a galaxy that spins so fast at its centre that it is thought to contain a black hole with a billion times more mass than the Sun*

Synoptic link

To remind yourself about the heavier elements and half-lives, look back at Topic P7.3 and Topic P7.5.

Study tip

Make sure you know how the heavier elements were formed and that they were *not* formed during the Big Bang.

Key points

- Stars become unstable when they have no more hydrogen nuclei that they can fuse together.
- Stars with about the same mass as the Sun: protostar → main-sequence star → red giant → white dwarf → black dwarf.
- Stars much more massive than the Sun: protostar → main-sequence star → red supergiant → supernova → neutron star (or black hole if enough mass).
- The Sun will eventually become a black dwarf.
- A supernova is the explosion of a red supergiant after it collapses.

P16.3 Planets, satellites, and orbits

Learning objectives

After this topic, you should know:

- what force keeps planets and satellites moving along their orbits
- Ⓗ the direction of the force on an orbiting body in a circular orbit
- Ⓗ how the velocity of a body in a circular orbit changes as the body moves around the orbit
- Ⓗ why an orbiting body needs to move at a particular speed for it to stay in a circular orbit.

The Earth orbits the Sun in an orbit that is almost circular. Most of the other planets orbit the Sun on orbits that are ellipses or slightly squashed circles. The Moon is a natural satellite that orbits the Earth on a circular orbit. Artificial satellites also, orbit the Earth. In each case, a body orbits a much bigger body. The force on the orbiting body is the force of gravitational attraction between it and the larger body.

Circular orbits

Figure 1 shows the force of gravity acting on a planet in a circular orbit around the Sun.

The force of gravity on the planet from the Sun acts towards the centre of the Sun. This force is the resultant force on the planet because no other forces act on it. The force is an example of a **centripetal force** because it acts towards the centre of the circle.

The direction of the planet's velocity (i.e., its direction of motion) is changed by this force. So it continues to orbit the Sun. The direction of motion of any planet (i.e., the direction of its velocity) in a circular orbit is *at right angles* to the direction of the force of gravity on it.

A planet in a circular orbit experiences acceleration towards the centre of the circle because the resultant force on it acts towards the centre of the circle. The acceleration is its change of velocity per second, and its change of velocity is directed towards the centre.

The speed of a planet in a circular orbit does not change, even though its velocity changes its direction. This is because the force on it is at right angles to its direction of motion. So no work is done by the force on the planet. So the kinetic energy and the speed of the planet do not change.

Into orbit

Satellites in orbits too close to the Earth gradually lose speed. This happens because of atmospheric drag if a satellite's orbit is in the Earth's upper atmosphere. If a satellite loses speed, it gradually spirals inwards until it hits the Earth's surface.

A satellite in a circular orbit above the Earth's atmosphere moves around the Earth at a constant height above the surface. The satellite is in a stable orbit. To stay in an orbit of a particular radius, the satellite has to move at a particular speed around the Earth. The same is true for a planet moving in a circular orbit around the Sun.

Imagine a satellite is launched from the top of a very tall mountain. Figure 2 shows what would happen if the satellite was launched too fast or too slow. If the launch speed is too slow, the satellite falls to the surface. If the launch speed is too high, the satellite flies off into space. At the correct speed, the satellite moves around the Earth in a circular orbit at a constant height and a constant speed.

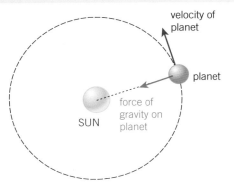

Figure 1 *A circular orbit*

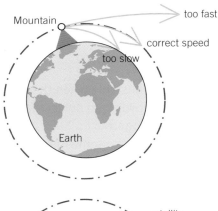

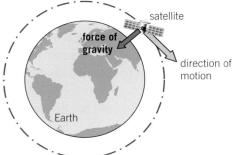

Figure 2 *Launching a satellite*

For any small body to stay in a circular orbit around another bigger body, the smaller body must move at a particular speed around the bigger body.

Using satellites

The further a satellite is from the Earth or a planet is from the Sun:

● the less the particular speed needed for it to stay in a circular orbit. The force of gravity on a satellite is weaker when the satellite is further from the Earth, and it does not need to travel as fast to stay in orbit. The same is true for a planet moving around the Sun.

● the longer the orbiting body takes to move around the orbit once. This is because the circumference of the orbit is bigger, and the orbiting body moves slower in order to remain in orbit. So the time for each complete orbit (= circumference ÷ speed) is longer.

If the speed of a satellite in a stable orbit changes, then the radius of the orbit has to change. For example, suppose a space vehicle above the Earth is in a circular orbit and its engines are used briefly to increase its speed. The vehicle moves out of its orbit and gains height. So its speed decreases as it moves into a higher orbit.

Communications satellites are usually in an orbit at about 36 000 kilometres above the equator with a period of 24 hours. They orbit the Earth in the same direction as the Earth's spin. So they stay above the same place on the Earth's surface as they go around the Earth. These kinds of orbits are described as geostationary.

Monitoring satellites are fitted with TV cameras pointing to the Earth. Their uses include weather forecasting, and monitoring the environment. Monitoring satellites are in much lower orbits than geostationary satellites and they orbit the Earth once every two or three hours.

1 A satellite is in a circular orbit around the Earth.
   a Write the direction of **i** the force of gravity on the satellie
      ii its acceleration. [2 marks]
   b Explain why its velocity continually changes even though its speed is constant. [3 marks]

2 **H** GPS satellites orbit the Earth about once every 12 hours. Write and explain whether or not a GPS satellite orbits the Earth above or below
   a a geostationary satellite [3 marks]
   b a weather satellite that has a period of two hours. [2 marks]

3 **H** Light from the Sun takes about 3 minutes to reach Mercury, about eight minutes to reach Earth, and about 40 minutes to reach Jupiter. Jupiter takes about 11 years to orbit the Sun, whereas Mercury takes about 3 months. Use this information to determine which of the three planets travels **a** slowest [5 marks]
   b fastest in its orbit. Explain your reasoning. **✓** [2 marks]

4 **H** Mercury orbits the Sun about 4 times each year. The Earth moves on its orbit at a speed of 30 km/s. Use the information here and in Q3 to estimate the speed of Mercury in km/s on its orbit. [3 marks]

Figure 3 *A satellite in orbit*

Key points

● The force of gravity between:
   ▪ a planet and the Sun keeps the planet moving along its orbit
   ▪ a satellite and the Earth keeps the satellite moving along its orbit.
● **H** The force of gravity on an orbiting body in a circular orbit is towards the centre of the circle.
● As a body in a circular orbit moves around the orbit:
   ▪ the magnitude of its velocity (its speed) does not change
   ▪ the direction of its velocity continually changes and is always at right angles to the direction of the force, so
   ▪ **H** it experiences an acceleration towards the centre of the circle.
● **H** To stay in orbit at a particular distance, a small body must move at a particular speed around a larger body.

P16.4 The expanding universe

Learning objectives

After this topic, you should know:

- what is meant by the red-shift of a light source
- how red-shift depends on speed
- how people know that the distant galaxies are moving away from Earth
- why people think the universe is expanding.

Red-shift

The Earth is the third planet from the Sun. The Sun is a star on the outskirts of the Milky Way galaxy. A galaxy is an enormous collection of stars that stay together because of the force of gravity between them. The Milky Way galaxy contains about 100 000 million stars. Its size is about 100 000 light years across. This means that light takes 100 000 years to travel across it. But it's just one of billions of galaxies in the universe. The furthest known galaxies are about 13 000 million light years away.

Figure 1 *In places where there is little light pollution you can see the Milky Way galaxy*

People can find out lots of things about stars and galaxies by studying the light from them. You can use a prism to split the light into a spectrum. The wavelength of light increases across the spectrum from blue to red. You can tell from its spectrum if a star or galaxy is moving towards Earth or away from Earth. This is because:

- the light waves are stretched out if the star or galaxy is moving away from you. The wavelength of the waves is increased. This is called a **red-shift** because the spectrum of light is shifted towards the red part of the spectrum.

- the light waves are squashed together if the star or galaxy is moving towards you. The wavelength of the waves is reduced. This is called a blue-shift because the spectrum of light is shifted towards the blue part of the spectrum.

The dark spectral lines shown in Figure 2 are caused by absorption of light by specific atoms such as hydrogen that make up a star or galaxy. The position of these lines tells you if there is a shift, and if there is a shift, whether it is a red-shift or a blue-shift.

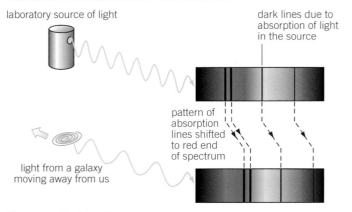

laboratory source of light

dark lines due to absorption of light in the source

pattern of absorption lines shifted to red end of spectrum

light from a galaxy moving away from us

Figure 2 *Red-shift*

The bigger the shift, the more the waves are squashed together or stretched out. So the faster the star or galaxy must be moving towards or away from you. In other words:

The faster a star or galaxy is moving (relative to you), the bigger the shift is.

Expanding universe

In 1929, the astronomer Edwin Hubble discovered that:

1 the light from distant galaxies was red-shifted

2 the further a galaxy is from Earth, the bigger its red-shift is.

He concluded that:

● the distant galaxies are moving away from Earth (i.e., receding)

● the greater the distance a galaxy is from Earth, the greater the speed at which it is moving away from Earth (its speed of recession).

Why should the distant galaxies be moving away from Earth? Humans have no special place in the universe, so all the distant galaxies must be moving away from each other. In other words, *the whole universe is expanding*.

1 a State whether each of the following is approaching the Earth or receding from the Earth:
  i a distant galaxy [1 mark]
  ii a galaxy that shows a blue-shift in its light. [1 mark]
 b The Sun is in the Milky Way galaxy. Astronomers think that the Andromeda Galaxy will eventually collide with the Milky Way galaxy. Write the evidence that astronomers have to support this prediction. [1 mark]

2 a Put these objects in order of increasing size:
  Andromeda galaxy   Earth   Sun   universe   [1 mark]
 b Quasars are astronomical objects much smaller in appearance than galaxies, and they have red-shifts of the same order of magnitude as the distant galaxies.
  i Write the part of the above statement that makes you conclude that quasars are much further away than nearby galaxies. [1 mark]
  ii Quasars can be as bright as a distant galaxy even though they are much smaller. Use this information to describe the power of the radiation emitted by a quasar. [2 marks]

3 Galaxy X has a larger red-shift than galaxy Y.
 a Explain what is meant by a red-shift. [2 marks]
 b Write which galaxy, X or Y, is:
  i nearer to Earth [1 mark] **ii** moving away faster. [1 mark]
4 Some of the nearest galaxies to Earth have different red shifts, and some have different blue shifts. Use this information to describe these galaxies. [3 marks]

Key points

● The red-shift of a distant galaxy is the shift to longer wavelengths (and lower frequencies) of the light from the galaxy because it is moving away from you.

● The faster a distant galaxy is moving away from you, the greater its red-shift is.

● All the distant galaxies show a red-shift. The further away a distant galaxy is from you, the greater its red-shift is.

● The distant galaxies are all moving away from you because the universe is expanding.

P16.5 The beginning and future of the Universe

Learning objectives

After this topic, you should know:

- what the Big Bang theory of the universe is
- why the universe is expanding
- what cosmic microwave background radiation is
- what evidence there is that the universe was created in a Big Bang.

Go further!

In 2016, physicists detected gravitational waves for the first time ever. These waves shake space and time and they were predicted by Einstein about 100 years ago. A new type of instrument was used to detect tiny gravitational wave ripples at two places on the Earth, 3000 km apart. When both instruments shook at the same time, scientists realised they had detected gravitational waves, which they then traced back to a distant merger of two black holes more than a billion light years away. Many discoveries about space were made by Galileo, and others after Galileo first used a telescope to observe light waves from the stars. Many more discoveries will be made now that astronomers can observe the Universe using gravitational waves.

The universe is expanding, but what is making it expand? The **Big Bang theory** was put forward as a model to explain the expansion. This says that:

- the universe is expanding after exploding suddenly (the Big Bang) from a very small and extremely hot and dense region
- space, time, and matter were created in the Big Bang.

Many scientists disagreed with the Big Bang theory. They put forward an alternative theory called the Steady State theory. These scientists said that the galaxies are being pushed apart. They thought that this is caused by matter entering the universe through 'white holes' (the opposite of black holes).

Which theory is weirder – everything starting from a Big Bang, or matter leaking into the universe from outside? Until 1965, most people supported the Steady State theory.

Evidence for the Big Bang

Scientists had two conflicting theories about the evolution of the universe: it was in a Steady State or it began at some point in the past with a Big Bang. Both theories could explain why *distant* galaxies are moving apart, so scientists needed to find some way of deciding which theory was correct. They worked out that if the universe began in a Big Bang, then high-energy electromagnetic radiation should have been produced very soon after the universe began. This radiation would have stretched as the universe expanded and become lower-energy radiation. Scientists thought up experiments to look for this trace energy as extra evidence for the Big Bang model.

It was in 1965 that scientists first detected microwaves coming from every direction in space. The existence of this **cosmic microwave background radiation (CMBR)** can be explained only by the Big Bang theory.

Cosmic microwave background radiation

Cosmic microwave background radiation was created as high-energy gamma radiation just after the Big Bang. It has been travelling through space since then. As the universe has expanded, it has stretched out to longer and longer wavelengths and is now microwave radiation. It has been mapped out using microwave detectors on the Earth and on satellites.

The future of the universe

Will the universe expand forever? Or will the force of gravity between the distant galaxies stop them moving away from each other? The answer to this question depends on the total mass of the galaxies, how much matter is between them, and how much space they take up – in other words, the density of the universe.

Astronomers think that the stars in a galaxy account for only a small percentage of the total mass of a galaxy. They know that galaxies would spin much faster if their stars were the only matter in galaxies. The missing mass is called **dark matter** because it can't be seen. Its presence means that the average density of the universe is much bigger than if dark matter didn't exist.

- If the density of the universe is less than a particular amount, it will expand forever. The stars will die out, and so will everything else as the universe heads for a Big Yawn!
- If the density of the universe is more than a particular amount, it will stop expanding and go into reverse. Everything will head for a Big Crunch!

Observations since 1998 of supernovae in distant galaxies suggest that the distant galaxies are accelerating away from each other. These observations have been checked and confirmed by other astronomers. So astronomers have concluded that the expansion of the universe is accelerating. It could be that the universe is in for a Big Ride followed by a Big Yawn.

The discovery that the distant galaxies are accelerating is puzzling astronomers. Scientists think some unknown source of energy, called dark energy, must be causing this accelerating motion. The only known force on the distant galaxies, the force of gravity, can't be used to explain dark energy, because it's an attractive force and so it acts against the outward motion of the distant galaxies away from each other.

There is still a lot about the universe, for example dark mass and dark energy, that astronomers don't understand. New telescopes and technologies will help to improve humans' understanding of the universe – and will certainly create more questions for scientists to investigate.

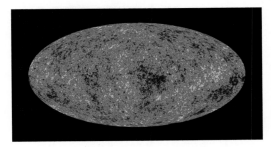

Figure 1 *A microwave image of the universe from the Cosmic Background Explorer satellite*

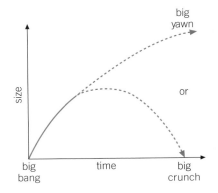

Figure 2 *The future of the universe?*

1 a Describe the Big Bang theory of the universe. [1 mark]
   b Explain why many astronomers did not support the Big Bang theory when it was first proposed. [2 marks]
   c Explain the significance of the discovery of cosmic microwave background radiation. [1 mark]

2 Put the following events **A–D** in the correct time sequence:
   A the distant galaxies were created
   B cosmic microwave background radiation was first detected
   C the Big Bang happened
   D the expansion of the universe began. [1 mark]

3 a Explain why astronomers think that the expansion of the universe is accelerating. [1 mark]
   b Describe what would have been the effect on the expansion of the universe if its density had been greater than a particular value. [2 marks]

4 a Hubble estimated that the speed of recession of a distant galaxy increases by about 22 km/s for every extra distance of one million light years. A distant galaxy is receding away at a speed of 150 000 km/s. Estimate how far away the galaxy is in light years. [2 marks]
   b The Milky Way galaxy is about 100 000 light years across. Make an order of magnitude estimate of the ratio of the distance to the galaxy in **a** to the distance across the Milky Way. [1 mark]

Study tip

Make sure you know more about the Big Bang theory than just 'the universe started with a big bang'.

Key points

- The universe started with the Big Bang, which was a massive explosion from a very small point.
- The universe has been expanding ever since the Big Bang.
- Cosmic microwave background radiation (CMBR) is electromagnetic radiation that was created just after the Big Bang.
- The red shifts of the distant galaxies provide evidence that the universe is expanding. CMBR can be explained only by the Big Bang theory.

Summary questions

1 **a** The stages in the development of the Sun are listed below. Put the stages in the correct sequence.
   A dust and gas   **D** red giant
   B main sequence   **E** white dwarf
   C protostar   [1 mark]
   b i Describe what will happen to the Sun after its present stage.   [3 marks]
   ii Describe what will happen to a star that has much more mass than the Sun.   [3 marks]
   c (H) The Earth moves around the Sun in a circular orbit at a constant speed. Explain why the velocity of the Earth changes and why it accelerates as it moves around the Sun.   [3 marks]

2 **a i** Define a supernova.   [1 mark]
   ii Explain how you could tell the difference between a supernova and a distant star like the Sun at present.   [1 mark]
   b i Define a black hole.   [1 mark]
   ii Describe what would happen to stars and planets near a black hole.   [1 mark]
   iii Define a neutron star, and describe how it is formed.   [2 marks]

3 **a i** Write the element that, as well as hydrogen, was formed in the early universe.   [1 mark]
   ii Write which of the two elements in part **i** is formed from the other one in a star.   [1 mark]
   b i Write which *two* of the elements listed below are *not* formed in a star that gives out radiation at a steady rate.   [1 mark]
       carbon   iron   lead   uranium
   ii Explain how the two elements given in your answer to part **i** would have been formed. [2 marks]
   iii Explain how you know that the Sun formed from the debris of a supernova.   [2 marks]

4 **a** Put these events in the correct sequence with the earliest event first:
   1 cosmic microwave background radiation was released
   2 hydrogen nuclei were first fused to form helium nuclei
   3 the Big Bang took place
   4 neutrons and protons formed.   [1 mark]
   b The stars were formed from clouds of dust and gas.
   i Write the force that can cause dust and gas particles to attract each other.   [1 mark]

   ii Write where the energy that heated the stars came from.   [2 marks]
   iii The stars in a galaxy revolve about the centre of the galaxy. Explain why the stars in a galaxy do not pull each other into a large single massive object at the centre.   [2 marks]

5 Light from a very distant galaxy has a change of wavelength because of the motion of the galaxy.
   a i Write whether this change of wavelength is an increase or a decrease.   [1 mark]
   ii Explain why this effect is called a red-shift. [3 marks]
   iii Explain what the change of wavelength tells you about the motion of the galaxy.   [2 marks]
   b Light from a particular nearby galaxy is found to have undergone a blue-shift because of the motion of the galaxy. Explain what this tells you about the motion of this galaxy.   [1 mark]
   c i Edwin Hubble discovered that the further a distant galaxy is from Earth, the greater the red-shift of the light from it. Explain what this tells you about the universe.   [1 mark]
   ii Give the crucial observational evidence that led scientists to accept the Big Bang theory of the universe.   [1 mark]

6 **a** Galaxy **A** is further from us than galaxy **B**.
   i Write which galaxy, **A** or **B**, produces light with a greater red-shift.   [1 mark]
   ii Galaxy **C** gives a bigger red-shift than galaxy **A**. Describe the distance to galaxy **C** compared with galaxy **A**.   [1 mark]
   b All the distant galaxies are moving away from each other.
   i Explain what this tells you about the universe.   [2 marks]
   ii Explain what it tells you about your place in the universe?   [1 mark]

7 Figure 1 shows two galaxies **X** and **Y**, which have the same diameter.
   a Write which galaxy, **X** or **Y**, is further from Earth. Give a reason for your answer.   [2 marks]
   b Write which galaxy, **X** or **Y**, produces the larger red-shift.   [3 marks]

Figure 1

Practice questions

01.1 One theory for the origin of the universe is that it began from a single point.
One of the boxes gives the correct name of the theory.
Choose the correct box. [1 mark]

Red Bang
Big Bang
Loud Bang

01.2 Observation of distant galaxies shows that the universe is expanding. Evidence for this is the light from the galaxies appears stretched. Give the name of this evidence. [1 mark]

01.3 Complete the sentence using the correct words from the box.

the Sun   all directions   the centre of the Earth

Cosmic background radiation has been found in the universe. It appears to come from _____. [1 mark]

02 In 1929 Edwin Hubble investigated the light from distant galaxies. He stated that distant galaxies were moving away from us. **Figure 1** shows a graph of the Hubble Data.

Figure 1

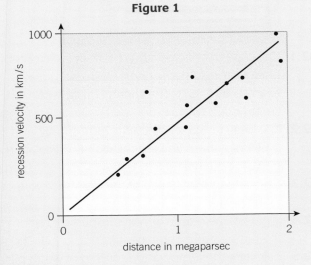

02.1 State what is meant by a distance of 1 light year. [1 mark]

02.2 Describe the relationship between the recession velocity of a galaxy and the distance from the Earth. [2 marks]

02.3 In 1929 Hubble used a large reflective telescope and calculated the Hubble Constant to determine the distance of galaxies from the Earth.
In 2013 NASA researchers surveyed more than 125 000 galaxies and claim to have measured the Hubble Constant with an uncertainty of less than 5%. Suggest two reasons why modern researchers can claim to be more accurate in measuring the Hubble Constant. [2 marks]

03.1 Describe how the Sun was formed and became a star. You should include details of how the Sun releases energy. [3 marks]

03.2 Give two reasons why the Sun is a main sequence star and is stable. [2 marks]

03.3 A star many times bigger than the Sun will eventually complete its life cycle and become a black hole. Explain what a black hole in space is. [2 marks]

03.4 Complete the sentence.
Elements heavier than iron are formed in a _____. The explosion of this massive star distributes _____ throughout the _____. [3 marks]

04 **Table 1** shows some information about the Earth, the Moon and satellites.

Table 1

Name	Mass in kg	Altitude in km	Orbit time in hours	Speed in km/h
Earth	5.98×10^{24}	0	0	0
Moon	7.45×10^{22}	384 000	672	3 600
GPS satellite	2.2×10^{2}	20 000	12	13 900
weather satellite	2.4×10^{2}	36 000	24	11 100
spy satellite	2.8×10^{2}	705	1.45	27 500

04.1 Evaluate the differences and similarities between planets, moons and artificial satellites. Use the information given in **Table 1**. [6 marks]

Paper 1 questions

01 A student read that all hot water tanks should be insulated to reduce the rate of energy transfer to the surroundings. She investigated the rate of energy transfer from a beaker of hot water using a series of tests. For each test, she used an insulation material of different thickness.

Figure 1

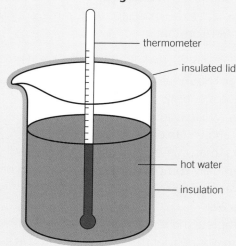

01.1 Give the independent and dependent variables in the investigation. [2 marks]

01.2 Suggest two things that should be kept constant. [2 marks]

01.3 State what resolution thermometer the student should choose. [1 mark]

01.4 The student plotted her results on graph paper. Copy the graph below and draw the line of best fit. [1 mark]

Figure 2

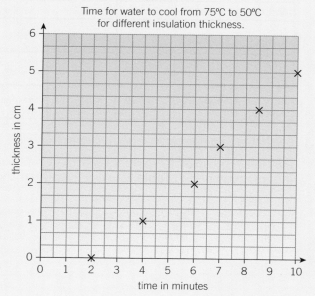

01.5 State the time for the water to cool from 75°C to 50°C without insulation. [1 mark]

01.6 Suggest the relationship between the thickness of insulation and the rate of energy transfer from a hot water tank. [2 marks]

01.7 Suggest one improvement that could be made to the investigation. [1 mark]

02 A warm plate stacker is used in a hotel. The two identical springs extend as plates are added, and they retract as plates are removed from the stack. The plates are kept warm by the electric heater at the bottom.

Figure 3

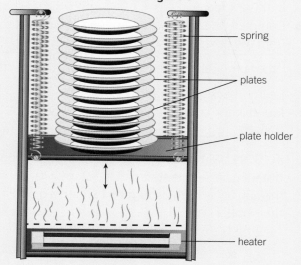

02.1 Calculate the energy stored in the elastic potential energy store of the springs when each spring extends by 0.35 m. The spring constant of each spring is 225 N/m. [2 marks]

02.2 The heater is connected to a 230 V supply, and has 2.17 A of current flowing through it. Calculate the power of the heater. [2 marks]

02.3 Calculate the resistance of the resistance wire used in the heater. [2 marks]

02.4 Suggest one reason why it is important to control the temperature of the heater. [1 mark]

02.5 Calculate the temperature of 15 kg of warm plates after 294 000 joules of energy are supplied to the heater. The plates start at a temperature of 15 °C. The specific heat capacity of the plate material is 980 J/kg°C. [3 marks]

03.1 The graph shows how the number of nuclei in a sample of the radioactive isotope plutonium-238 changes with time.

Figure 4

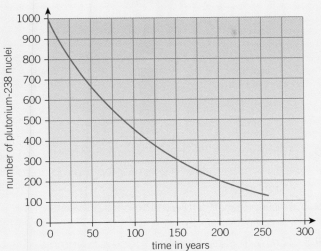

Copy and use the graph to find the half-life of plutonium-238.

Show clearly on the graph how you obtain your answer. [2 marks]

03.2 Plutonium-238 decays by emitting alpha particles. What is an alpha particle? [1 mark]

03.3 Plutonium-238 is highly dangerous. A tiny amount taken into the body is enough to kill a human. Plutonium-238 is unlikely to cause any harm if it is outside the body but is likely to kill if it is inside the body.

Explain why. [3 marks]

AQA, 2009

04 Some elements have radioactive isotopes.

04.1 Describe the similarities and differences between the structure of a radioactive isotope and a stable isotope of the same element. [3 marks]

04.2 Copy and complete the diagram below to show the radioactive decay of plutonium-238. [3 marks]

238   Y   4

  Plutonium ⟶ Uranium + Z

X   92   2

05 A hill farmer has used a mountain stream to build a system that generates electricity.

05.1 Give the name of the type of energy resource the farmer is using. Choose one word from the box. [1 mark]

| Geothermal   Water   Hydroelectric   Uphill |

05.2 Evaluate the different methods of generating electricity during the day. You should include environmental issues, reliability and costs. [6 marks]

	Coal	Gas	Nuclear	Wind	Tidal
Cost (£/MWh)	100–155	60–130	80–105	150–210	155–390
Start up Time (hours)	18 hours	3 hours	70 hours	variable	12 hours

06 Energy can be stored or transferred.

06.1 Complete the sentences below using words from the box.

| gravitational potential   kinetic   thermal chemical   elastic potential |

An aeroplane's _____ energy store increases as it gains height.

A train's _____ energy store increases as it accelerates.

The _____ energy store of the brakes in a car increases as the car slows down.

A gas boiler uses energy from its _____ energy store to boil water. [4 marks]

06.2 A child sits on a swing and is given a push.

Figure 5

Describe the energy transfers and changes to energy stores as the swing moves from right to left [3 marks]

06.3 Suggest one reason why the swing eventually comes to a stop [1 mark]

07 A group of students build a bird feeder of length 0.2 m and mass 1.3 kg and attach it to a rope. The rope is used to pull the feeder up a tree and then lower the feeder to refill it with seeds.

Figure 6

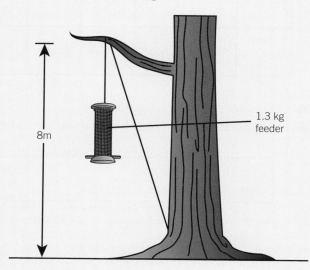

1.3 kg feeder

8m

07.1 Calculate the weight of the bird feeder. [2 marks]
Gravitational field strength on Earth is 9.8 N/kg.

07.2 Calculate the increase of the gravitational potential energy store of the bird feeder when it is pulled up to the branch. [2 marks]

07.3 Ⓗ A student lets go of the rope as the bird feeder reaches the branch. Calculate the velocity of the bird feeder as it hits the ground. [4 marks]

07.4 The table shows the frequency of refilling the feeder when the feeder is suspended at different heights.

Refills per week	2	3	5	5	2
Height of feeder	2 m	4 m	6 m	7 m	8 m

Draw conclusions about the relationship between the height of the feeder and the number of refills needed. [3 marks]

08 Explain in terms of particles how a bicycle pump increases the temperature of the air being pumped into a tyre and why the tyre pressure increases. [5 marks]

09 A gardener replaces the cable on a plastic electric lawnmower. The new cable from the lawnmower is connected to 3-pin plug.

Figure 7

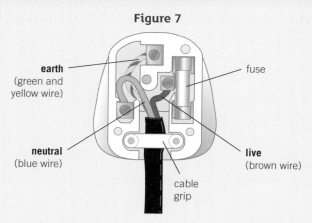

earth
(green and yellow wire)

fuse

neutral
(blue wire)

live
(brown wire)

cable grip

09.1 Comment on whether the gardener needed to use a three-core cable to connect the 3-pin plug to the lawnmower. [1 mark]

09.2 Explain how the fuse in the plug prevents the cables from overheating. [2 marks]

09.3 The instruction handbook for the lawnmower states that the lawn mower is double insulated. Describe what is meant by the term 'double insulated'. [2 marks]

09.4 Explain how the gardener may get an electric shock if the cable is cut whilst she is using the lawnmower. [3 marks]

10 A thermistor is used in a circuit to maintain the temperature of a room used to keep tropical plants.

10.1 Draw the circuit symbol for a thermistor. [1 mark]

10.2 Describe what happens to the resistance of the thermistor as the temperature decreases below 20 °C. [2 marks]

10.3 Sketch a graph of current (y-axis) against potential difference (x-axis) for a thermistor that has a constant temperature. [2 marks]

11 Athletes can suffer from tendon and ligament injuries. Recent scientific advances mean that doctors can inject bone marrow stem cells into these injuries to promote healing.

A radioactive tracer is inserted into the stem cells to quantify the number of cells retained in the tendon and the distribution of stem cells around the body. Technetium-99 has a half- life of 6 hours and emits gamma radiation.

Radon–224 has a half- life of 3.6 days and emits alpha radiation.

11.1 Which radioactive tracer would you choose? Explain your answer. [3 marks]

11.2 Ⓗ A sample of cobalt– 60 is left to decay for 27 days. Calculate what percentage of the original sample will remain. [2 marks]

Cobalt-60 has a half life of 5.27 days.

11.3 When radioactive materials are used in industry or medicine, the technicians must be aware that radioactive contamination can occur.

Describe what is meant by radioactive contamination. [2 marks]

11.4 Suggest **one** reason why it is important for scientists to publish the results of any research on cancer treatment. [1 mark]

12 Figure 8 shows the nuclear fission reaction of Uranium-235.

Figure 8

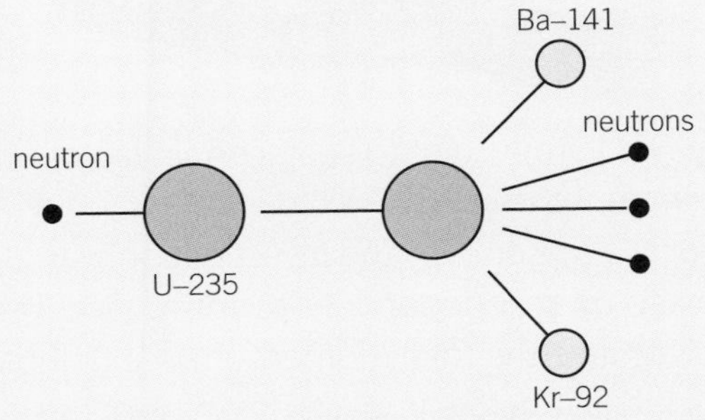

Ba–141

neutrons

neutron

U–235

Kr–92

12.1 Copy and complete the diagram to show the next stage in the chain reaction. [3 marks]

12.2 Complete the sentences using words from the box. Use Figure 8 to help you.

released	protons	energy	absorbed
neutrons	depth	uranium-236	
plutonium-244	weight	chain	

In the nuclear fission of uranium-235, a neutron is _____ by the Uranium-235.

The uranium-235 then becomes a new unstable isotope called _____ .

This is a nuclear reaction, and it releases _____ .

To keep the reaction under control, control rods in the reactor core absorb surplus _____ .

The _____ of the rods in the reactor core is adjusted to maintain a steady chain reaction. [5 marks]

13.1 What is meant by the term nuclear fusion? [2 marks]

This is an article from a science journal.

Fusion energy may soon be used in small-scale laser-fired fusion reactors with no radiation. The reactors are fueled by heavy hydrogen (deuterium) found abundantly in sea water. The reactors have already been shown to produce more energy than is needed to start them.

13.2 Evaluate the advantages and disadvantages of the new nuclear fusion reactors compared to traditional nuclear fission reactors. [6 marks]

14 A light sensing circuit is used to maintain the level of light in a cabinet used to grow plants.

Figure 9

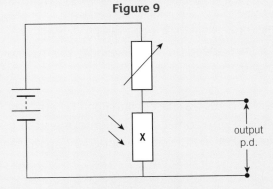

output p.d.

X

14.1 Name the component labelled X. [1 mark]

14.2 Give **one** reason why a variable resistor is included in the circuit. [1 mark]

Paper 2 questions

01 A deep-sea diver is lowered gradually below the surface of the sea.

01.1 **H** Explain why the diver cannot reduce the pressure of the sea water. [2 marks]

01.2 **H** Calculate the pressure of the sea water on the diver at a depth of 30 m.
The density of sea water is 1024 kg/m³. The gravitational field strength is 9.8 N/kg. [2 marks]

01.3 The diver breathes underwater using gas in a cylinder. Explain why divers often need to wear additional weight belts to stop them floating to the surface. [2 marks]

01.4 Calculate the pressure of a 30 N sample of sea water on the base of a glass jar. The cross sectional area of the base of the jar is 360 cm². [3 marks]

02.1 Copy and complete the diagram of the life cycle of a star. [3 marks]

Figure 1

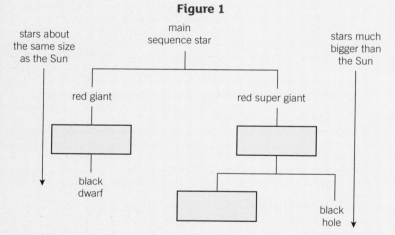

02.2 Describe how elements, including those much heavier than iron, are formed during the life cycle of a star. [4 marks]

02.3 **H** When a satellite orbits the Earth it moves at a constant speed but with a velocity that is always changing. Explain why. [2 marks]

03 In March 2004, a spacecraft named Rosetta set off to reach a comet deep in our solar system. In November 2014, Rosetta dropped the spacecraft Philae on to the comet. Philae was the first spacecraft to land on a comet. The gravity on the comet is approximately 1/200 000 of the gravity on Earth.

03.1 Suggest **one** reason why the spacecraft took so long to reach the comet. [1 mark]

03.2 Calculate the weight of Rosetta and Philae just before leaving Earth.
Their combined mass was 2900 kg. Gravitational field strength on Earth is 9.8 N/kg. [2 marks]

03.3 Explain why Philae took 7 hours to reach the surface of the comet after leaving the Rosetta spacecraft. [2 marks]

03.4 Calculate the average speed of descent of Philae. Give your answer in m/s. [3 marks]

03.5 Some people think that space travel is a waste of money. Do you agree? Give a reason for your answer. [1 mark]

04 A manufacturer of golf clubs investigated the distance a golf ball would travel using a new type of golf club. They used an automatic golf swing machine and carried out the investigation on the same day at the same location.

Figure 2

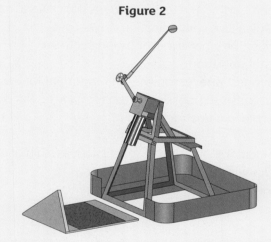

04.1 Why was it important to keep certain factors in the experiment the same? [1 mark]

04.2 Suggest a reason why the investigation was checked by an independent scientist. [1 mark]

04.3 **H** Calculate the total momentum of a golf ball and club just before impact. The mass of the club head is 0.195 kg and velocity of the club at impact is 54 m/s. [2 marks]

04.4 **H** Calculate the velocity of a 46 g golf ball just after being hit by the club. The velocity of the club slows to 42 m/s. [3 marks]

04.5 A laser range finder contains a transmitter and receiver of laser light. Describe how a laser range finder could be used to determine the distance to an object on a golf course. [3 marks]

04.6 Suggest a safety precaution that should be undertaken when laser light is used. [2 marks]

05.1 The normal speed limit in built up areas is 30 mph (13.4 m/s). Give **two** reasons why this speed limit is reduced to 20 mph (8.9 m/s) near schools. [2 marks]

05.2 A road safety council official wants drivers who exceed the 20 mph speed limit near schools to be penalised more harshly than other speeding drivers. Do you agree? Give reasons for your answer. [2 marks]

05.3 Give **two** road safety features, other than speed reduction, that can be found near schools. [2 marks]

05.4 A 1000 kg car hits a stationary object when it is accelerating at 2 m/s². Calculate the impact force. [2 marks]

06 A student investigated how changing the grade of sand paper affected the friction force of the sand paper.

Figure 3

200 g block
newton-meter
sand paper

The student pulled on the newton-meter and measured the force needed to start moving the block of wood. The student recorded the results in a table.

Grade of sand paper	1	2	3	4	5
Force in N	7.5	3.5	2.0	1.5	0.5

06.1 Name a variable that the student controlled. [1 mark]

06.2 What is the resolution of the newton-meter? [1 mark]

06.3 Identify any trends or patterns in the results of the investigation. [2 marks]

06.4 Describe how the results the student obtained can be made reproducible. [2 marks]

06.5 The student wanted to draw a line graph of her results. Give **one** reason why she had to draw a bar chart. [1 mark]

07 The table gives some information about the International Space Station (ISS), which is a manned spacecraft in orbit around the Earth.

Orbital height above the Earth	400 km
Speed of space station	7.6 km/s
Time of one complete orbit	92.6 mins

07.1 Calculate how many complete orbits the ISS will make in one day. [2 marks]

07.2 Calculate the total distance the ISS will travel in one day. [2 marks]

07.3 Explain why the ISS stays at constant speed as it orbits the Earth. You should refer to Newton's 1st law of motion. [2 marks]

08 A converging lens of focal length 5 cm is used as a magnifying glass.

08.1 An object 1.6 cm tall is placed 2.4 cm from the lens. Copy the diagram below and draw to scale. [1 mark]

Figure 4

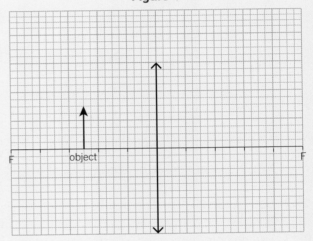

F

object

F

08.2 Construct a ray diagram to show how the light travels through the lens and then forms an image. [3 marks]

08.3 State whether the image is real or virtual. Explain your answer. [3 marks]

08.4 Calculate the magnification of the image. [2 marks]

09 A student used the lens shown to investigate how the distance of the object from the lens affected the magnification of the image.

Figure 5

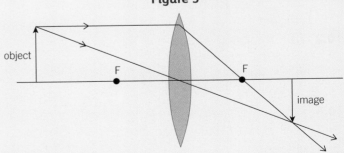

object

F

F

image

09.1 The results were plotted on a graph.
Student A drew a line of best fit as a curve, Student B said that was incorrect because a line of best fit is always a straight line.

Figure 6

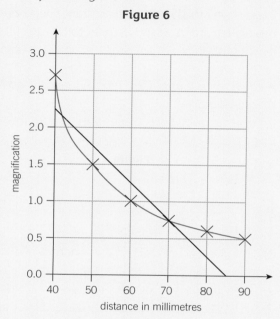

distance in millimetres

Which student drew the correct line of best fit?
[1 mark]

09.2 Both students drew line graphs and not bar charts. Suggest the reason why. [1 mark]

09.3 Use the information in the graphs to determine the magnification when the distance to the lens is 50 mm. [1 mark]

10 A power-station alternator generates an alternating potential difference of 25 kV and is designed to deliver up to 800 kW of electrical power through the National Grid.
A transformer in the national grid is used to step-up the potential difference to 130 kV.

10.1 Ⓗ Calculate the current in the primary coil of the transformer, assuming the transformer is 100% efficient. [2 marks]

10.2 Ⓗ Calculate the number of turns in the secondary coil.
The number of turns in the primary coil is 12 000 turns. [2 marks]

10.3 Explain why stepping the potential difference up from 25 kV to 130 kV makes the grid system more efficient. [2 marks]

11 A large crate on a platform is being pulled to the left by a horizontal rope. The force of friction acts in the opposite direction to the horizontal force of the rope.

Figure 7

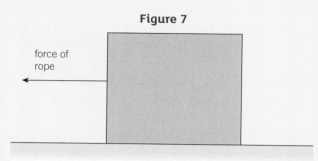

force of rope

11.1 Copy and complete the free body diagram to show the weight of the crate, the reaction force from the ground, and the friction force acting against the force of the rope. [4 marks]

11.2 The crate is placed on a very low friction surface and it begins to slide to the left. There is a horizontal force of 240 N on the crate. There is a vertical force of 80 N from the weight of the crate.
Draw a vector diagram to determine the magnitude and direction of the resultant force on the crate. Use graph paper. [4 marks]

12 A toy rocket launcher propels a rocket into the air using a strong elastic band. A parachute opens as the rocket begins to fall to Earth.

12.1 Name the main energy store of the rocket launcher. [1 mark]

12.2 Describe the velocity and forces on the rocket as it falls to Earth. [3 marks]

12.3 Two rockets **A** and **B** are identical in shape and size and have the same size parachute.
Rocket **A** is heavier than rocket **B**. Both rockets are launched to the same height.
Choose the correct sentence from the list below. Give reasons for your choice.
Rocket **A** will land first
Rocket **B** will land first
Rocket **A** and rocket **B** will land together. [2 marks]

13 ⓗ Ultrasound can be used to detect flaws hidden beneath the surface in metals. This is called non-destructive testing (NDT).

13.1 Describe what is meant by ultrasound.　[2 marks]

13.2 An ultrasound wave is transmitted by a receiver on the surface of a steel metal plate. The reflected wave is returned to a receiver on the surface in 3.5×10^{-6} s. The speed of ultrasound waves in steel metal is 3.2×10^3 m/s
Calculate the depth of the flaw in the metal.
[3 marks]

13.3 Suggest **one** reason why it is important to use NDT on metals that are going to be used in the nuclear industry.　[2 marks]

13.4 Give **one** use of ultrasound for medical imaging.
[1 mark]

14 A rugby payer uses a board attached to a spring to strengthen her calf muscles.
The unstretched spring is 25 cm long, and when the board is pushed down fully the spring depresses to 8 cm long. The spring constant is 5200 N/m.

Figure 8

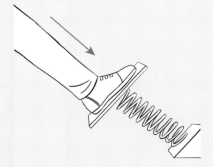

14.1 Calculate the force needed to depress the spring fully.　[3 marks]

14.2 Name the main energy store of the spring.　[1 mark]

14.3 Calculate the work done needed to fully depress the spring. Give your answer to two significant figures.　[3 marks]

$$\text{elastic potential energy} = 0.5 \times \text{spring constant} \times \text{extension}^2$$

14.4 Manufacturers always test the springs up to the elastic limit.
What is meant by the term elastic limit?　[1 mark]

14.5 Give **one** reason why the data for the apparatus should be checked by an independent investigator.
[1 mark]

15 ⓗ This information is from a physics textbook. *Ultrasound waves can be produced by electronic systems. They have a frequency higher than the upper limit for hearing in humans. When they meet a boundary between two different media, ultrasound waves are partially reflected.*

15.1 Give the upper frequency for the human hearing range. Choose **one** of the options in the box below.
[1 mark]

| 200 Hz | 2000 Hz | 20 000 Hz | 200 000 Hz |

15.2 Describe what happens to the ultrasound that reaches the boundary between two different media and is **not** reflected.　[2 marks]

15.3 Explain why ultrasound, and not x-rays, is used for pre-natal scans.　[2 marks]

16 ⓗ A father and son ride the dodgem cars in a fairground.
The mass of each dodgem car is 150 kg.
The mass of the father is 100 kg.
The mass of the son is 40 kg.
Both dodgem cars travel in the same direction.

16.1 Calculate the speed and direction of the father's dodgem car after the two dodgem cars bump into each other.
Before the collision the speed of the son's dodgem car was 7 m/s, and the speed of the father's car was 4 m/s.
After the collision the son's dodgem car continues to travel at 3 m/s in the same direction.　[4 marks]

16.2 Calculate the impact force on the son's dodgem car. The collision time was 0.4 s.　[2 marks]

16.3 Explain how wearing a seat belt reduces the impact force on the drivers in the dodgem car.　[3 marks]

16.4 Suggest another design feature of a dodgem car that reduces the impact force.　[1 mark]

Maths skills for Physics
MS1 Arithmetic and numerical computation

Learning objectives

After this topic, you should know how to:

- recognise and use expressions in decimal form
- recognise and use expressions in standard form
- use ratios, fractions and percentages
- make estimates of the results of simple calculations.

Figure 1 *How far away is the Moon?*

Figure 2 *The air pressure at the summit of Mount Everest is significantly lower than at sea level*

Study tip

Always remember to add a unit, if appropriate, when quoting a number.

What is the speed of a radio wave? How far away is the Moon? What is the difference in air pressure between sea level and the summit of Mount Everest?

Scientists use maths all the time – when collecting data, looking for patterns, and making conclusions. This chapter includes all the maths for your GCSE Physics course. The rest of the book gives you many opportunities to practise using maths in physics.

1a Decimal form

There will always be a whole number of protons, neutrons, or electrons in an atom.

When you make measurements in science the numbers may *not* be whole numbers, but numbers *in between* whole numbers. These are numbers in decimal form, for example 3.2 cm, or 4.5 g.

The value of each digit in a number is called its place value.

Thousands	Hundreds	Tens	Units	·	Tenths	Hundredths	Thousandths
4	5	1	2	·	3	4	5

1b Standard form

Place value can help you to understand the size of a number, however some numbers in science are too large or too small to understand when they are written as ordinary numbers. For example, distance from the Earth to the Sun, 150 000 000 000 m or the diameter of the nucleus of a hydrogen atom, 0.000 000 000 000 001 75 m.

We use standard form to show very large or very small numbers more easily.

In standard form, a number is written as $A \times 10^n$.

- A is a decimal number between 1 and 10 (but not including 10), for example 1.5.
- n is a whole number. The power of ten can be positive or negative, for example 10^{11}.

This gives you a number in standard form, for example, 1.5×10^{11} m.

Table 1 explains how you convert numbers to standard form.

Table 1 *How to convert numbers into standard form*

The number	The number in standard form	What you did to get to the decimal number	...so the power of ten is...	What the *sign* of the power of ten tells you
1000 m	1.0×10^3 m	You moved the decimal point 3 places to the *left* to get the decimal number	+3	The positive power shows the number is *greater* than one.
0.01 s	1.0×10^{-2} s	You moved the decimal point 2 places to the *right* to get the decimal number	−2	The negative power shows the number is *less* than one.

It is much easier to write some of the very big or very small numbers that you find in real life using standard form. For example

- the distance from the Earth to the Sun is 150 000 000 000 m = 1.5×10^{11} m
- the diameter of an atom is 0.000 000 000 1 m = 1.0×10^{-10} m
- the wavelength of light of is around 0.000 007 m = 7×10^{-7} m
- the speed of light is 300 000 000 m/s = 3×10^8 m/s.

Multiplying numbers in standard form

You can use a scientific calculator to calculate with numbers written in standard form. You should work out which button you need to use on your own calculator (it could be EE , EXP , 10ˣ , or ×10ˣ).

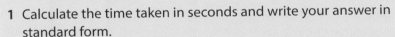

> **Worked example: Standard form**
> A train travelled a distance of 180 km at a constant speed in a time of 235 minutes
>
> 1 Calculate the time taken in seconds and write your answer in standard form.
>
> 2 Calculate the distance travelled by the train each second. Write your answer in standard form to 3 significant figures.
>
> **Solution**
> 1 **Step 1:** Because there are 60 seconds in 1 minute, multiply the time in minutes by 60 to give the time in seconds.
>
>   Time in seconds = 235 × 60 = 14 100 s.
>
>   **Step 2:** Convert the numbers to standard form
>
>   14 100 s = 1.41×10^4 s.
>
> 2 **Step 1:** Distance travelled each second = $\dfrac{180 \text{ km}}{1.41 \times 10^4 \text{ s}}$
>
>   = 0.012 766 km
>
>   **Step 2:** Convert the numbers to standard form
>
>   0.012 766 km = 1.2766×10^{-2} km
>
>   **Step 3:** Write the numbers to 3 significant figures
>
> 1.2766×10^{-2} km = 1.28×10^{-2} km

Figure 3 *The Sun is 1.5×10^{11} m from the Earth*

Figure 4 *Uranium isotopes are used in nuclear power plants. The atomic radius of a uranium atom is 1.75×10^{-10} m*

> **Study tip**
> Check that you understand the power of ten, and the sign of the power.

Figure 5 *You can use a scientific calculator to do calculations involving standard form*

> **Study tip**
> In step 3, the third significant figure is rounded up to 8 because the 4th significant figure is greater than or equal to 5.

1c Ratios, fractions, and percentages
Ratios

A ratio compares two quantities. A ratio of $2:4$ of radioactive atoms to non-radioactive atoms in a radioactive sample means for every two radioactive atoms there are four non-radioactive atoms.

You can compare ratios by changing them to the form $1:n$ or $n:1$.

$1:n$ Divide both numbers by the *first* number.
For every one radioactive atom there are two non-radioactive atoms.

$$\div 2 \left(\begin{array}{c} 2:4 \\ 1:2 \end{array} \right) \div 2$$

$n:1$ Divide both numbers by the *second* number.
For every half a radioactive atom there is one non-radioactive atom (even though you can't really get 'half a radioactive atom').

$$\div 4 \left(\begin{array}{c} 2:4 \\ 0.5:1 \end{array} \right) \div 4$$

You can describe the number of radioactive atoms in relation to the number of non-radioactive atoms using three different ratios $2:4$, $1:2$ and $0.5:1$. All of the ratios are equivalent – they mean the same thing.

You can simplify a ratio so that both numbers are the lowest whole numbers possible.

Worked example: Simplifying ratios

A student draws a vector arrow of length 120 mm to represent a force X of 60 N on a scale diagram. She then needs to draw a second vector arrow to represent a force Y of 36 N on the same diagram.

1 Calculate the ratio of force Y to force X.

2 Use your answer to **1** to calculate the required length of the vector arrow for Y.

Solution

1 Divide force Y by force X

$$\frac{\text{force Y}}{\text{force X}} = \frac{36\,\text{N}}{60\,\text{N}} = 0.60$$

2 Because the diagram is a scale diagram, the ratio of the vector arrow lengths is the same as the ratio of the force.

The length of the vector arrow for Y = 0.60 × the length of the vector arrow for X

$$= 0.60 \times 120\,\text{mm}$$

$$= \mathbf{72\,mm}$$

Fractions

A fraction is a part of a whole.

$\frac{1}{3}$ ⟶ The numerator tells you how many parts of the whole you have.

⟶ The denominator tells you how many equal parts the whole has been divided into.

To convert a fraction into a decimal, divide the numerator by the denominator.

$\frac{1}{3} = 1 \div 3 = 0.33333\ldots = 0.\dot{3}$ (the dot shows that the number 3 recurs, or repeats over and over again).

To convert a decimal to a fraction, use the place value of the digits, then simplify. $0.045 = \frac{45}{1000} = \frac{9}{200}$

Take care if you are asked to use ratios to find a fraction. For example in a fitness run in which the runner walks then jogs alternately, a ratio of $2:4$ in the walking to jogging distance does not mean that the walking distance is $\frac{2}{4}$ (or half) of the total distance. If the total distance was 6 km, the walking distance would be 2 km and the jogging distance would be 4 km. So the walking distance would be $\frac{1}{3}\left(=\frac{2\ km}{6\ km}\right)$ of the total distance.

Figure 6 *One square of this chocolate bar represents $\frac{1}{24}$. A column of four squares represents $\frac{4}{24}$, which can be simplified to $\frac{1}{6}$*

Worked example: Calculating the fraction of a quantity

A satellite in a circular orbit above the Earth's equator at a certain height takes 24 hours to orbit the Earth once. During that time, it takes six hours to go over the Atlantic Ocean.

Figure 7 shows a satellite in a circular orbit.

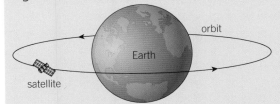

orbit

Earth

satellite

Figure 7 *A satellite in a circular orbit*

1 Calculate the fraction of one complete orbit which is over the Atlantic Ocean and write this down as a decimal.

2 The distance travelled by the satellite in one orbit = 4.0×10^4 km. Calculate the distance travelled by the satellite when it is over the Atlantic Ocean.

Solution

1 Divide the time taken to go over the Atlantic Ocean by the time taken for one complete orbit:

$$\frac{\text{time taken over the Atlantic Ocean}}{\text{time taken for one compete orbit}} = \frac{6}{24} = 0.25$$

2 Distance travelled by the satellite above the ocean

$$= 0.25 \times 4.0 \times 10^4\,\text{km} = \mathbf{1.0 \times 10^4\,km}$$

Percentages

A percentage is a number expressed as a fraction of 100. For example:

$$77\% = \frac{77}{100} = 0.77$$

Worked example: Calculating a percentage

A car accelerates from rest for 4.5 s then travels at a constant speed for 13.5 s before braking to a standstill in 7.0 s. Calculate the percentage of time the car travels at constant speed.

Solution

Step 1: Calculate for the total time taken.

Total time taken = 4.5 + 13.5 + 7.0 = 25.0 s

Step 2: Divide the time taken at constant speed by the total time taken and express the answer as a decimal.

$$\frac{\text{time taken at constant speed}}{\text{total time taken}} = \frac{13.5}{25} = 0.54$$

Step 3: Multiply the answer above by 100

0.54 × 100 = **54%**

Figure 8 *What percentage of the time the car travelled was at a constant speed?*

Worked example: Calculating a percentage change

A spring has an unstretched length of 300 mm. Calculate the percentage change in its length when it is stretched to a length of 396 mm.

Solution

Step 1: Calculate the actual change in length.

396 mm – 300 mm = 96 mm

Step 2: Divide the actual change in length by the unstretched length.

$$\frac{96\,\text{mm}}{300\,\text{mm}} = 0.32$$

Step 3: Multiply the answer to Step 2 by 100%.

0.32 × 100% = **32%**

You may need to calculate a percentage of a quantity. To do this, convert the percentage to a decimal and multiply the quantity by the decimal.

Worked example: Using percentage change

The spring is damaged if it is stretched by more than 35% of its unstretched length. Calculate the maximum length of the spring if it is not to be damaged.

Solution

Step 1: Convert 35% to decimal

$35\% = 0.35$

Step 2: Multiply the unstretched length (300 nm) by 0.35 to give the extension of the spring for no damage.

extension $= 0.35 \times 300\,mm = 105\,mm$

maximum length = original length + extension
$= 300\,mm + 105\,mm = \textbf{405 mm}$

Synoptic link

To see more examples of the use of percentages, see Maths skill MS1c

1d Estimating the result of a calculation

When you use your calculator to work out the answer to a calculation you can sometimes press the wrong button and get the wrong answer. The best way to make sure that your answer is correct is to estimate it in your head first.

Worked example: Estimating an answer

You want to calculate distance travelled and you need to find $34\,m/s \times 8\,s$. Estimate the answer and then calculate it.

Solution

Step 1: Round each number up or down to get a whole number multiple of 10.

34 m/s is about 30 m/s

8 s is about 10 s

Step 2: Multiply the numbers in your head.

$30\,m/s \times 10\,s = 300\,m$

Step 3: Do the calculation and check it is close to your estimate.

Distance $= 34\,m/s \times 8\,s = 272\,m$

This is quite close to 300 so it is probably correct.

Notice that you could do other things with the numbers:

$34 + 8 = 42$ $\qquad$ $\dfrac{34}{8} = 4.3$ $\qquad$ $34 - 8 = 26$

Not one of these numbers is close to 300. If you got any of these numbers, you would know that you needed to repeat the calculation.

Sometimes the calculations involve more complicated equations, or standard form.

When carrying out multiplications or divisions using standard form, you should add or subtract the powers of ten to work out roughly what you expect the answer to be. This will help you to avoid mistakes.

1 A sheet of card has a length of 297 mm, a width of 210 mm and a mass of 25.2 grams. A student cut a strip of even width and length 297 mm from the card. The mass of the strip was 5.6 grams.
   a Calculate the ratio of the mass of the strip to the mass of the original card. [1 mark]
   b The ratio of the width of the strip to the width of the original card is equal to mass ratio calculated in part a. Use mass ratio and the width of the original card to calculate the width of the strip of card. Give your answer to the nearest millimetre. [2 marks]

2 A car and a coach join a motorway at the same junction and travel in opposite directions at constant speed. The car travels at a speed of 30 m/s and the coach travels at a speed of 24 m/s.
   a Calculate the ratio of the speed of the car to the speed of the coach. [2 marks]
   b Calculate how far the car has travelled from the motorway junction when the coach has travelled a distance of 12 km from the junction. [2 marks]

MS2 Handling data

2a Significant figures

Numbers are rounded when it is not appropriate to give an answer that is too precise.

When rounding to significant figures, count from the first non-zero digit.

These masses each have three significant figures (s.f.). The significant figures are underlined in each case.

<u>153</u> g   0.<u>153</u> g   0.00<u>153</u> g

Table 1 below shows some more examples of measurements given to different numbers of s.f.

Table 1 *The number of significant figures – the significant figures in each case are underlined*

Number	0.0<u>5</u> s	<u>5.1</u> nm	0.<u>775</u> g/s	<u>23.50</u> cm³
Number of significant figures	1	2	3	4

In general, you should give your answer to the same number of s.f. as the as the data in the question that has the lowest number of s.f.

Remember that rounding to s.f. are **not** the same as decimal places. When rounding to decimal places, count the number of digits that follow the decimal point.

Learning objectives

After this topic, you should know how to:

- use an appropriate number of significant figure
- find arithmetic means
- construct and interpret frequency tables and bar charts
- make order of magnitude calculations.

Figure 1 *What is the average speed of the eagle?*

2b Finding arithmetic means

To calculate the mean value of a series of values:

- add together all the values in the series to get a total
- divide the total by the number of values in the data series

Worked example: Calculating a mean

A student's reaction time was measured and recorded five times. Her results were as follows:

$$0.46\,s \qquad 0.48\,s \qquad 0.52\,s \qquad 0.44\,s \qquad 0.56\,s$$

Calculate the mean value of these measurements.

Solution

Step 1: Add together the recorded values.

$$0.46\,s + 0.48\,s + 0.52\,s + 0.44\,s + 0.56\,s = 2.46\,s$$

Step 2: Divide by the number of recorded values (in this case, five measurements were measured).

$$\frac{2.46}{5} = \mathbf{0.49\,s}\ (2\ \text{s.f.})$$

The mean reaction time for the student was 0.49 s (2 s.f.)

2c Frequency tables and bar charts
Tables

Tables are used to present data from observations and measurements that are made during experiments or from surveys. Data can be

- qualitative (descriptive, but with no numerical measurements)
- quantitative (including numerical measurements).

Qualitative data includes categoric variables, such as types of power stations. The values of categoric variables are names not numbers.

Quantitative data includes

- continuous variables – can take any value (usually collected by measuring), such as length or time.
- discrete variables – can only take exact values (usually collected by counting), such as number of paper clips picked up by an electromagnet, or complete swings of a pendulum.

Table 2 below shows the energy output in 2010 of different types of UK power stations as a percentage of the total UK energy output.

- The first column is a categoric variable as it shows the different types of power stations.
- The second column is a continuous variable as it shows the energy output of each type of power station as a percentage of the total energy output.

Table 2 *The percentage energy output of different types of electricity power stations in the UK in 2010*

Type of power station	% energy output
Gas	40
Coal	31
Oil	1
Nuclear	18
Wind	3
Hydroelectric	2
Bio-energy	3
Other	2

Sometimes data is grouped into classes, as in Table 3, which shows the ages of the winners of the Nobel Prize for Physics. If you need to group data into classes:

● make sure the values in each class do not overlap

● aim for a sensible number of classes.

Table 3 *Ages of the winners of the Nobel Prize for Physics*

Ages, years	Number of physicists
21–30	1
31–40	28
41–50	55
51–60	51
61–70	35
71–80	23
81–90	8

Figure 2 *William Lawrence Bragg is the youngest winner of the Nobel Prize for Physics. He was 25 when he was jointly award the prize with his father, William Henry Bragg, in 1915*

Bar charts

You can use a bar chart to display data.

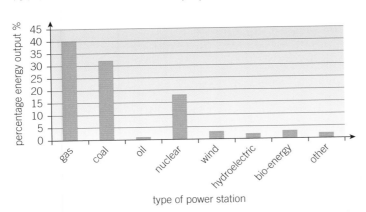

Figure 3 *The percentage energy output of different types of electricity power stations in the UK in 2010*

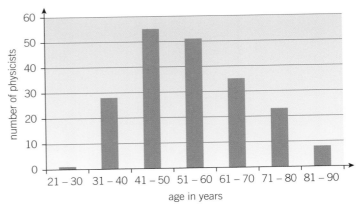

Figure 4 *Ages of the winners of the Nobel Prize for Physics*

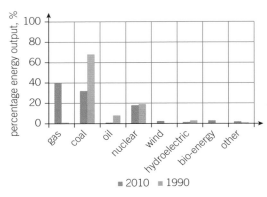

■ 2010 ■ 1990

Figure 5 *The different types of electricity power stations in the UK and their percentage power output in 2010 and 1990*

Figure 6 *What is the median speed of a group of horses?*

You can also use bar charts to compare two or more sets of data (Figure 5)

2f Averages

When you collect data, it is sometimes useful to calculate an average. There are three ways you can calculate an average – the mean, the mode, and the median.

You saw how to calculate a mean earlier in this topic (section 2b).

How to calculate a median

When you put the values of a series in order from smallest to biggest the middle value is called the median. When the series of values has two central values, the median is the mean of these two values.

> **Worked example: Calculating a median (odd number of values)**
> The speed of seven different horses galloping is shown below.
>
>   11.1 m/s 13.2 m/s 11.9 m/s 12.5 m/s 12.8 m/s 11.2 ms 12.3 m/s
>
> Calculate the median speed of the horses.
>
> **Solution**
> **Step 1:** Place the values in order from smallest to largest:
>
>   11.1 m/s 11.2 m/s 11.9 m/s 12.3 m/s 12.5 m/s 12.8 m/s 13.2 m/s
>
> **Step 2:** Select the middle value – this is the median value.
>
>   median value = 12.3 m/s

If you have an even number of values, you select the middle pair of values and calculate a mean. That then becomes your median value.

How to calculate a mode

The mode is the value that occurs most often in a series of results. If there are two values that are equally common, then the data is bimodal.

> **Worked example: Calculating a mode**
> A student measured the time it took for 12 different cars to travel a 100 m stretch of road.
>
>   12.9 m/s 13.1 m/s 17.3 m/s 14.5 m/s 13.0 m/s 13.1 m/s
>
>   12.9 m/s 13.2 m/s 13.4 m/s 12.5 m/s 14.1 m/s 12.9 m/s
>
> Calculate the modal time for the cars to travel the stretch of road.
>
> **Solution**
> **Step 1:** Place the values in order from smallest to largest.
>
>   12.5 m/s 12.9 m/s 12.9 m/s 12.9 m/s 13.0 m/s 13.1 m/s
>   13.1 m/s 13.2 m/s 13.4 m/s 14.1 m/s 14.5 m/s 17.3 m/s
>
> **Step 2:** Select the value which occurs the most often.
>
>   mode = 12.9 m/s

2g Scatter diagrams and correlations

You may collect data and plot a scatter graph (see Topic M4). You can add a line to show the trend of the data, called a line of best fit. The line of best fit is a line that goes through as many points as possible and has the same number of points above and below it.

If the gradient of the line of best fit is:

positive it means as the independent variable *increases* the dependent variable *increases*

negative it means as the independent variable *increases* the dependent variable *decreases*

zero it means changing the independent variable has no effect on the dependent variable.

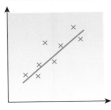
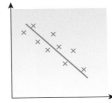

A relationship where there happens to be a link is called a correlational relationship, or correlation. You say that the relationship between the variables is positive, negative, or that there is no relationship.

For example:

- As you increase the force on an object, the acceleration of the object will increases – a positive correlation.

- As you increase the density of a liquid, the speed with which an object falls through the liquid will decrease – a negative correlation.

- The intensity of a light beam has no effect of the angle of refraction in a glass block – no relationship.

The presence of a relationship does not always mean that changing the independent variable *causes* the change in the dependent variable. In order to claim a causal relationship, you must use science to predict or explain *why* changing one variable affects the other.

Often there is a third factor that is common to both so it looks as if they are related. You could collect data for shark attacks and ice cream sales. A graph shows a positive correlation but shark attacks do not make people buy ice cream. Both are more likely to happen in the summer.

2h Order of magnitude calculations

Being able to make a rough estimate is helpful. It can help you to check that a calculation is correct by knowing roughly what you expect the answer to be. A simple estimate is an order of magnitude estimate, which is an estimate to the nearest power of 10.

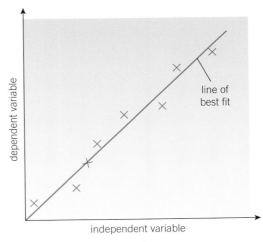

Figure 7 *A scatter graph*

Study tip

Use a transparent ruler to help you draw the line of best fit so you make sure that there are the same number of points on either side of the line.

For example, to the nearest power of 10, you are probably 1 m tall and can walk at a speed of about 1 m/s.

You, your desk, and your chair are all of the order of 1 m tall. The diameter of a molecule is of the order of 1×10^{-9} m, or 1 nanometre.

1 A student used an electronic stopwatch to measure how long a ball took to roll a measured distance down a slope after being released. She repeated the measurement 10 times. Her results were as follows:

1.70 s	1.65 s	1.82 s	1.61 s	1.74 s
1.61 s	1.63 s	1.71 s	1.67 s	1.65 s

   a Calculate the mean value of these measurements. Give your answer to 3 significant figures. [1 mark]

   b Calculate the median value of these measurements. [1 mark]

   c Calculate the mode value of these measurements. [1 mark]

2 How many significant figures are the following numbers quoted to?

   a 0.001 [1 mark]   b 85.4 [1 mark]

   c 70.0 [1 mark]   d 8.314 [1 mark]

   e 3.08×10^8 [1 mark]   f 3.07×10^{-3} [1 mark]

3 Look back at Figure 5. The bar chart tells you about the types of power stations used to generate electricity in the United Kingdom.

   a For each non-renewable power station, write down one conclusion you draw about their contribution to UK electricity generation in 2010 compared with 1990. [4 marks]

   b State two conclusions you draw from the chart about the use of renewable energy sources compared with the use of fossil fuel in

     i 1990 [2 marks]   ii 2010? [2 marks]

4 Write down an order of magnitude value for the speed of an ant and estimate the distance it could travel in 1 minute. [2 marks]

MS3 Algebra

3a Mathematical symbols

You have used lots of different symbols in maths, such as +, −, ×, ÷. There are other symbols that you might meet in science. These are shown in Table 1.

Table 1 *The symbols you will meet whilst studying physics*

Symbol	Meaning	Example
=	Equal to	$2\,m/s \times 2\,s = 4\,m$
<	Is less than	The mean weight of a child in a family < the mean weight of an adult in a family
<<	Is very much less than	The diameter of an atom << the diameter of an apple
>>	Is very much bigger than	The diameter of the Earth >> the diameter of a football
>	Is greater than	The density of lead > the density of steel
∝	Is proportional to	F(force) ∝ x (extension) for a spring
≈	Is approximately equal to	$272\,m \sim 300\,m$ (see estimating an answer in the worked example above)
~	Is the same order of magnitude as	The diameter of an atom $\sim 10^{-10}\,m$

3b Changing the subject of an equation

An equation shows the relationship between two or more variables. You can change an equation to make *any* of the variables become the subject of the equation.

To change the subject of an equation, you can do an opposite (inverse) operation to both sides of the equation to get the variable that you want on its own.

This means that:

- subtracting is the opposite of adding (and adding is the opposite of subtracting)
- dividing is the opposite of multiplying (and multiplying is the opposite of dividing)
- taking the square root is the opposite of squaring (and squaring is the opposite of taking the square root)

You can use these steps to change the subject of an equation, such as the equation for kinetic energy.

Learning objectives

After this topic, you should know how to:

- understand and use the symbols: =, <, <<, >>, >, ∝, ~
- change the subject of an equation
- substitute numerical values into algebraic equations using appropriate units for quantities.

Synoptic link

To see more examples of rearranging equations, see Topic P3.1 and Topic P3.3.

Figure 1 *If you know the kinetic energy and mass of this roller coaster car you can work out the speed*

Worked example: Kinetic energy

Change the equation $KE = \frac{1}{2}mv^2$ to make v the subject.

Solution

Step 1: Multiply both sides of the equation by 2.

$$2 \times KE = 2 \times \frac{1}{2}mv^2$$

Remove the fraction from the right hand side of the equation: $2 \times KE = mv^2$

Step 2: Divide by m to get the v^2 on its own.

$$\frac{2 \times KE}{m} = \frac{mv^2}{m} = \frac{2 \times KE}{m} = v^2$$

Step 3: Take the square root of both sides.

$$\sqrt{v^2} = \sqrt{\frac{2 \times KE}{m}} = v = \sqrt{\frac{2 \times KE}{m}}$$

3c Quantities and units
SI Units

When you make a measurement in science you need to include a number *and* a unit.

When you do a calculation your answer should also include both a number *and* a unit. There are some special cases where the units cancel, but usually they do not.

Everyone doing science, including you, needs to use the SI system of units.

Table 2 *Some quantities and their units you will meet during your physics GCSE*

Quantity and symbol	Unit
distance s	metre m
mass m	kilogram kg
time t	second s
current I	ampere A
temperature t	kelvin K
frequency f	hertz Hz
force F	newton N
energy E	joule J
power P	watt W
pressure p	pascal Pa
charge Q	coulomb C
potential difference V	volt V
electric resistance R	ohm Ω

For example, 1.5 N is a *measurement*. The number 1.5 is not a measurement because it does not have a unit.

Some quantities that you *calculate* do not have a unit because they are a ratio.

Using units in equations

When you put quantities into an equation it is best to write the number *and* the unit. This helps you to work out the unit of the quantity that you are calculating.

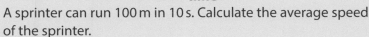

Worked example: Speed = $\dfrac{\text{distance}}{\text{time}}$

A sprinter can run 100 m in 10 s. Calculate the average speed of the sprinter.

Solution

Step 1: Write down what you know.

distance = 100 m

time = 10 s

Step 2: Write down the equation you need.

average speed (m/s) = $\dfrac{\text{distance (m)}}{\text{time (s)}}$

Step 3: Do the calculation and include the units.

average speed = $\dfrac{100 \text{ m}}{10 \text{ s}}$ = **10 m/s**

m/s are the units of speed.

Figure 2 *If you know the distance a sprinter ran, and the time it took him, you can calculate his average speed*

Metric prefixes

You can use metric prefixes to show large or small multiples of a particular unit. Adding a prefix to a unit means putting a letter in front of the unit, for example km. It shows you that you should multiply your value by a particular power of 10 for it to be shown in an SI unit.

For example, 3 millimetres = 3 mm = 3×10^{-3} m. Most of the prefixes that you will use in physics involve multiples of 10^3. However, when dealing with volumes in density calculations, we often deal with cubic centimetres (cm^3), where 1 cm is one hundredth of a metre (or 1×10^{-2} m).

Table 3 *Common prefixes you will use in your units*

Prefix	giga	mega	kilo	deci	centi	milli	micro	nano
Symbol	G	M	k	d	c	m	μ	n
Multiplying factor	10^9	10^6	10^3	10^{-1}	10^{-2}	10^{-3}	10^{-6}	10^{-9}

When carrying out multiplications or divisions using standard form, you should add or subtract the powers of ten to work out roughly what you expect the answer to be. This will help you to avoid mistakes.

Converting between units

It is helpful to use standard form when you are converting between units. To do this, it's best to consider how many of the 'smaller' units are contained within one of the 'bigger' units. For example

- There are 1000 mm in 1 m. So $1\,mm = \frac{1}{1000}\,m = 10^{-3}\,m$

- There are 1000 m in 1 km. So $1\,km = 1000\,m = 10^3\,m$.

1 a Write down the SI symbol for the SI unit of each of the following quantities
  i length [1 mark]   **ii** volume [1 mark]
  iii mass [1 mark]   **iv** force [1 mark]
  b Write down each of the following amount in standard form without the prefix
  i 72 mm [1 mark]   **ii** 16 cm³ [1 mark]
  iii 385 MJ [1 mark]   **iv** 56 kN [1 mark]

2 How would you read the following expressions as a sentence?
  a circumference of the Earth << circumference of the Sun [1 mark]
  b atmospheric pressure ∝ altitude [1 mark]
  c 4.5 kJ ~ 4.61 kJ [1 mark]

3 The force F (N) needed to extend a spring by extension e (m) is given by the equation $F = ke$ where k is the spring constant of the spring.
  a Rearrange the equation to make k the subject. [1 mark]
  b Show that the unit of k is N/m [1 mark]

4 The energy E stored in a stretched spring is given by the equation
  $E = \frac{1}{2}ke^2$ where k is the spring constant of the spring and e is its extension.
  a Rearrange the above equation to make e the subject of the equation. [1 mark]
  b What is the unit of e? [1 mark]

5 An object accelerates uniformly for a distance s with an acceleration a. its final velocity v is given by the equation $v^2 = u^2 + 2as$ where $u =$ its initial velocity is u.
  a Rearrange the equation $v^2 = u^2 + 2as$ to make s the subject of the equation. [1 mark]
  b A car accelerates for a certain time with a constant acceleration of 3.4 m/s² from an initial speed of 4.0 m/s to a speed on 22 m//s. Calculate the distance it travels in this time. [1 mark]

MS4 Graphs

During your GCSE course you will collect data in different experiments or investigations. In investigations, the data will either be:

- from an experiment where you have changed *one* independent variable (or allowed time to change) and measured the effect on a dependent variable
- from an investigation where you have collected data about *two* independent variables to see if they are related.

4a Collecting data by changing a variable

In many investigations you change one variable (the independent variable) and measure the effect on another variable (the dependent variable). In a fair test, the other variables are kept constant.

For example, you can vary the amount of force you apply to an object (independent variable) and measure the effect on the acceleration of the object (dependent variable).

A scatter diagram lets you show the relationship between two numerical values.

- The independent variable is plotted on the horizontal axis – the *x*-axis.
- The dependent variable is plotted on the vertical axis – the *y*-axis.

The line of best fit is a line that goes roughly through the middle of all point on the scatter graph. The line of best fit is drawn so that the points are evenly distributed on either side of the line.

Sometimes, graphs are plotted with the independent variable on the *y*-axis because the gradient represents a physical quantity. For example, in an investigation of the motion of an object, the time taken by the object to fall through different distances can be measured. The distance fallen is the independent variable and the time taken is the dependent variable. Plotting the distance on the *y*-axis and time taken on the *x*-axis gives a graph where the gradient of the line represents the speed of the object.

4b Straight line graphs

When people say 'plot a graph' they usually mean plot the points then draw a line of best fit. This is a smooth line that passes through or near each plotted point. If the line of best fit is a straight line, the gradient of the line is constant and there is a linear relationship between the two variables.

After this topic, you should know how to:

- translate information between graphical and numeric form
- explain that $y = mx + c$ represents a linear relationship
- plot two variables from experimental or other data
- determine the slope and intercept of a linear graph
- draw and use the slope of a tangent to a curve as a measure of rate of change
- describe the physical significance of area between a curve and the *x*-axis and measure it by counting squares as appropriate.

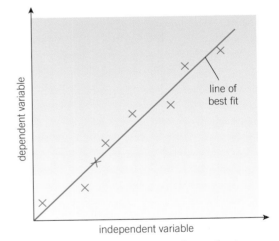

Figure 1 *A graph with the independent variable plotted on the x-axis and the dependent variable plotted on the y-axis*

Study tip

Look back at section M2g to remind yourself about interpreting scatter graphs and correlations.

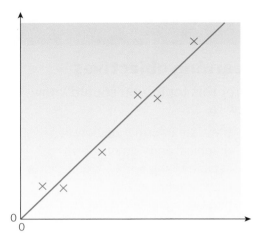

Figure 2 *A line of best fit which passes through the origin*

The equation for a straight line graph

A straight line graph tells you about the mathematical relationship between variables but there are other things that you can calculate from a graph.

The mathematical equation of a straight line is $y = mx + c$, where m is the **gradient** and c is the point on the y-axis where the graph intercepts, called the y-intercept.

Straight line graphs that go through the origin (0,0) are special (Figure 2). For these graphs, y is directly proportional to x, and the mathematical equation is $y = mx$ (because the y-intercept is zero).

When you describe the relationship between two *physical* quantities, you should think about the reason why the graph might (or might not) go through (0,0).

Worked example: Hooke's Law

A spring is being stretched by a force F and experiences an extension x.

A line of best fit for a graph of force (y-axis) against extension (x-axis) is a straight line through (0, 0). Explain what the gradient shows in this context, and state why the graph goes through (0,0).

Solution

Step 1: Match the equation to $y = mx$ to work out what the gradient means.

$y = mx$, and Hooke's Law says $F = kx$.

So the gradient gives us the value of k, the spring constant.

Step 2: Think about what happens to x when the y quantity is zero.

The line goes through (0,0) because when the force is zero, there is no extension in the spring.

4c Plotting data

When you draw a graph you choose a scale for each axis.

- The scale on the x-axis should be *the same* all the way along the x-axis but it can be *different* to the scale on the y-axis.

- Similarly, the scale on the y-axis should be *the same* all the way along the y-axis but it can be *different* to the scale on the x-axis.

- Each axis should have a label and a unit, such as time / s.

4d Determining the slope and intercept of a straight line

You often need to calculate a gradient. These may represent important physical quantities. Their units can give you a clue as to what they represent. The gradient of a straight line is calculated using the equation:

$$\text{gradient} = \frac{\text{change in } y}{\text{change in } x}$$

For all graphs where the quantity on the x-axis is time, the gradient will tell you the rate of change of the quantity on the y-axis with time.

You can also find the y-intercept of a graph. This is the value of the quantity on the y-axis when the value of the quantity on the x-axis = 0.

For all graphs where the quantity on the x-axis is time, the gradient will tell you the rate of change of the quantity on the y-axis with time. For example,

● the rate of change of distance (y-axis) with time (x-axis) is speed,

● the rate of change of velocity (y-axis) with time (x-axis) is acceleration

If the line is straight, its gradient is constant so the rate of change is constant.

4e Using tangents with curved graphs

When you plot a graph of the relationship between certain variables, you may not get a straight line – the relationship is non-linear.

To find the gradient at a point **T** you need to draw a **tangent** to the curve at that point (Figure 3):

● Draw a tangent to the curve – the line should pass through point **T** and have the same slope as the curve at that point.

● Make a right-angled triangle with your line as the hypotenuse. The triangle can be as big as you like , but make sure that the triangle is large enough for you to calculate sensible changes in values.

● Calculate the slope of the tangent using $\dfrac{\text{change in } y}{\text{change in } x}$

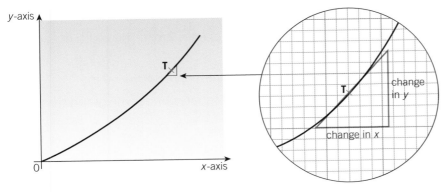

Figure 3 *You find the gradient by drawing a tangent*

If the *x*-axis is time, the gradient will be equal to the rate of change of the variable on the *y*-axis.

On a distance-time graph this is $\frac{\Delta s}{\Delta t}$, or the rate of change of distance with time, or speed.

On a speed-time graph of an object moving in a constant direction, this is $\frac{\Delta v}{\Delta t}$, or the rate of change of speed with time, or acceleration.

4f Measuring the area of graphs

You may also need to find the area of the graph. To find the area you can find the value of one square and then count the squares.

Worked example: Finding the distance travelled

Figure 4 shows a small section of a cyclist's journey.

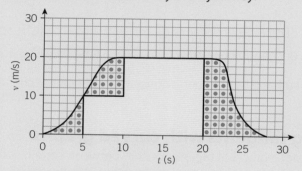

Figure 4 *The cyclist's journey*

Use the graph to calculate an estimate of the distance travelled by the cyclist in this section of her journey.

Solution

Step 1: Calculate the area of any large squares or rectangles contained under the graph (the shaded area).

Small shaded square (from *t* = 5 to *t* = 10), distance represented = 10 m/s × 5 s = 50 m.

Large shaded square (from *t* = 10 to *t* = 20), distance represented = 20 m/s × 10 s = 200 m.

Step 2: Calculate the area of one small square.

The height of each small square is 2 m/s. The width of each small square is 1 s.

So each small square represents 2 m/s × 1 s = 2 m.

Step 3: Count the remaining small squares under the graph that we didn't include in Step 1.

Add half squares together.

Total squares = 64.

Step 4: Multiply by the distance represented by each small square.

Distance = 64 squares × 2 m/square

= 128 m.

Step 5: Add together the distances to find the total distance.

Total distance = 50 m + 200 m + 128 m

= **378 m**

Synoptic link

To see more examples of the use of graphs in physics, see Topics P9.1 and P9.4.

1 a The mathematical equation for a straight line graph is $y = mx + c$. What is represented in this equation by:

  i m [1 mark]   **ii** c [1 mark]

  b Sketch a straight line graph with:

  i   a positive gradient and y-intercept (0,0).   [2 marks]

  ii a smaller positive gradient than your graph from **i** and a negative y-intercept.   [1 mark]

2 The equation $F = k\,e$ gives the force F applied to stretch a spring so its extension is e. The spring constant of the spring is k. Match this equation to the straight line graph equation $y = mx + c$ [2 marks]

3 For an object that accelerates uniformly from an initial velocity u, its velocity v after time t is given by the equation $v = u + at$. Match the equation to the straight line graph equation $y = mx + c$.   [2 marks]

4 Figure 5 shows a distance–time graph.

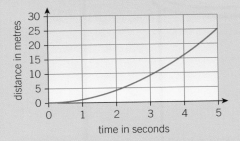

Figure 5 *Distance–time graph for a sprinter*

Calculate the gradient of the line at 4 seconds.   [3 marks]

MS5 Geometry and trigonometry

Learning objectives

After this topic, you should know how to:

- use angular measures in degrees
- visualise and represent 2D and 3D forms including two dimensional representations of 3D objects
- calculate areas of triangles and rectangles, surface areas and volumes of cubes.

5a Measuring and using angles

You measure angles with a protractor. Angles are measured in degrees (°). The angle shown in Figure 1 is 45°.

There are 360° in a circle, 180° in a half circle, and 90° in a quarter of a circle.

The three angles of a triangle always add up to 180°. In Figure 1, the angle in the bottom right corner of the triangle is 90°. This is also called a right angle.

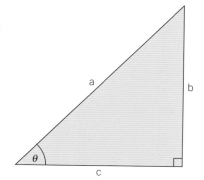

Figure 1 *The symbol for an angle is usually θ*

5b Representation of 3D objects

Three-dimensional (3D) objects have width, height, and depth. Two-dimensional (2D) objects only have width and height. You cannot accurately show all aspects of a 3D object in a diagram. As such, it is easier to draw laboratory apparatus in 2D. Figure 2 and Figure 3 show a beaker drawn in different ways.

Figure 2 *A perspective drawing to give a sense of three dimensions*

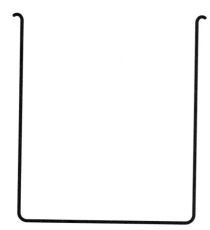

Figure 3 *A cross-section drawing with no attempt to represent three dimensions*

Other 2D representations of 3D objects include:

- nets – 2D shapes that can be cut out and folded to make 3D shapes (Figure 4)
- elevation views (showing objects from a side)
- plan views (showing objects from above or below).

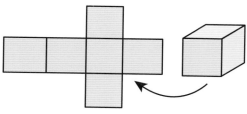

Figure 4 *A cube can be represented as a net*

5c Calculating areas and volumes
Areas of rectangles and triangles

The SI unit of area is the square metre m².

Because a distance of 1 metre = 100 cm = 1000 mm, an area of 1 square metre (m²) = 10 000 cm² = 1 000 000 mm²

A rectangle is a flat shape with four straight sides and a right angle at each of its four corners.

$$\text{area of a rectangle} = \text{height } h \times \text{base } b$$

A triangle is a flat shape with three straight sides.

$$\text{area of a triangle} = \frac{1}{2} \times \text{height } h \times \text{base } b$$

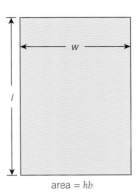

area = hb

Figure 5 *Calculating an area of a rectangle*

Worked example: Area of a triangle
Calculate the area of a triangle that has a base of length 0.35 m and a height of 0.12 m. Give your answer in standard form.

Solution
area = $\frac{1}{2}$ × its height × its base = 0.5 × 0.12 m × 0.35 m = 0.021 m²
    = **2.1 × 10⁻² m²**

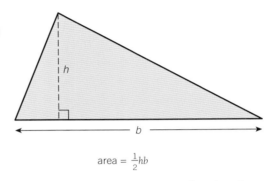

area = $\frac{1}{2}hb$

Figure 6 *Calculating an area of a triangle*

Volumes of cubes

The SI unit of volume is the cubic metre (m³).

Because a distance of 1 metre = 100 cm = 1000 mm,

a volume of 1 cubic metre (m³) = 1 000 000 cm³ = 1 000 000 000 mm³

A cuboid is a rectangular box with sides of unequal length (Figure 7).

$$\text{volume of a cuboid} = \text{length } l \times \text{width } w \times \text{height } h$$

Synoptic link

To see more examples of calculating areas, see Maths skill MS5c.

Worked example: Volume of a cuboid
Calculate the volume of a cuboid that has sides of length 8.0 cm, height 5.0 cm and width 3.0 cm. Give your answer in cubic metres in standard form.

Solution
volume of cuboid = length × width × height
    = (8.0 cm × 5.0 cm × 3.0 cm)³
    = (0.080 m × 0.050 m × 0.030 m)³
    = 0.000120 m³
    = **1.20 × 10⁻⁴ m³**

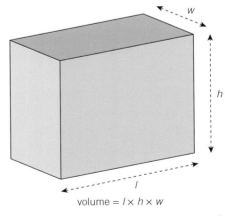

volume = l × h × w

Figure 7 *Calculating the volume of a cuboid*

A cube is a type of cuboid where the length, height, and width are all the same. The volume of a cube can therefore be simplified to:

$$\text{volume} = \text{length}^3$$

1 Calculate the area of
   a a rectangle of length 2.3 m and width 1.9 m   [1 mark]
   b a triangle which has a base of 0.35 m and a height
      of 1.2 m.   [1 mark]
   Give your answers to 2 significant figures.

2 Each side of a cube has a length of 1.5 cm.
   a Calculate the volume of the cube in cubic metres.   [1 mark]
   b Calculate the surface area of the cube in square metres. [1 mark]
   Give your answers to 2 significant figures.

3 The kite in Figure 8 has a length of 1.60 m and a width of 1.15 m.

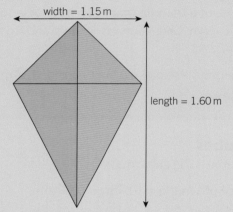

Figure 8

By dividing the area of the kite as a red and a blue triangle, show
that the area of the kite is 1.84 square metres.   [3 marks]

Working scientifically

WS1 Development of scientific thinking

Science works for us all day, every day. Working as a scientist you will have knowledge of the world around you, particularly about the subject you are working with. You will observe the world around you. An enquiring mind will then lead you to start asking questions about what you have observed.

Science usually moves forward by slow steady steps. Each small step is important in its own way. It builds on the body of knowledge that we already have. In this book you can find out about:

- how scientific methods and theories change over time (Topics P7.1, P7.2, P7.8, P16.1, P16.4, P16.5)
- the models that help us to understand theories (Chapter 7)
- the limitations of science, and the personal, social, economic, ethical and environmental issues that arise (Topics P2.5, P3.4, P3.5, P7.8, P7.9, P10.7)
- the importance of peer review in publishing scientific results (Topic P7.2)
- evaluating risks in practical work and in technological applications (Topics WS2, P3.4, P3.5, P7.4, P7.6, P7.9).

The rest of this section will help you to 'work scientifically' when planning, carrying out, analysing and evaluating your own investigations.

Figure 1 *All around you, everyday, there are many observations you can make. Studying science can give you the understanding to explain and make predictions about some of what you observe*

WS2 Experimental skills and strategies

Deciding on what to measure

Variables can be one of two different types:

- A categoric variable is one that is best described by a label (usually a word). The type of material used in an experiment is a categoric variable (e.g., copper).
- A continuous variable is one that you measure, so its value could be any number. Temperature (as measured by a thermometer or temperature sensor) is a continuous variable. Continuous variables have values (called quantities). These are found by taking measurements (e.g., mass, volume, etc.) and S.I. units such as grams (g), metres (m), and joules(J) should be used.

Making your data repeatable and reproducible

When you are designing an investigation you must make sure that you, and others, can trust the data you plan to collect. You should ensure that each measurement is *repeatable*. You can do this by getting consistent sets of repeat measurements and taking their mean. You can also have more confidence in your data if similar results are obtained by different investigators using different equipment, making your measurements *reproducible*.

You must also make sure you are measuring the actual thing you want to measure. If you don't, your data can't be used to answer your original question. This seems very obvious, but it is not always easy to set up. You need to make sure that you have controlled as many other variables as you can. Then no-one can say that your investigation, and hence the data you collect and any conclusions drawn from the data, is not valid.

How might an independent variable be linked to a dependent variable?

- The independent variable is the one you choose to vary in your investigation.
- The dependent variable is used to judge the effect of varying the independent variable.

These variables may be linked together. If there is a pattern to be seen (e.g., as one thing gets bigger the other also gets bigger), it may be that:

- changing one has caused the other to change
- the two are related (there is a correlation between them), but one is not necessarily the cause of the other.

Starting an investigation

As scientists, we use observations to ask questions. We can only ask useful questions if we know something about the observed event. We will not have all of the answers, but we will know enough to start asking the correct questions.

When you are designing an investigation you have to observe carefully which variables are likely to have an effect.

An investigation starts with a question and is followed by a prediction, and backed up by scientific reasoning. This forms a hypothesis that can be tested against the results of your investigation. You, as the scientist, predict that there is a relationship between two variables.

You should think about carrying out a preliminary investigation to find the most suitable range and interval for the independent variable.

Making your investigation safe

Remember that when you design your investigation, you must:

- look for any potential hazards
- decide how you will reduce any **risk**.

You will need to write these down in your plan:

- write down your plan
- make a risk assessment
- make a prediction and hypothesis
- draw a blank table ready for the results.

Different types of investigation

A fair test is one in which only the independent variable affects the dependent variable. All other variables are controlled and kept constant.

Figure 2 *Safety precautions should be appropriate for the risk. The wires in electrical circuits may become warm, but you do not need to wear safety gloves. You should, however, let your teacher know if circuit wires begin to heat up*

This is easy to set up in the laboratory, but it can be difficult in outdoor experiments (e.g., measuring the speed of sound in air), and is almost impossible in fieldwork. Investigations in the environment are not that simple and easy to control. There are complex variables that are changing constantly.

So how can we set up the fieldwork investigations? The best you can do is to make sure that all of the many variables change in much the same way, except for the one you are investigating. For example, if you are monitoring the effects of aircraft noise on people living near an airport, they should all be experiencing the same noise from other outdoor sources – even if it is constantly changing.

If you are investigating two variables in a large population then you will need to do a survey. Again, it is impossible to control all of the variables. For example, imagine scientists investigating the effect of overhead electricity cables on the health of people living at different distances from the cables. They would have to choose people of the same age and same family history to test. Remember that the larger the sample size tested, the more valid the results will be.

Control groups are used in these investigations to try to make sure you are measuring the variable you intend to measure. For example, when investigating the effect of aircraft noise on people living near an airport, the control groups would use people not living near an airport, but still experiencing the same noise from other outdoor sources as the people living near the airport. The control groups would need to be in similar areas to the airport groups, with similar traffic and other non-airport sources of noise. In this way, the effect on people of living near an airport could be compared with the effect on the control groups.

Designing an investigation
Accuracy
Your investigation must provide **accurate** data. Accurate data is essential if your results are going to have any meaning.

How do you know if you have accurate data?
It is very difficult to be certain. Accurate results are very close to the true value. However, it is not always possible to know what the true value is.

- Sometimes you can calculate a theoretical value and check it against the experimental evidence. Close agreement between these two values could indicate accurate data.

- You can draw a graph of your results and see how close each result is to the line of best fit.

- Try repeating your measurements and check the spread or range within sets of repeat data. Large differences in a repeated measurement suggest inaccuracy. Or try again with a different measuring instrument and see if you get the same readings.

Precision
Your investigation must provide data with sufficient **precision** (i.e., close agreement within sets of repeat measurements). If it doesn't then you will not be able to make a valid conclusion.

Figure 3 *Imagine you wanted to investigate the effect of overhead electricity cables on the health of people living at different distances from the cables. You would need to choose a control group using people far away enough from the cables to not be affected by them, but close enough to be still experiencing similar environmental conditions*

> **Study tip**
>
> Trial runs will tell you a lot about how your investigation might work out. They should get you to ask yourself:
> - do I have the correct conditions?
> - have I chosen a sensible range?
> - have I got sufficient readings that are close enough together? The minimum number of points to draw a line graph is generally taken as five.
> - will I need to repeat my readings?

> **Study tip**
>
> A word of *caution*.
>
> Just because your results show precision it does not mean your results are accurate.
>
> Imagine you carry out an investigation into the specific heat capacity of a substance. You get readings of the temperature change in the substance that are all the same. This means that your data will have precision, but it doesn't mean that they are necessarily accurate.

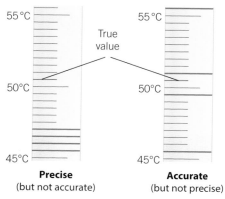

Figure 4 *The green line shows the true value and the pink lines show the readings two different groups of students measured. Precise results are not necessarily accurate results*

Figure 5 *Despite the fact that a stopwatch has a high resolution, it is not always the most appropriate instrument to use for measuring time*

Precision versus accuracy

Imagine measuring the temperature after a set time when a fuel is used to heat a fixed volume of water. Two students repeated this experiment, four times each. Their results are marked on the thermometer scales in Figure 4:

● A precise set of results is grouped closely together.

● An accurate set of results will have a mean (average) close to the true value.

How do you get precise, repeatable data?

● You have to repeat your tests as often as necessary to improve repeatability.

● You have to repeat your tests in exactly the same way each time.

● You should use measuring instruments that have the appropriate scale divisions needed for a particular investigation. Smaller scale divisions have better resolution.

Making measurements
Using measuring instruments

There will always be some degree of uncertainty in any measurements made (WS3). You cannot expect perfect results. When you choose an instrument you need to know that it will give you the accuracy that you want (i.e., it will give you a true reading). You also need to know how to use an instrument properly.

Some instruments have smaller scale divisions than others. Instruments that measure the same thing, such as mass, can have different resolutions. The **resolution** of an instrument refers to the smallest change in a value that can be detected (e.g., a ruler with centimetre increments compared to a ruler with millimetre increments). Choosing an instrument with an inappropriate resolution can cause you to miss important data or make silly conclusions.

But selecting measuring instruments with high resolution might not be appropriate in some cases where the degree of uncertainty in a measurement is high, for example, judging when a wave in a ripple tank reaches the end of the tank (Topic P12.4). In this case an electronic timer measuring to within one thousandth of a second isn't going to improve the precision of the data collected.

WS3 Analysis and evaluation

Errors

Even when an instrument is used correctly, the results can still show differences. Results will differ because of a random error. This can be a result of poor measurements being made. It could also be due to not carrying out the method consistently in each test. Random errors are minimised by taking the mean of precise repeat readings, looking out for any outliers (measurements that differ significantly from the others within a set of repeats) to check again, or omit from calculations of the mean.

The **error** may be a **systematic error**. This means that the method or measurement was carried out consistently incorrectly so that an error was being repeated. An example could be a balance that is not set at zero correctly. Systematic errors will be consistently above, or below, the accurate value.

Presenting data
Tables
Tables are really good for recording your results quickly and clearly as you are carrying out an investigation. You should design your table before you start your investigation.

The range of the data
Pick out the maximum and the minimum values and you have the range. You should always quote these two numbers when asked for a range. For example, the range is between the lowest value in a data set, and the highest value. *Don't forget to include the units.*

The mean of the data
Add up all of the measurements and divide by how many there are. You can ignore outliers in a set of repeat readings when calculating the mean, if found to be the result of poor measurement.

Bar charts
If you have a categoric independent variable and a continuous dependent variable then you should use a bar chart.

Line graphs
If you have a continuous independent and a continuous dependent variable then use a line graph.

Scatter graphs
These are used in much the same way as a line graph, but you might not expect to be able to draw such a clear line of best fit. For example, to find out if the melting point of an element is related to its density you might draw a scatter graph of your results.

Using data to draw conclusions
Identifying patterns and relationships
Now you have a bar chart or a line graph of your results you can begin looking for patterns. You must have an open mind at this point.

Firstly, there could still be some anomalous results. You might not have picked these out earlier. How do you spot an anomaly? It must be a significant distance away from the pattern, not just within normal variation.

A line of best fit will help to identify any anomalies at this stage. Ask yourself – 'do the anomalies represent something important or were they just a mistake?'

Secondly, remember a line of best fit can be a straight line or it can be a curve – you have to decide from your results.

The line of best fit will also lead you into thinking what the relationship is between your two variables. You need to consider whether the points you have plotted show a linear relationship. If so, you can draw a straight

Figure 6 *How you record your results will depend upon the type of measurements you are taking*

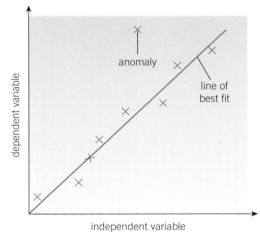

Figure 7 *A line of best fit can help to identify anomalies*

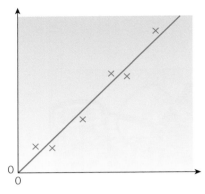

Figure 8 *When a straight line of best fit goes through the origin (0, 0) the relationship between the variables is directly proportional*

line of best fit on your graph (with as many points above the line as below it, producing a 'mean' line. Then consider if this line has a positive or negative gradient.

A **directly proportional** relationship is shown by a positive straight line that goes through the origin (0, 0).

Your results might also show a curved line of best fit. These can be predictable, complex or very complex. Carrying out more tests with a smaller interval near the area where a line changes its gradient will help reduce the error in drawing the line (in this case a curve) of best fit.

Drawing conclusions

Your graphs are designed to show the relationship between your two chosen variables. You need to consider what that relationship means for your conclusion. You must also take into account the repeatability and the reproducibility of the data you are considering.

You will continue to have an open mind about your conclusion.

You will have made a prediction. This could be supported by your results, it might not be supported, or it could be partly supported. It might suggest some other hypothesis to you.

You must be willing to think carefully about your results. Remember it is quite rare for a set of results to completely support a prediction or be completely repeatable.

Look for possible links between variables, remembering that a positive relationship does not always mean a causal link between the two variables.

Your conclusion must go no further than the evidence that you have. Any patterns you spot are only strictly valid in the range of values you tested. Further tests are needed to check whether the pattern continues beyond this range.

The purpose of the prediction was to test a hypothesis. The hypothesis can:

● be supported,

● be refuted, or

● lead to another hypothesis.

You have to decide which it is on the evidence available.

Making estimates of uncertainty

You can use the range of a set of repeat measurements about their mean to estimate the degree of uncertainty in the data collected.

For example, in a test to look at the descent of a parachute over a measured vertical distance, a student got the following results:

4.5 s, 4.9 s, 4.4 s, and 4.8 s.

The mean result = (4.5 + 4.9 + 4.4 + 4.8) ÷ 4 = 4.65 s

The range of these readings is 4.4 s to 4.9 s = 0.5 s.

So, a reasonable estimate of the uncertainty in the mean value would be half the range. In this case, the time taken would be ± 0.25 s.

You can include a final column in your table of results to record the 'estimated uncertainty' in your mean measurements.

The level of uncertainty can also be shown on a graph (Figure 9).

As well as this, there will be some uncertainty associated with readings from any measuring instrument. You can usually take this as:

- half the smallest scale division. For example, 0.5 cm on a centimetre ruler, or
- on a digital instrument, half the last figure shown on its display. For example, on a balance reading to 0.01 g the uncertainty would be ± 0.005 g.

Anomalous results

Anomalies (or outliers) are results that are clearly out of line compared with others. They are not those that are due to the natural variation that you get from any measurement. Anomalous results should be looked at carefully. There might be a very interesting reason why they are so different.

If anomalies can be identified while you are doing an investigation, then it is best to repeat that part of the investigation. If you find that an anomaly is due to poor measurement, then it should be ignored.

Evaluation

If you are still uncertain about a conclusion, it might be down to the repeatability, reproducibility and uncertainty in your measurements. You could check reproducibility by: looking for other similar work on the Internet or from others in your class, getting somebody else, using different equipment, to redo your investigation (this occurs in peer review of data presented in articles in scientific journals), trying an alternative method to see if it results in you reaching the same conclusion.

When suggesting improvement that could be made in your investigation, always give your reasoning.

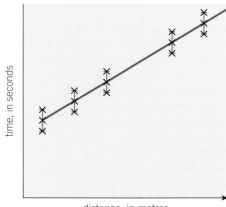

Figure 9 *Indicating levels of uncertainty. These are the results for the time it took for a parachute to fall a measured distance. For each distance, the test was repeated three times*

Study tip

The method chosen for an investigation can only be evaluated as being valid if it actually collects data that can answer your original question. The data should be repeatable and reproducible, and the control variables should have been kept constant (or taken into account if they couldn't be directly manipulated).

Synoptic link

See the Maths skills section to learn how to use SI units, prefixes and powers of ten for orders of magnitude, significant figures, and scientific quantities.

Glossary

Acceleration Change of velocity per second (in metres per second per second, m/s^2).

Activity The number of unstable atoms that decay per second in a radioactive source.

Alpha radiation Alpha particles, each composed of two protons and two neutrons, emitted by unstable nuclei.

Alternating current Electric current in a circuit that repeatedly reverses its direction.

Alternator An alternating current generator.

Amplitude The height of a wave crest or trough of a transverse wave from the rest position. For oscillating motion, the amplitude is the maximum distance moved by an oscillating object from its equilibrium position.

Angle of incidence Angle between the incident ray and the normal.

Angle of reflection Angle between the reflected ray and the normal.

Atomic number The number of protons (which equals the number of electrons) in an atom. It is sometimes called the proton number.

Beta radiation Beta particles that are high-energy electrons created in, and emitted from, unstable nuclei.

Big Bang theory The theory that the universe was created in a massive explosion (the Big Bang), and that the universe has been expanding ever since.

Biofuel Any fuel taken from living or recently living materials, such as animal waste.

Black body radiation The radiation emitted by a perfect black body (a body that absorbs all the radiation that hits it).

Black dwarf A star that has faded out and gone cold.

Black hole An object in space that has so much mass that nothing, not even light, can escape from its gravitational field.

Boiling point Temperature at which a pure substance boils or condenses.

Boyle's Law For a fixed mass of gas at constant temperature, its pressure multiplied by its volume is constant.

Braking distance The distance travelled by a vehicle during the time it takes for its brakes to act.

Carbon neutral A biofuel from a living organism that takes in as much carbon dioxide from the atmosphere as is released when the fuel is burned.

Carrier waves Waves used to carry any type of signal.

Centripetal force The resultant force towards the centre of a circle acting on an object moving in a circular path.

Chain reaction Reactions in which one reaction causes further reactions, which in turn cause further reactions, etc.

Charge-coupled device (CCD) An electronic device that creates an electronic signal from an optical image formed on the CCD's array of pixels

Circuit breaker An electromagnetic switch that opens and cuts off current if too much current passes through it.

Compression Squeezing together.

Concave lens A lens that makes parallel rays diverge (spread out).

Condense Turn from vapour into liquid.

Conservation of energy Energy cannot be created or destroyed.

Conservation of momentum In a closed system, the total momentum before an event is equal to the total momentum after the event. Momentum is conserved in any collision or explosion, provided no external forces act on the objects that collide or explode.

Contrast medium An X-ray absorbing substance used to fill a body organ so the organ can be seen on a radiograph.

Control variable A variable that may, in addition to the independent variable, affect the outcome of the investigation, and therefore has to be kept constant or at least monitored.

Convection Circulation of a liquid or gas caused by increasing its thermal energy.

Converging (convex) lens A lens that makes light rays parallel to the principal axis converge (meet) at a point.

Cosmic microwave background radiation Electromagnetic radiation that has been travelling through space ever since it was created shortly after the Big Bang.

Count rate The number of counts per second detected by a Geiger counter.

Dark matter Matter in a galaxy that cannot be seen. Its presence is deduced because galaxies would spin much faster if their stars were their only matter.

Deceleration Change of velocity per second when an object slows down.

Density Mass per unit volume of a substance.

Diffuse reflection Reflection from a rough surface - the light rays are scattered in different directions.

Diffusion The Spreading out of particles away from each other.

Diode A non-ohmic conductor that has a much higher resistance in one direction (its reverse direction) than in the other direction (its forward direction).

Direct current Electric current in a circuit that is in one direction only.

Directly proportional A graph will show this if the line of best fit is a straight line through the origin.

Dispersion The splitting of white light into the colours of the spectrum.

Displacement Distance in a given direction.

Dissipation of energy The energy that is not usefully transferred and stored in less useful ways.

Diverging (concave) lens A lens that makes light rays parallel to the axis diverge (spread out) as if from a single point – also referred to as a concave lens.

Driving force Force of a vehicle that makes it move (sometimes referred to as motive force).

Dynamo A direct-current generator.

Earth wire The wire in a mains cable used to connect the metal case of an appliance to earth.

Echo Reflection of sound that can be heard.

Efficiency Useful energy transferred by a device ÷ total energy supplied to the device.

Effort The force applied to a device used to raise a weight or move an object.

Elastic A material is elastic if it is able to regain its shape after it has been squashed or stretched.

Electric current Flow of electric charge. The size of an electric current (in amperes, A) is the rate of flow of charge.

Electric field A charged object (X) creates an electric field around itself, which causes a non-contact force on any other charged object in the field.

Electromagnet An insulated wire wrapped round an iron bar that becomes magnetic when there is a current in the wire.

Electromagnetic induction The process of inducing a potential difference in a wire by moving the wire so it cuts across the lines of force of a magnetic field.

Electromagnetic spectrum The continuous spectrum of electromagnetic waves.

Electromagnetic waves Electric and magnetic disturbances that transfer energy from one place to another.

Electrons Tiny negatively charged particles that move around the nucleus of an atom.

Endoscope A medical instrument that uses optical fibres to see inside the body.

Energy levels Specific energy values of electrons in an atom.

Errors Sometimes called uncertainties.

Evaporate Turn from liquid into vapour.

Extension The increase in length of a spring (or a strip of material) from its original length.

Field lines See lines of force.

Fleming's left-hand rule A rule that gives the direction of the force on a current-carrying wire in a magnetic field according to the directions of the current and the field.

Fluid A liquid or a gas.

Focal length The distance from the centre of a lens to the point where light rays parallel to the principal axis are focused (or, in the case of a diverging lens, appear to diverge from).

Force A force (in newtons, N) can change the motion of an object.

Force diagram A diagram showing the forces on an object.

Force multiplier A lever used so that a weight or force can be moved by a smaller force.

Free electron Electron that moves about freely inside a metal and is not held inside an atom.

Free body force diagram A diagram that shows the forces acting on an object without any other objects or forces shown.

Freezing point The temperature at which a pure substance freezes.

Frequency The number of wave crests passing a fixed point every second.

Frequency of an alternating current The number of complete cycles an alternating current passes through each second. The unit of frequency is the hertz (Hz).

Frequency of oscillating motion Number of complete cycles of oscillation per second, equal to 1 ÷ the time period. The unit of frequency is the hertz (Hz).

Friction The force opposing the relative motion of two solid surfaces in contact.

Fuse A fuse contains a thin wire that melts and cuts the current off if too much current passes through it.

Gamma radiation Electromagnetic radiation emitted from unstable nuclei in radioactive substances.

Generator effect The production of a potential difference using a magnetic field.

Geothermal Energy that comes from energy released by radioactive substances deep within the Earth.

Gradient (of a straight line graph) Change of the quantity plotted on the y-axis divided by the change of the quantity plotted on the x-axis.

Gravitational field strength, g The force of gravity on an object of mass 1 kg (in newtons per kilogram, N/kg). It is also the acceleration of free fall.

Half-life Average time taken for the number of nuclei of the isotope (or mass of the isotope) in a sample to halve.

Hooke's law The extension of a spring is directly proportional to the force applied, as long as its limit of proportionality is not exceeded.

Independent variable The variable for which values are changed or selected by the investigator.

Induced magnetism Magnetisation of an unmagnetised magnetic material by placing it in a magnetic field.

Inertia The tendency of an object to stay at rest or to continue in uniform motion.

Infrared radiation Electromagnetic waves between visible light and microwaves in the electromagnetic spectrum.

Input energy Energy supplied to a device.

Internal energy The energy of the particles of a substance due to their individual motion and positions

Inverse proportionality This is where two variables are related such that making one variable n times bigger causes the other one to become n times smaller (e.g. doubling one quantity causes the other to halve).

Ion A charged atom.

Ionisation Any process in which atoms become charged.

Irradiated An object that has been exposed to ionising radiation.

Isotopes Atoms with the same number of protons and different numbers of neutrons.

Kilowatt-hour (kWh) The energy in electricity supplied to a 1 kW electrical device in 1 hour.

Latent heat The energy transferred to or from a substance when it changes its state

Light-depending resistor (LDR) A resistor whose resistance depends on the intensity of the light incident on it.

Light-emitting diode (LED) A diode that emits light when it conducts.

Limit of proportionality The limit for Hooke's law applied to the extension of a stretched spring.

Line of force Line in a magnetic field along which a magnetic compass points – also called a magnetic field line.

Line of force in an electric field Line along which a free positive charge moves along in an electric field.

Live wire The mains wire that has a voltage that alternates in voltage (between +325 V and 325 V in Europe).

Load The weight of an object raised by a device used to lift the object, or the force applied by a device when it is used to shift an object.

Longitudinal waves Waves in which the vibrations are parallel to the direction of energy transfer.

Magnetic field The space around a magnet or a current-carrying wire.

Magnetic field line Line in a magnetic field along which a magnetic compass points – also called a line of force.

Magnetic flux density A measure of the strength of the magnetic field defined in terms of the force on a current-carrying conductor at right angles to the field lines.

Magnification The image height ÷ the object height.

Magnifying glass A converging lens used to magnify a small object which must be placed between the lens and its focal point.

Magnitude The size or amount of a physical quantity.

Main sequence The main sequence is the life stage of a star during which it radiates energy because of fusion of hydrogen nuclei in its core.

Mass The quantity of matter in an object – a measure of the difficulty of changing the motion of an object (in kilograms, kg).

Mass number The number of proton and neutrons in a nucleus.

Mechanical wave Vibration that travels through a substance.

Melting point Temperature at which a pure substance melts or freezes (solidifies).

Microwaves Electromagnetic waves between infrared radiation and radio waves in the electromagnetic spectrum.

Moderator Substance in a nuclear reactor that slows down fission neutrons.

Moment The turning effect of a force defined by the equation: moment of a force (in newton metres, Nm) = force (in newtons, N) × perpendicular distance from the pivot to the line of action of the force (in metres, m).

Momentum This equals mass (in kg) × velocity (in m/s).

Motor effect When a current is passed along a wire in a magnetic field, and the wire is not parallel to the lines of the magnetic field, a force is exerted on the wire by the magnetic field.

National Grid The network of cables and transformers used to transfer electricity from power stations to consumers (i.e., homes, shops, offices, factories, etc.).

Neutral wire The wire of a mains circuit that is earthed at the local substation so its potential is close to zero.

Neutron star The highly compressed core of a massive star that remains after a supernova explosion.

Neutrons Uncharged particles of the same mass as protons. The nucleus of an atom consists of protons and neutrons.

Newton's First Law of motion If the resultant force on an object is zero, the object stays at rest if it is stationary, or it keeps moving with the same speed in the same direction.

Newton's Second Law of motion The acceleration of an object is proportional to the resultant force on the object, and inversely proportional to the mass of the object.

Newton's Third Law When two objects interact with each other, they exert equal and opposite forces on each other.

Normal Straight line through a surface or boundary perpendicular to the surface or boundary.

Nuclear fission The process in which certain nuclei (uranium-235 and plutonium-239) split into two fragments, releasing energy and two or three neutrons as a result.

Nuclear fission reactors Reactors that release energy steadily due to the fission of a suitable isotope such as uranium-235.

Nuclear fuel Substance used in nuclear reactors that releases energy due to nuclear fission.

Nuclear fusion The process where small nuclei are forced together to fuse and form a larger nucleus.

Nucleus Tiny positively charged object composed of protons and neutrons at the centre of every atom.

Ohm's law The current through a resistor at constant temperature is directly proportional to the potential difference across the resistor.

Opaque object An object that light cannot pass through.

Optical fibre Thin glass fibre used to transmit light signals.

Oscillate Move to and fro about a certain position along a line.

Oscilloscope A device used to display the shape of an electrical wave.

Parallel Components connected in a circuit so that the potential difference is the same across each one.

Parallelogram of forces A geometrical method used to find the resultant of two forces that do not act along the same line.

Perpendicular At right angles.

Physical change A change in which no new substances are produced.

Plugs A plug has an insulated case and is used to connect the cable from an appliance to a socket.

Potential difference A measure of the work done or energy transferred to the lamp by each coulomb of charge that passes through it. The unit of potential difference is the volt (V).

Power The energy transformed or transferred per second. The unit of power is the watt (W).

Pressure Force per unit cross-sectional area for a force acting on a surface at right angles to the surface. The unit of pressure is the pascal (Pa) or newton per square metre (N/m^2).

Primary seismic wave Longitudinal waves that push or pull on the material that they move through as they travel through the Earth.

Principal focus The point where light rays parallel to the principal axis of a lens are focused (or, in the case of a diverging lens, appear to diverge from).

Principle of moments For an object in equilibrium, the sum of all the clockwise moments about any point = the sum of all the anti-clockwise moments about that point.

Protons Positively charged particles with an equal and opposite charge to that of an electron.

Protostar The concentration of dust clouds and gas in space that forms a star.

Radiation dose Amount of ionising radiation a person receives.

Radio waves Electromagnetic waves of wavelengths greater than 0.10 m.

Radioactive contamination The unwanted presence of materials containing radioactive atoms on other materials.

Rarefaction Stretched apart.

Reactor core The thick steel vessel used to contain fuel rods, control rods and the moderator in a nuclear fission reactor.

Real image An image formed by a lens that can be projected on a screen.

Red giant A star that has expanded and cooled, resulting in it becoming red and much larger and cooler than it was before it expanded.

Red supergiant A star much more massive than the Sun will swell out after the main sequence stage to become a red supergiant before it collapses.

Redshift Increase in the wavelength of electromagnetic waves emitted by a star or galaxy due to its motion away from us. The faster the speed of the star or galaxy, the greater the redshift is.

Reflection The change of direction of a light ray or wave at a boundary when the ray or wave stays in the incident medium.

Refraction The change of direction of a light ray when it passes across a boundary between two transparent substances (including air).

Relay A switch opened or closed by an iron armature that is attracted to the relay's electromagnet when a current is in the electromagnet

Renewable energy Energy from natural sources that is always being replenished so it never runs out

Repeatable A measurement is repeatable if the original experimenter repeats the investigation using the same method and equipment and obtains the same results.

Reproducible A measurement is reproducible if the investigation is repeated by another person, or by using different equipment or techniques, and the same results are obtained.

Resistance Resistance (in ohms, Ω) = potential difference (in volts, V) ÷ current (in amperes, A).

Resistive forces Forces such as friction and air resistance that oppose the motion of an object.

Resolution of forces The process of considering a force in terms of two perpendicular components, which together have the same effect on an object as the force.

Resultant force A single force that has the same effect as all the forces acting on the object.

Risk The likelihood that a hazard will actually cause harm.

Scalar A physical quantity, such as mass or energy, that has magnitude only (unlike a vector which has magnitude and direction).

Secondary seismic waves Transverse waves that shake the Earth from side to side as they pass through.

Seismic waves Shock waves that travel through the Earth and across its surface as a result of an earthquake

Series Components connected in a circuit in such a way that the same current passes through them.

Simple pendulum A pendulum consisting of a small spherical bob suspended by a thin string from a fixed point.

Solenoid A long coil of wire that produces a magnetic field in and around the coil when there is a current in the coil

Specific heat capacity Energy needed to raise the temperature of 1 kg of a substance by 1 °C.

Specific latent heat of fusion Energy needed to melt 1 kg of a substance with no change of temperature.

Specific latent heat of vaporisation Energy needed to boil away 1 kg of a substance with no change of temperature.

Specular reflection Reflection from a smooth surface. Each light ray is reflected in a single direction.

Speed The speed of an object (metres per second) = distance moved by the object (metres) ÷ time taken to move the distance travelled (seconds).

Split-ring commutator Metal contacts on the coil of a direct current motor that connects the rotating coil continuously to its electrical power supply.

Spring constant Force per unit extension of a spring.

Static electricity Electric charge stored on insulated objects.

Step-down transformer Electrical device that is used to step-down the size of an alternating potential difference.

Step-up transformer Electrical device used to step-up the size of an alternating potential difference.

Stopping distance The distance travelled by the vehicle in the time it takes for the driver to think and brake.

Supernova The explosion of a massive star after fusion in its core ceases and the matter surrounding its core collapses on to the core and rebounds.

Systematic errors Cause readings to be spread a value other than the true value, due to results differing from the true value by a consistent amount each time a measurement is made.

Tangent A straight line drawn to touch a point on a curve so it has the same gradient as the curve at that point.

Terminal velocity The velocity reached by an object when the drag force on it is equal and opposite to the force making it move.

Thermal conductivity Property of a material that determines the energy transfer through it by conduction.

Thermistor A resistor whose resistance depends on the temperature of the thermistor.

Thinking distance The distance travelled by the vehicle in the time it takes the driver to react.

Three-pin plug A three-pin plug has a live pin, a neutral pin, and an earth pin.

Total internal reflection The total reflection of a light ray in a transparent substance when it reaches a boundary with air or another transparent substance.

Transformer Electrical device used to change an (alternating) voltage. See also Step-up transformer and Step-down transformer.

Translucent object An object that allows light to pass through, but the light is scattered or refracted.

Transmission A wave passing through a substance.

Transparent object Object that transmits all the incident light that enters the object.

Transverse wave A wave where the vibration is perpendicular to the direction of energy transfer.

Ultrasound wave Sound wave at frequency greater than 20 000 Hz (the upper frequency limit of the human ear).

Ultraviolet radiation Electromagnetic waves between visible light and X-rays in the electromagnetic spectrum.

Upthrust The upward force that acts on a body partly or completely submerged in a fluid.

Useful energy Energy transferred to where it is wanted in the way that is wanted.

Vector A vector is a physical quantity, such as displacement or velocity, that has a magnitude and a direction (unlike a scalar which has magnitude only).

Velocity Speed in a given direction (in metres/second, m/s).

Vibrate Oscillate (move to and fro) rapidly about a certain position.

Virtual image An image, seen in a lens or a mirror, from which light rays appear to come after being refracted by a lens or reflected by a mirror.

Wasted energy Energy that is not usefully transferred.

Wave speed The distance travelled per second by a wave crest or trough.

Wavelength The distance from one wave crest to the next.

Weight The force of gravity on an object (in newtons, N).

White dwarf A star that has collapsed from the red giant stage to become much hotter and denser.

White light Light that includes all the colours of the spectrum.

Work The energy transferred by a force. Work done (joules, J) = force (newtons, N) × distance moved in the direction of the force (metres, m).

X-rays Electromagnetic waves smaller in wavelength than ultraviolet radiation produced by X-ray tubes.

Index

Appendix 1: Physics equations

You should be able to remember and apply the following equations, using SI units, for your assessments.

Word equation	Symbol equation
weight = mass × gravitational field strength	$W = mg$
force applied to a spring = spring constant × extension	$F = ke$
acceleration = $\dfrac{\text{change in velocity}}{\text{time taken}}$	$a = \dfrac{\Delta v}{t}$
Ⓗ momentum = mass × velocity	$p = mv$
gravitational potential energy = mass × gravitational field strength × height	$E_p = mgh$
power = $\dfrac{\text{work done}}{\text{time}}$	$P = \dfrac{W}{t}$
efficiency = useful power output × total power output	
charge flow = current × time	$Q = It$
power = potential difference × current	$P = VI$
energy transferred = power × time	$E = Pt$
density = $\dfrac{\text{mass}}{\text{volume}}$	$\rho = \dfrac{m}{V}$
work done = force × distance (along the line of action of the force)	$W = Fs$
distance travelled = speed × time	$s = vt$
resultant force = mass × acceleration	$F = ma$
kinetic energy = 0.5 × mass × (speed)²	$E_k = \dfrac{1}{2}mv^2$
power = $\dfrac{\text{energy transferred}}{\text{time}}$	$P = \dfrac{E}{t}$
efficiency = $\dfrac{\text{useful output energy transfer}}{\text{useful input energy transfer}}$	
wave speed = frequency × wavelength	$v = f\lambda$
potential difference = current × resistance	$V = IR$
power = current² × resistance	$P = I^2R$
energy transferred = charge flow × potential difference	$E = QV$

The following equations are GCSE Physics only and will not be required for the GCSE Combined Science: Trilogy course.

GCSE Physics only

pressure = $\dfrac{\text{force normal to a surface}}{\text{area of that surface}}$	$p = \dfrac{F}{A}$
moment of a force = force × distance (normal to direction of force)	$M = Fd$

You should be able to select and apply the following equations from the Physics equation sheet.

Word equation	Symbol equation
(final velocity)^2 − (initial velocity)^2 = 2 × acceleration × distance	$v^2 - u^2 = 2\,a\,s$
elastic potential energy = 0.5 × spring constant × extension^2	$E_e = \frac{1}{2}k\,e^2$
period = $\dfrac{1}{\text{frequency}}$	
Ⓗ force on a conductor (at right angles to a magnetic field) carrying a current = magnetic flux density × current × length	$F = B\,I\,l$
change in thermal energy = mass × specific heat capacity × temperature change	$\Delta E = m\,c\,\Delta\theta$
thermal energy for a change of state = mass × specific latent heat	$E = m\,L$
Ⓗ potential difference across primary coil × current in primary coil = potential difference across secondary coil × current in secondary coil	$V_s\,I_s = V_p\,I_p$

The following equations are GCSE Physics only and will not be required for the GCSE Combined Science: Trilogy course.

GCSE Physics only

Ⓗ pressure due to a column of liquid = height of column × density of liquid × gravitational field strength	$p = h\rho g$
Ⓗ $\dfrac{\text{potential difference across primary coil}}{\text{potential difference across secondary coil}}$ $= \dfrac{\text{number of turns in primary coil}}{\text{number of turns in secondary coil}}$	$\dfrac{V_p}{V_s} = \dfrac{n_p}{n_s}$
For gases: pressure × volume = constant	$p\,V = \text{constant}$
Ⓗ force = $\dfrac{\text{change in momentum}}{\text{time taken}}$	$F = \dfrac{m\,\Delta v}{\Delta t}$
magnification = $\dfrac{\text{image height}}{\text{object height}}$	

CODEX™

ELDAR®

BY GAVIN THORPE

Additional Text: Andy Chambers,
Jervis Johnson & Tuomas Pirinen
Book Cover Art: Geoff Taylor
Internal Art: John Blanche, David Gallagher,
Jes Goodwin, Des Hanley, Neil Hodgson,
Nuala Kennedy & Paul Smith
Miniatures Designers: Tim Adcock, Chris Fitzpatrick,
Jes Goodwin, Mike McVey & Gary Morley

PRODUCED BY GAMES WORKSHOP

Citadel & the Citadel logo, 'Eavy Metal, Eldar, Games Workshop & the Games Workshop logo, Space Marine and Warhammer are trademarks of Games Workshop Ltd registered in the UK and elsewhere in the world.

Aspect Warrior, Asurmen, Avatar, Baharroth, Codex, Dark Reaper, Defender Squad, Dire Avenger, Eldrad Ulthran, Exarch, Falcon, Farseer, Fire Dragon, Fire Prism, Fuegan, Guardian, Howling Banshee, Iyanna Arienal, Jain Zar, Jetbike, Karandras, Maugan Ra, Nuadhu 'Fireheart', Shining Spears, Storm Squad, Striking Scorpions, Swooping Hawk, Vyper, Warlock, Warp Spider, War Walker, Wave Serpent, Wild Rider, Wraithguard and Wraithlord are all trademarks of Games Workshop Ltd.

British Cataloguing-in-Publication Data. A catalogue record for this book is available from the British Library.

Second Printing.

UK	US	AUSTRALIA	CANADA	JAPAN
GAMES WORKSHOP LTD.	GAMES WORKSHOP INC.	GAMES WORKSHOP,	2679 BRISTOL CIRCLE,	GAMES WORKSHOP,
WILLOW RD, LENTON,	6721 BAYMEADOW DRIVE,	23 LIVERPOOL ST,	UNITS 2&3, OAKVILLE,	WILLOW RD, LENTON,
NOTTINGHAM,	GLEN BURNIE,	INGLEBURN,	ONTARIO	NOTTINGHAM,
NG7 2WS	MARYLAND, 21060 6401	NSW 2565	L6H 6Z8	NG7 2WS

PRODUCT CODE: 60 03 01 04 001   Games Workshop World Wide Web site: http://www.games-workshop.com   ISBN: 1 869893 39

INTRODUCTION

Welcome, revered leader, star of the firmament to Codex: Eldar, a book dedicated to helping you collect, paint and fight battles with the warriors of the Craftworld Eldar in Warhammer 40,000.

OVERVIEW OF THE ELDAR

The Eldar are an incredibly ancient race, who once ruled a vast empire across the stars. Then came the hideous times of the Fall, when the Eldar were consumed by their own decadence and fell from power. The few who survived were scattered across the stars.

Though the Eldar are few in number, they are one of the most technologically advanced races in the galaxy. This advantage is combined with the prodigious abilities of their Farseers, who scry the future and guide their kin along the most favourable paths of fate.

WHY COLLECT AN ELDAR ARMY?

The Eldar are a very powerful army with many specialist squads, in the form of the Aspect Warriors. Their vehicles are highly effective – fast skimmers without exception, heavily armed for their size. With their Falcons and Wave Serpents, the Eldar excel at swift

attacks and flexible defence, enabling them to move the right units to where they are needed. Although generally not as well armoured and tough as Space Marines, for example, Craftworld Eldar do have access to reasonably heavily armoured troops, such as the Wraithguard, Striking Scorpions and Warp Spiders.

Craftworld Eldar also have the most powerful battlefield psykers in the Warhammer 40,000 game, and in a greater ratio when compared to other races. Farseers and Warlocks can provide excellent support with their psychic powers, confounding the enemy's plans and bolstering your own attack.

To summarise, a Craftworld army has many different tools at their disposal: psykers, specialist troops and excellent vehicles. However, they tend to be more fragile than other races and if used unwisely they will not fare well. Many of the troops and vehicles are relatively expensive in terms of their points cost, so it is essential that you maximise their strengths at every opportunity. If your long range support units get engaged in close combat, or your assault units have to spend most of the battle trekking across the tabletop, you'll get a poor return for your investment of points.

Craftworld Eldar armies look very striking on the battlefield. Each craftworld has an overall colour scheme which unifies the army, broken by the ritual colours of the different Aspect Warrior squads. With the Avatar, Wraithlords and War Walkers there are plenty of centre-piece models you can really go to town on, and the Falcon and its variants are possibly the slickest-looking vehicles in Warhammer 40,000!

WHAT'S IN THIS BOOK?

Codex: Eldar is split into the following sections, each of which highlights a different aspect of using a Craftworld army in Warhammer 40,000:

The Army List. You can use this revised army list to put together your battle-winning force.

The Painting and Collecting Guide. Here you will find advice on how to collect a Craftworld Eldar army, ways it can be used in battle and, of course, lots of painting information to help you get your army painted and ready for battle.

The Craftworld Eldar. This section contains full rules for all of the various weapons and items of wargear that are used by the Craftworld Eldar, as well as the powers of Farseers and Warlocks. In addition, there is also a mixture of background text, stories and Imperial reports; each of which illuminates some part of the Eldar psyche. Also included is a selection of famous Eldar characters that you may use in your battles if you want to.

"Your understanding is not required mon-keigh, merely your surrender..."

Message to Colonel Brand at the Third Battle of Belafon I

To: Magos Xenologis Frantix
Transmitted: Onchestus
Received: Sherilax
Date: 4543848.M38
Duct: Astropath-terminus Lumir
Author: Tech-Engineer Pilamist
Purity Seal: Inquisitor Abhorrun

Thought for the day: Study the alien, the better to kill it

Our preliminary findings of the artefact found in the Trojan Sector can be summarised thus:

+ The edifice [illustrated in the attached schematic] is some 5.36 metres high, constructed of an unknown material, possibly some form of organic-based polymer.

+ The surface of the obelisk is decorated with numerous Eldar runes, with character height varying from .23 metres to 1.2 metres.

+ There is no external mechanism, wiring or other evidence of any mechanical operation.

+ Intricate internal crystal circuitry bears a resemblance to the designs on the surface, which may be some kind of psycho-activated mechanism for operating the device.

+ Within a central cavity, there is a small warp core. This rates at approximately 1.02 microtheres — extremely dense for its volume.

The warp aura generated by the core is of a different signature to our own, containing an alternating fluctuation wave-form. It appears that the warp breach is made into a self-contained domain, a part of, but separated from, warp space itself. Such a portal could be opened by releasing the potential energy of the collapsed warp core, enabling sizeable objects to pass through [our estimate is that a stable warp breach of 10-13.5 metres could be sustained indefinitely and expanded to 40-50 metres for short periods of time]. Inquisitor Abhorrun also made such a comment on the >>>Purity Censored<<< If this is true, it would explain the Eldar's ability to deploy from warp space from within a high gravity field, as the gravimetric forces would not affect the sub-containment field. Truly these creatures are one of the most advanced races in our galaxy.

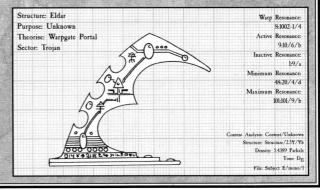

Structure: Eldar
Purpose: Unknown
Theorise: Warpgate Portal
Sector: Trojan

Warp Resonance: 8:1002-1/4
Active Resonance: 9:10/6/b
Inactive Resonance: 1:9/a
Minimum Resonance: 48:20/4/d
Maximum Resonance: 101:101/9/b

Content Analysis: Content/Unknown
Structure: Structure/2.5Y/Yb
Density: 5.4389 Parkels
Tone: D_II
File: Subject E/mono/1

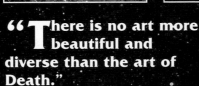

> ❝ **There is no art more beautiful and diverse than the art of Death.** ❞
>
> **Laconfir of Biel-tan**

CRAFTWORLD ELDAR ARMY LIST

The following pages contain an army list that enables you to field a Craftworld Eldar army and fight battles using the scenarios included in the Warhammer 40,000 rulebook. It also provides you with the basic information you'll need in order to field a Craftworld Eldar army in scenarios you've devised yourself, or that form part of a campaign.

The army list is split into five sections. All the squads, vehicles and characters in the army are placed into one of these depending upon their role on the battlefield. Each model is also given a points value, which varies depending on how effective that model is in battle. Before you choose an army, you will need to agree with your opponent upon a scenario and the total number of points each of you will spend. Then you can proceed to pick your army as described below.

USING A FORCE ORGANISATION CHART

The army lists are used in conjunction with the force organisation chart from a scenario. Each chart is split into five categories that correspond to the sections in the army list, and each category has one or more boxes. Each box indicates that you **may** make one choice from that section of the army list, while a dark-toned box means that you **must** make a choice from that section.

Note that unless a model or vehicle forms part of a squad or a squadron, it is a single choice from what is available to your army.

STANDARD MISSIONS

COMPULSORY	OPTIONAL
1 HQ	1 HQ
2 Troops	4 Troops
	3 Elites
	3 Fast Attack
	3 Heavy Support

The Standard Missions force organisation chart is a good example of how to choose an army. To begin with you will need at least one HQ unit and two Troop units (dark shaded boxes indicate units that must be taken for the mission). This leaves the following for you to choose from to make up your army's total points value: up to 1 HQ unit, 0-3 additional Elite units, 0-4 additional Troop units, 0-3 additional Fast Attack units or 0-3 additional Heavy Support units.

USING THE ARMY LISTS

To make a choice, look in the relevant section of the army list and decide what unit you want to have in your army, how many models there will be in it, and which upgrades you want (if any). Remember that you cannot field models that are equipped with weapons or wargear not shown on the model. Once this is done subtract the points value of the unit from your total points, and then go back and make another choice. Continue doing this until you have spent all your points. Then you're ready to do battle!

Army List Entries

Each army list entry consists of the following:

Unit Name: The type of unit, which may also show a limitation on the maximum number of choices you can make of that unit type (0-1, for example).

Profile: These are the characteristics of that unit type, including its points cost. Where the unit has different warriors, there may be more than one profile.

Number/Squad: This shows the number of models in the unit, or the number of models you may take for one choice from the force organisation chart. Often this is a variable amount, in which case it shows the minimum and maximum unit size.

Weapons: These are the unit's standard weapons.

Options: This lists the different weapon and equipment options for the unit and any additional points for taking these options. It may also include an option to upgrade one squad member to a character.

Special Rules: This is where you'll find any special rules that apply to the unit.

SPECIAL RULES

Fleet of Foot

The Eldar are noted for their grace and agility, and are able to move more swiftly than other races when necessary. In the shooting phase, you may declare that an Eldar unit is going to run instead of shoot. Roll a D6. The score is the distance in inches the models in the unit may move in that shooting phase. This move is not affected by difficult terrain.

The following models may not run: Avatar; Eldar vehicles or jetbikes; any model with a saving throw better than 4+, except Dire Avengers, Howling Banshees and Fire Dragon Exarchs; any unit containing an anti-grav platform or support weapon; Dark Reapers.

CRAFTWORLD ELDAR ARMOURY

A Farseer or Warlock may be given up to two weapons from the list below. In addition, a Farseer may be given any items from the Wargear list. No piece of wargear may be taken more than once by a single character and all weapons and wargear must be represented on the model.

A Farseer must take between one to four Farseer psychic powers and a Warlock may be given a single Warlock power.

ELDAR WEAPONS

Close combat weapon . . . . .1 pt
Shuriken pistol . . . . .1 pt
Singing spear . . . . .18 pts
Witch blade . . . . .15 pts

WARLOCK POWERS

Conceal . . . . .20 pts
Destructor . . . . .15 pts
Embolden . . . . .10 pts
Enhance . . . . .15 pts

WARGEAR

Eldar jetbike (twin-linked shuriken catapults) . . . . .30 pts
Ghosthelm . . . . .5 pts
Runes of warding . . . . .10 pts
Runes of witnessing . . . . .5 pts
Spirit stones . . . . .40 pts

VEHICLE UPGRADES

Eldar vehicles may be given certain vehicle upgrades, as noted in their army list entry, for the points costs indicated below. A vehicle may not be given the same upgrade more than once. As with wargear, all vehicle upgrades must be represented on the model.

Crystal targeting matrix . . . . .30 pts
Holo-field . . . . .25 pts
Scythes . . . . .10 pts
Spirit stone . . . . .10 pts
Star engines . . . . .15 pts
Vectored engines . . . . .5 pts

FARSEER PSYCHIC POWERS

Eldritch storm . . . . .35 pts
Fortune . . . . .20 pts
Guide . . . . .30 pts
Mind war . . . . .15 pts

ELDAR SUMMARY

	WS	BS	S	T	W	I	A	Ld	Sv
Avatar	10	0	6	6	4	5	3	10	5+
Farseer	5	5	3	4	3	5	1	10	4+
Warlock	4	4	3	3	1	4	1	8	4+
Exarch	5	5	3	3	1	6	2	9	3+
Warp Spider	4	4	3	3	1	5	1	9	3+
Striking Scorpion	4	4	4	3	1	5	1	9	3+
Howling Banshee	4	4	3	3	1	5	1	9	4+
Fire Dragon	4	4	3	3	1	5	1	9	4+
Wraithguard	4	4	5	5	1	4	1	10	3+
Ranger	3	4	3	3	1	4	1	8	5+
Guardian	3	3	3	3	1	4	1	8	5+
Dire Avenger	4	4	3	3	1	5	1	9	4+
Guardian J/Bike	3	3	3	3(4)	1	4	1	8	3+
Shining Spear	4	4	3	3(4)	1	5	1	9	3+
S/Spear Exarch	5	5	3	3(4)	1	6	2	9	3+
Swooping Hawk	4	4	3	3	1	5	1	9	4+
Wraithlord	4	4	5(10)	8	3	4	2(3)	10	3+
Dark Reaper	4	4	3	3	1	5	1	9	4+

RANGED WEAPONS

Weapon	Range	Str.	AP	Type
Shuriken pistol	12"	4	5	Pistol
Shuriken catapult	12"	4	5	Assault 2
Shuriken cannon	24"	6	5	Heavy 3
Death spinner	12"	6	–	Rapid fire
Exarch d/spinner	12"	6	–	Assault 2
Fusion gun	12"	6	1	Assault 1*
Firepike	18"	8	1	Assault 1*
Wraithcannon	12"	X	1	Assault 1*
Ranger long rifle	36"	X	6	Heavy 1*
Flamer	Template	4	5	Assault 1
Scatter laser	36"	6	6	Heavy D6*
M. Launcher (krak)	48"	8	3	Heavy 1*
M. Launcher (plasma)	48"	4	4	Heavy 1 Blast*
Bright lance	36"	8	2	Assault 1
Starcannon	36"	6	2	Heavy 3
Lasblaster	24"	3	6	Assault 2
Hawks Talon	24"	4	6	Assault 3
Prism cannon	60"	9	2	Heavy 1 Blast
D-cannon	Guess 24"	10	2	Heavy 1 Blast*
Shadow weaver	Guess 48"	6	–	Heavy 1 Blast
Vibro cannon	36"	4	–	Heavy 1*
Reaper launcher	48"	5	3	Heavy 2

*These weapons have additional special rules, see the Weapons section of the Warhammer 40,000 rulebook or the Eldar Wargear section for details.

	Armour			
	Front	Side	Rear	BS
Wave Serpent	12	12	10	3
Falcon	12	12	10	3
Vyper	10	10	10	3
Fire Prism	12	12	10	3

	WS	BS	S	Armour Front	Side	Rear	I	A
War Walker	3	3	5	10	10	10	4	2

'The mon-keigh do not understand their peril,' thought Faeruithir. 'We sent them warnings, messengers to tell them that they must not delve into any past mysteries and terrors, and they ignored us. Well, they will not be able to ignore our weapons, they cannot dare turn a blind eye to our magnificent warriors.

The Warp Spider Exarch felt the mind of Farseer Durell touch his own, hearing the Farseer's message in the instant of contact. Shifting his position slightly so that he could survey the battle, he spoke to the others of his squad.

"The time for our attack is nigh. Prepare to move with me along the skein, my kin."

The armoured warrior-women of the mon-keigh were holding out against the Eldar attack for now. However, Durell had informed the Exarch that Silvanol and his Shining Spears were currently outflanking the enemy to the west, and that the Warp Spiders were to move around the enemy line to the east. When the time was ready, the two squads would attack from behind at the same time, confusing the humans and allowing the Striking Scorpions and Guardians to finish them off with a frontal attack.

Faeruithir activated his jump generator. His stomach lurched momentarily as his body was shifted into the warp. Visions of the immaterial realm swept across his eyes for a split-second and his mind was filled with an anarchic wailing as he could sense a great hungering maw tugging at his spirit. Then the jump generator deposited him back into the real universe.

For a moment after re-appearing, he thought that he could hear a distant heartbeat, thundering across the universe like the pulse of a god, and then the transition was fully complete. Faeruithir found himself in the grounds of some ancient edifice, the remains of its curving walls and strange arches clearly showing that it had not been built by human hands, but by beings far older than the men of Terra.

As the Eldar attack continued, Faeruithir saw the mighty form of the Wraithlord Kuladan, heroine from the Battle of a Thousand Blades and Saviour of the Flame, striding through the ruins, her starcannon sending flares of energy into the foe. A Falcon swept past the towering Wraithlord, pulse laser picking out its target with unerring accuracy, sending plumes of smoke rising from one of the crude mon-keigh transports.

As he watched, the Warp Spider Exarch saw a group of male humans charge across the debris of the ancient human settlement. Kyli and her Storm Squad reacted quickly, pulling their blades from ornamented sheathes. They swiftly intercepted the barbaric aliens before they could reach the hill where Durell stood with his Warlocks, surveying the tide of battle. Faeruithir gasped in horror at what happened next. As Kyli's Guardians encircled the humans, the savages seemed to rip at their clothing. It was only when the first detonation flung four Guardians into the air that the Exarch realised that the mon-keigh had been wearing bandoliers of explosives. As more explosions tore through the Guardians, Faeruithir felt disgust seep through every cell in his body – only humans would put so little value on life that they would gladly commit suicide.

A sudden noise attracted Faeruithir's attention, the sound of stone rolling against stone. Spinning around, he noticed a group of humans trying to sneak past the Eldar lines. This could not be allowed, Durell's orders were that all humans were to be exterminated, none were to be allowed to continue their delving into the Hrudian catacombs.

"Kill them!" Faeruithir told his squad, thrusting an elegant finger at the humans as they clambered amongst the fallen masonry. The Warp Spiders readied their death spinners and a moment later the air was full of a cloud of monofilament wire. As the humans became enmeshed in the tangle, Faeruithir could see them becoming more terrified, the strands slicing at skin and flesh. Their fear grew and they thrashed wildly, lopping off their own limbs as they struggled against the constricting mass. Faeruithir noticed the look in the eyes of one of the mon-keigh – it reminded him of the wild stare of a food animal that knows it is to be slaughtered. Within a few more heartbeats, nothing was left except a red pile of unidentifiable fleshy ruin. Then Durell's voice flashed in the Exarch's mind again, ordering him to move onward.

Activating his warp jump again, the Exarch felt the pull of the warp even more strongly than before. The heartbeat resounded through his ears, the craving for life swept around him, almost overwhelming him. He felt the spirit stone at his chest burning with golden fire against his own heart. Then it was over once again, as their jump generators set him and the squad a hundred yards behind the nearest of their adversaries, within the shelter of the ruins of an old human building; its walls long since crumbled with time, its bricks held together by the moss and vines.

Durell passed on the telepathic command to wait while the trap was set. When Silvanol was in position, that trap would be sprung and the humans would be doomed…

HEADQUARTERS

It is said that when the Great Enemy was born into the universe, the war god Kaela Mensha Khaine fought her. Khaine was defeated, but rather than being destroyed, his substance was scattered across the material realm. It is also said that the Avatars of the Bloody-handed God were found where these fragments came to rest, in the middle of the wraithbone core of the craftworlds that had fled the Fall. They are fighters without equal, with skin of the toughest metals and molten magma for blood. Each carries a Wailing Doom, a weapon of immense power that may take the form of a vicious spear, a mighty sword or a many-bladed axe. An Avatar cannot be wholly killed; if its body is destroyed, its spirit will return to the inner sanctum on the craftworld until it has grown a new form.

0-1 AVATAR OF THE BLOODY-HANDED GOD	Points/Model	WS	BS	S	T	W	I	A	Ld	Sv
Avatar	80	10	0	6	6	4	5	3	10	5+

Weapons: The Wailing Doom.

SPECIAL RULES

Fearless: The Avatar is the living incarnation of a god. It will never fall back, even if attacked by a weapon that would normally make the enemy fall back without a Morale check, and cannot be pinned.

Inspiring: When led by their Avatar, the Craftworld Eldar are filled with thoughts of bloodshed, and its presence inspires them to the greatest acts of valour. Any Eldar unit with a model within 12" of an Avatar becomes fearless in close combat. This means that when the unit is fighting in an assault, it will automatically pass any Morale checks it is required to make. Also, if the Avatar itself is in close combat, all Eldar units with a model within 12" add +1 to their score when working out who has won a round of close combat (in effect they count as having inflicted 1 more wound).

Independent Character: The Avatar is an independent character and follows the Independent Character special rules as given in the Warhammer 40,000 rulebook.

Monstrous Creature: The Avatar is a huge and fearsomely strong opponent. It is treated as a monstrous creature and therefore rolls 2D6 for armour penetration and ignores opponents' armour saves in close combat.

Invulnerable: The Avatar is a supernatural creature whose physical vessel is very difficult to destroy. It is treated as being Invulnerable and therefore may make an armour save against any and all wounds it takes, even those that would normally pierce its armour or that allow no save to be made.

Daemon: To all intents and purposes, an Avatar is a Daemon and will be affected by weapons or abilities that use special rules against Daemons.

Farseers are potent psykers, whose prodigious powers allow them to see the future. By casting their runes, they can travel the tangled skeins of probability to divine which course of action should be taken.

FARSEER	Points/Model	WS	BS	S	T	W	I	A	Ld	Sv
Farseer	40	5	5	3	4	3	5	1	10	4+

Options: A Farseer may be given any equipment allowed by the Craftworld Eldar armoury.

Bodyguard: The Farseer may be accompanied by up to 5 Warlocks (see separate entry).

Transport: The Farseer and Warlocks may be mounted in a Wave Serpent for +110 pts.

SPECIAL RULES

Independent Character: Unless accompanied by one or more Warlocks, the Farseer is an independent character and follows the Independent Character special rules as given in the Warhammer 40,000 rulebook.

Psychic Powers: The Farseer must choose between 1 and 4 psychic powers for the points cost listed in the Craftworld Eldar armoury.

Rune Armour: See the Wargear section for details.

> "What do humans know of our pain? We have sung songs of lament since before your ancestors crawled on their bellies from the sea."
>
> Farseer Eldrad Ulthran

WARLOCK BODYGUARD

	Points/Model	WS	BS	S	T	W	I	A	Ld	Sv
Warlock	11	4	4	3	3	1	4	1	8	4+
Jetbike Warlock	+25	4	4	3	3(4)	1	4	1	8	3+

Number: Each Farseer allows you to field up to 5 Warlocks.

Options: A Warlock may be given any equipment allowed by the Craftworld Eldar armoury.

Character: Warlocks are characters, but may not move on their own. They must either remain in a unit with the Farseer, or they may be assigned to join a Wraithguard or Guardian squad as indicated in the appropriate entries in the army list. With the exception of the Farseer's bodyguard, you may not have more than one Warlock in a unit.

Transport: Warlocks assigned to Guardian Jetbike squadrons or in a unit with a Farseer with a jetbike must be mounted on an Eldar jetbike at an additional cost of +25 pts each. The jetbike is armed with twin-linked shuriken catapults.

Warlock Powers: Each Warlock may be given a single Warlock power at the points cost listed in the Craftworld Eldar armoury.

Rune Armour: See the Wargear section for details.

ELITES

WARP SPIDERS

	Points/Model	WS	BS	S	T	W	I	A	Ld	Sv
Warp Spider	22	4	4	3	3	1	5	1	9	3+
Exarch	+12	5	5	3	3	1	6	2	9	3+

Squad: The squad consists of between 5 and 10 Warp Spiders.

Weapons: Death spinner.

Character: One model in the squad may be upgraded to an Exarch for +12 pts. The Exarch may be armed with an additional death spinner for +10 pts, turning his death spinner into an Assault 2 weapon instead of rapid fire. The Exarch may also be armed with powerblades at +15 pts.

The Exarch may have the following warrior powers: Surprise Assault at +30 pts; Withdraw at +15 pts.

SPECIAL RULES

Warp Jump Generators: See the Wargear section for details.

The Eldar god of war is Kaela Mensha Khaine – the Bloody-handed God. The Aspect Warriors each represent a different facet of Khaine's existence, a different 'aspect' of death and destruction. When an Eldar treads the Path of the Warrior, he or she will choose an Aspect Shrine in which to study the arts of war.

The Aspect Shrines are tended by the Exarchs, who are also responsible for passing on their deadly skills to the Aspect Warriors who attend their shrine. Exarchs are Eldar who have become trapped on the Path of the Warrior, unable to suppress their love of war and their desire for combat. Exarchs wear the finest armour and carry ancient and exotic weapons that have been maintained since the founding of the Shrine. They are examples of what can happen to an Eldar if they stray from the path, and are held with a mixture of fear and awe by other Eldar.

Warp Spiders are named after the tiny crystalline creatures that roam a craftworld's infinity matrix, purging it of non-Eldar psychic presences. The Warp Spiders epitomise the aggressive defence of these creatures, using their warp jump generators to materialise next to their foes and attack, slipping away again before the enemy can retaliate. As Warp Spiders use their jump generators to travel, they are able to wear bulkier and heavier armour, which could constrict the movements of other Eldar.

Striking Scorpions are close assault specialists who excel in dense terrain. They use every nook and crevice to get close to the enemy before springing an attack. The sting comes in the form of the deadly mandiblaster, used to attack a foe from a few paces away. Only the toughest Eldar can become Striking Scorpions, as strong physique is needed to wear the heavy armour and swing the chainswords to smash through armour and bone.

STRIKING SCORPIONS

	Points/Model	WS	BS	S	T	W	I	A	Ld	Sv
Scorpion	16	4	4	4	3	1	5	1	9	3+
Exarch	+12	5	5	4	3	1	6	2	9	3+

Squad: The squad consists of between 5 and 10 Striking Scorpions.

Weapons: Shuriken pistol, chainsword and mandiblaster.

Options: The squad may be equipped with Plasma and Haywire grenades for +3 pts per model.

Character: One model in the squad may be upgraded to an Exarch at an additional cost of +12 pts. The Exarch may exchange his chainsword for one of the following weapons: Biting Blade at +5 pts; Scorpion's Claw at +15 pts.

The Exarch may be given the following warrior powers: Crushing Blow at +10 pts; Stealth at +20 pts.

Transport: The Striking Scorpions may be mounted in a Wave Serpent for +110 pts.

"There can be no peace while alien feet still tread upon Ath-Ethon."

Response to the surrender of the Fourth Imperial Garrison, Rigal IV

The banshee is a harbinger of woe and death in Eldar mythology, whose cry is said to herald ill fate and can tempt a soul from its spirit stone. It is fitting, therefore, that perhaps the most feared of all the Aspect Warriors draw their inspiration from this unearthly creature. A female Aspect, like the banshee of legend, Howling Banshees are fearsome close combat opponents, whose Banshee masks and gleaming power weapons have meant the doom of countless foes over the millennia.

HOWLING BANSHEES

	Points/Model	WS	BS	S	T	W	I	A	Ld	Sv
Banshee	16	4	4	3	3	1	5	1	9	4+
Exarch	+16	5	5	3	3	1	6	2	9	3+

Squad: The squad consists of between 5 and 10 Howling Banshees.

Weapons: Shuriken pistol and power weapon.

Character: One model in the squad may be upgraded to an Exarch for +16 pts. The Exarch may exchange her power weapon for an Executioner for +5 pts or be equipped with powerblades for +5 pts.

The Exarch may be given the following warrior powers: War Shout at +20 pts; Acrobatic at +8 pts.

Wave Serpent: The Howling Banshees may be mounted in a Wave Serpent for +110 pts.

SPECIAL RULES

Banshee Masks: See the Wargear section for details.

The Fire Dragon Aspect is based upon the writhing, sinewy dragon of Eldar myth; an incarnation of destruction and devastation. Fire Dragons are experts at close quarter fighting, where their fusion guns and melta bombs can destroy almost any foe, vehicle or fortification, no matter how well armoured.

FIRE DRAGONS

	Points/Model	WS	BS	S	T	W	I	A	Ld	Sv
Fire Dragon	17	4	4	3	3	1	5	1	9	4+
Exarch	+11	5	5	3	3	1	6	2	9	3+

Squad: The squad consists of between 5 and 10 Fire Dragons.

Weapons: Fusion gun and Melta bombs.

Options: The squad may be equipped with Plasma grenades for +2 pts per model.

Character: One model in the squad may be upgraded to an Exarch for +11 pts. The Exarch may exchange his fusion gun for a Firepike for +18 pts.

The Exarch may be given the following warriors powers: Burning Fist for +20 pts; Tank Hunter for +15 pts.

Transport: The Fire Dragons may be mounted in a Wave Serpent for +110 pts.

WRAITHGUARD

	Points/Model	WS	BS	S	T	W	I	A	Ld	Sv
Wraithguard	35	4	4	5	5	1	4	1	10	3+

Squad: The squad consists of between 5 and 10 Wraithguard.

Weapons: Wraithcannon.

Character: The Wraithguard may be joined by a Warlock from the Farseer's retinue. See the Farseer entry in the HQ section of the army list.

Transport: If there are 5 Wraithguard and up to 1 Warlock in the squad, it may be mounted in a Wave Serpent for +110 pts.

SPECIAL RULES

Fearless: Wraithguard are not living creatures and are therefore not affected by emotions such as dread and urges of self-preservation. A Wraithguard unit never falls back and cannot be pinned. Even attacks which normally cause the enemy to automatically fall back have no effect on Wraithguard. If a character joins the Wraithguard, then he also becomes fearless.

Wraithsight: Wraithguard do not see the world as mortals do, but instead witness an ever-shifting image of spirits, which makes them slow to react to changes on the battlefield. At the start of every Eldar turn, roll a D6 for each Wraithguard unit that is not led by a Farseer or Warlock. On a roll of a 1, the Wraithguard may do nothing that turn.

TROOPS

RANGERS

	Points/Model	WS	BS	S	T	W	I	A	Ld	Sv
Ranger	19	3	4	3	3	1	4	1	8	5+

Squad: The squad consists of between 3 and 10 Rangers.

Weapons: Ranger long rifle, shuriken pistol.

SPECIAL RULES

Infiltrators: In the right circumstances, Rangers have the ability to work their way into a forward position on the battlefield. To represent this, they may set up using the Infiltrators rule, but only if the mission allows for Infiltrators to be used. If the mission does not allow use of the Infiltrators rule then the Rangers must set up normally with the rest of the army.

Difficult Terrain: Rangers are famed for their ability to slip unseen through the most rugged terrain, causing no more disturbance than the passing of a breeze. A Ranger squad moving through difficult terrain can roll one dice more than usual, choosing the highest roll as their movement as normal.

Cameleoline Cloaks: Rangers are swathed in cloaks and robes that make them all but invisible to the naked eye. A Ranger squad adds +1 to any cover saves it is allowed (eg, a 5+ cover save becomes a 4+ cover save). If they are not in cover then they have a 6+ cover save.

> "We used to think of them as wandering vagrants. Well, those vagrants held up my whole platoon for five days."
>
> Lieutenant Pharàik on Eldar Rangers

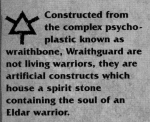
Constructed from the complex psycho-plastic known as wraithbone, Wraithguard are not living warriors, they are artificial constructs which house a spirit stone containing the soul of an Eldar warrior.

Life in the craftworld is strict and disciplined, and there have been many Eldar over the millennia who have tired of the Eldar Path. They leave their craftworld seeking excitement and adventure, and tread the Path of the Outcast. Many die, alone and forgotten. Some fall from grace and become consumed by their dark passions, while others manage to exorcise their wanderlust and eventually return to their craftworld.

Guardian Storm squads are formed from Eldar who were once warriors from one of the close combat Aspect Shrines such as the Striking Scorpions. Some Guardians carry a close-range weapon, such as a fusion gun, which enables them to fire a powerful and deadly blast of energy at any enemy that get too close.

GUARDIAN STORM SQUAD

	Points/Model	WS	BS	S	T	W	I	A	Ld	Sv
Guardian	8	3	3	3	3	1	4	1	8	5+
Warlock	11	4	4	3	3	1	4	1	8	4+

Squad: The squad consists of between 5 and 20 Guardians.

Weapons: Shuriken pistol and close combat weapon.

Options: The Guardians may be armed with Plasma and Krak grenades at +2 pts per model and Haywire grenades for +3 pts per model.

Up to two models in the squad may exchange their weapons for either a fusion gun at +4 pts each or a flamer at +3 pts each.

Character: The unit may be joined by a Warlock from the Farseer's bodyguard. See the entry in the HQ section of the army list.

Transport: A Guardian Storm squad numbering 10 or less models may be mounted in a Wave Serpent for +110 pts.

Highly skilled with their shuriken catapults, Dire Avengers are deadly on the attack and immovable in defence.

DIRE AVENGERS

	Points/Model	WS	BS	S	T	W	I	A	Ld	Sv
Dire Avenger	12	4	4	3	3	1	5	1	9	4+
Exarch	+14	5	5	3	3	1	6	2	9	3+

Squad: The squad consists of between 5 and 10 Dire Avengers.

Weapons: Shuriken catapult.

Character: One model in the squad may be upgraded to an Exarch for +14 pts. The Exarch may exchange his shuriken catapult for a shuriken pistol and a power weapon for +10 pts; or a shuriken pistol and Diresword for +20 pts.

The Exarch may be given the following warrior powers: Distract for +8 pts; Defend for +12 pts.

Transport: The Dire Avengers may be mounted in a Wave Serpent for +110 pts.

> "Some call the Eldar decadent. If that is true, the Imperial Army could do with that kind of decadence."
>
> Last words of Colonel Brin, executed Heretic 463.M38

The Wave Serpent is the main troop carrier of a craftworld's army. Protected inside its hull and force field, Guardians and Aspect Warriors can be transported in safety to any part of the battlefield. Its powerful anti-grav engines give it great speed, making it possibly the best troop transport in the galaxy.

Transport: WAVE SERPENT

	Points	Front Armour	Side Armour	Rear Armour	BS
Wave Serpent	110	12	12	10	3

Energy Field: The prow is protected by an energy field to ward off enemy shots. Any ranged attack against the Wave Serpent from the front or side arc with a Strength greater than 8 counts as Strength 8. In addition, any attacks against a Wave Serpent never roll more than +1D6 for their armour penetration (for example, melta weapons at half range or ordnance only roll one dice). Attacks in close combat, or from the rear, are unaffected by the energy field and do not suffer any of these penalties.

Type: Skimmer, Tank, Fast.

Weapons: Twin-linked shuriken catapults, twin-linked shuriken cannons.

Options: You may upgrade the twin-linked shuriken cannons with one of the following: twin-linked scatter lasers at +5 pts; twin-linked Eldar missile launchers at +20 pts; twin-linked bright lances at +15 pts; twin-linked starcannons at +15 pts. The twin-linked catapults can be upgraded to a single shuriken cannon for +20 pts. A Wave Serpent may be given the following vehicle upgrades: crystal targeting matrix, spirit stone, vectored engines, star engines, scythes.

Transport: The Serpent can carry up to 10 models, or 5 Wraithguard and a Warlock. It may not carry an Avatar or Wraithlord or a squad that has an anti-grav platform.

GUARDIAN DEFENDER SQUAD

	Points/Model	WS	BS	S	T	W	I	A	Ld	Sv
Guardian	8	3	3	3	3	1	4	1	8	5+
Warlock	11	4	4	3	3	1	4	1	8	4+

Squad: The squad consists of between 5 and 20 Guardians.

Weapons: Shuriken catapult.

Options: The squad may be armed with Plasma grenades at a cost of +2 pts per model.

The Guardian squad may be joined by a heavy weapon platform at the additional points cost listed: shuriken cannon +35 pts; scatter laser +40 pts; Eldar missile launcher +55 pts; bright lance +50 pts; starcannon +50 pts. The heavy weapon platform has two Guardians as crew, each armed with a shuriken catapult or shuriken pistol and close combat weapon. These do not count towards the maximum or minimum squad size. It requires one crewman to fire the platform, the other may shoot with their own weapon freely. If one crewman is killed the platform operates as normal; if both crew are killed the platform is useless (the platform itself can't be hit). The heavy weapon platform can move and fire with a heavy weapon.

Character: The unit may be joined by a Warlock from the Farseer's bodyguard. See the entry in the HQ section of the army list.

Transport: Unless accompanied by an anti-grav platform, a Guardian Defender squad numbering 10 or fewer models may be mounted in a Wave Serpent for +110 pts.

In times of need, those taught the ways of war form squads of Guardian Defenders. Anti-grav platforms allow them to provide mobile heavy firepower when on the advance.

FAST ATTACK

GUARDIAN JETBIKE SQUADRON

	Points/Model	WS	BS	S	T	W	I	A	Ld	Sv
Jetbike	35	3	3	3	3(4)	1	4	1	8	3+
Jetbike Warlock	36	4	4	3	3(4)	1	4	1	8	3+

Squadron: The squadron consists of between 3 and 10 Guardian jetbikes.

Type: Eldar jetbike.

Weapons: The jetbikes are armed with twin-linked shuriken catapults. The riders are armed with a shuriken pistol.

Options: Up to one in three jetbikes may replace their shuriken catapults with a single shuriken cannon at +20 pts per model.

Character: The unit may be joined by a Warlock from the Farseer's bodyguard. See the entry in the HQ section of the army list.

Eldar anti-gravitic technology is a source of constant amazement and jealousy to the Adeptus Mechanicus. It is this that enables the Eldar to create anti-grav vehicles ranging from the huge Scorpion super heavy tanks down to the one-man jetbikes. Through subtle manipulation of the gravity field, jetbikes combine high speed with incredible manoeuvrability, making them an ideal craft for launching rapid hit-and-run attacks against the enemy.

SHINING SPEARS

	Points/Model	WS	BS	S	T	W	I	A	Ld	Sv
Shining Spear	50	4	4	3	3(4)	1	5	1	9	3+
Exarch	+20	5	5	3	3(4)	1	6	2	9	3+

Squadron: The squadron consists of between 3 and 5 Shining Spears.

Type: Eldar jetbike.

Weapons: The jetbikes are armed with twin-linked shuriken catapults. The riders are armed with a laser lance.

Character: One model in the squad may be upgraded to an Exarch for +20 pts. The Exarch may exchange his laser lance for a bright lance for +25 pts or a power weapon for +5 pts.

The Exarch may be given the following warrior powers: Skilful Rider for +5 pts; Evade for +10 pts.

The Shining Spears are one of the rarest, most specialised Aspects. They embody the spear of Kaela Mensha Khaine, that struck like lightning and could kill any foe with a single blow. The Shining Spears ride jetbikes so that they can strike without warning, pouncing on their enemy in an instant and dealing death with blasts from their laser lances.

Vypers are highly mobile weapons platforms, capable of laying down a withering hail of fire even at high speed. Although not heavily armoured, their ability to skim quickly through the air provides them with as sure a defence as any amount of thick armoured plating.

Swooping Hawks specialise in bringing death to anyone, no matter who they are. Their wings allow them to swiftly move to anywhere on the battlefield, picking off the enemy with a hail of energy bolts.

VYPER SQUADRON

	Points/Model	Front Armour	Side Armour	Rear Armour	BS
Vyper	50	10	10	10	3

Squadron: The squadron consists of 1-3 Vypers.

Type: Fast, Skimmer, Open-topped.

Weapons: The Vyper is armed with twin-linked shuriken catapults and a shuriken cannon.

Options: The shuriken cannon may be replaced with one of the following heavy weapons: scatter laser at +5 pts; Eldar missile launcher at +20 pts; bright lance at +15 pts; starcannon at +15 pts.

The shuriken catapults can be upgraded to a shuriken cannon for +20 pts.

A Vyper may be given the following vehicle upgrades: crystal targeting matrix, spirit stone, holo-field, vectored engines, star engines, scythes.

SWOOPING HAWKS

	Points/Model	WS	BS	S	T	W	I	A	Ld	Sv
Swooping Hawk	21	4	4	3	3	1	5	1	9	4+
Exarch	+12	5	5	3	3	1	6	2	9	3+

Squad: The squad consists of between 5 and 10 Swooping Hawks.

Weapons: Lasblaster, Plasma grenades, Swooping Hawk grenade pack.

Character: One model in the squad may be upgraded to an Exarch for +12 pts. The Exarch may be armed with a power weapon for +10 pts or exchange his lasblaster for a Web of Skulls and shuriken pistol for +20 pts or a Hawk's Talon for +15 pts.

The Exarch may be given the following warrior powers: Bounding Leap for +5 pts; Sustained Assault for +20 pts.

SPECIAL RULES
Swooping Hawk Wings: See the Wargear section.

HEAVY SUPPORT

During the war in heaven, it was Falcon, consort of the Great Hawk, who retrieved Vaul's mighty sword, Anaris, and gave it to the Eldar hero Eldanesh to continue the battle with Khaine. It is this principle of deliverance which is behind the design of the Falcon grav-tank. With its potent armament and ability to carry a small squad of fighters, the Falcon is designed to take the fight to the enemy, or to extricate the warriors should resistance prove too fierce for them.

FALCON

	Points/Model	Front Armour	Side Armour	Rear Armour	BS
Falcon	120	12	12	10	3

Type: Skimmer, Tank, Fast.

Weapons: Twin-linked shuriken catapults, pulse laser and one weapon from the following list: shuriken cannon at +20 pts; scatter laser at +25 pts; Eldar missile launcher at +40 pts; bright lance at +35 pts; starcannon at +35 pts.

> "Ask not the Eldar a question, for they will give you three answers; all of which are true and horrifying to know."
>
> Inquisitor Czevak

Options: The shuriken catapults can be upgraded to a single shuriken cannon for +20 pts.

A Falcon may be given the following vehicle upgrades: crystal targeting matrix, spirit stone, holo-field, vectored engines, star engines, scythes.

Transport: The Falcon can carry 6 models. It may not carry an Avatar, Wraithlord, Wraithguard or a squad containing an anti-grav platform.

FIRE PRISM

	Points	Front Armour	Side Armour	Rear Armour	BS
Fire Prism	115	12	12	10	3

Type: Skimmer, Tank, Fast.

Weapons: Twin-linked shuriken catapults and prism cannon.

Options: The shuriken catapults can be upgraded to a shuriken cannon for +20 pts.

A Fire Prism may be given the following vehicle upgrades: crystal targeting matrix, spirit stone, holo-field, vectored engines, star engines, scythes.

WRAITHLORD

	Points/Model	WS	BS	S	T	W	I	A	Ld	Sv
Wraithlord	75	4	4	5(10)	8	3	4	2(3)	10	3+

Weapons: Two Dreadnought close combat weapons. Each fist also incorporates a flamer or a shuriken catapult. Note that the Wraithlord's profile already includes the extra Attack for having two close combat weapons.

Options: The Wraithlord must be armed with one of the following heavy weapons: shuriken cannon at +25 pts; scatter laser at +30 pts; Eldar missile launcher at +50 pts; bright lance at +45 pts; starcannon at +45 pts.

SPECIAL RULES

Implacable Advance: The Wraithlord is a towering construct capable of laying down a curtain of fire as it advances into combat. It can shoot up to two weapons if it moves, and all of its weapons if it stays stationary. It can even fire heavy weapons whilst moving.

Fearless: The Wraithlord carries the spirit of a mighty warrior who has witnessed many battles over the centuries. The Wraithlord cannot be pinned and never falls back, even from attacks which would normally cause the enemy to automatically fall back without a Morale check.

WAR WALKER SQUADRON

	Points/Model	WS	BS	S	Front	Side	Rear	I	A
War Walker	30	3	3	5	10	10	10	4	2

Number: A squadron consists of between 1 and 3 War Walkers.

Type: Walker, Open-topped.

Weapons: The War Walker is armed with two of the following weapons: shuriken cannon at +20 pts; scatter laser at +25 pts; Eldar missile launcher at +45 pts; bright lance at +35 pts; starcannon at +35 pts. Note that the points cost for weapons is not included in the profile, but must be added to the basic cost of 30 pts.

SPECIAL RULES

Energy Field: A War Walker's pilot is encased in a powerful force field. Any ranged attack against a War Walker from the front arc with a Strength greater than 8 counts as Strength 8. In addition, such attacks never roll more than +1D6 for their armour penetration (for example, melta weapons at half range or ordnance only roll one dice). Attacks from the side or rear, or in close combat, are not affected by the energy field and so do not suffer any of these penalties.

> "We warned you of the price of your actions, now you must pay it in full – in blood."
> Message received prior to the Assyri Devastation

This dedicated anti-tank vehicle uses the most advanced forms of laser technology ever seen. Its powerful prism cannon can blast apart armoured vehicles and cut through swathes of infantry. With its ability to move at high speed over almost any obstacle, this lethal attack can be made anywhere on the battlefield.

Towering over its foes, the Wraithlord is controlled by the essence of one of the craftworld's mightiest warriors. Only the most worthy are installed into its armoured shell.

War Walkers are used in rough terrain, to scout out enemy positions. Its two heavy weapons provide a considerable arsenal for its size, but it lacks the heavy armour that would allow it to fight at the centre of an attack.

The Eldar use many forms of technology that cannot be matched by other races. These exotic weapons are used by the Craftworld armies to support their advance or to form a solid defence.

The Dark Reapers represent the war god in his role as the Destroyer. They are perhaps the most sinister and lethal of all the Aspect Warriors and their dark armour is adorned with symbols of death and destruction. They excel at long range support and carry the deadly reaper launcher – a fast-firing heavy weapon that shoots a hail of armour-piercing rockets.

SUPPORT WEAPON BATTERY

	Points/Model	WS	BS	S	T	W	I	A	Ld	Sv
Support Weapon	20	3	3	3	3	1	4	1	8	5+

Battery: Each unit consists of 1-3 support weapons.

Weapons: All of the support weapons must be armed with the same type of weapon from the following list: D-cannon at +30 pts per model; Vibro-cannon at +40 pts per model; Shadow Weaver at +25 pts per model.

Crew: Two Guardians armed with shuriken catapults or shuriken pistol and close combat weapon.

SPECIAL RULES

Weapons Platform: The support weapon has two Guardians as crew. One crewman is required to fire the support weapon, the other is free to fire his own weapon. If one crewman is killed the platform operates as normal, if both crew are killed the platform is useless (the platform itself can't be hit).

Unlike other anti-grav platforms, the special nature of a support weapon means that it must be absolutely still when it is fired and so it may not fire if it was moved in the movement phase.

Character: The unit may be joined by a Warlock from the Farseer's retinue. See the entry in the HQ section of the army list.

DARK REAPERS

	Points/Model	WS	BS	S	T	W	I	A	Ld	Sv
Dark Reaper	37	4	4	3	3	1	5	1	9	4+
Exarch	+18	5	5	3	3	1	6	2	9	3+

Squad: The squad consists of between 3 and 5 Dark Reapers.

Weapons: Reaper launcher.

Character: One model in the squad may be upgraded to an Exarch for +18 pts. The Exarch may exchange his reaper launcher for one of the following weapons: shuriken cannon for +5 pts; Eldar missile launcher for +10 pts.

The Exarch may be given the following warrior powers: Fast Shot for +20 pts; Crack Shot for +10 pts.

Transport: A Dark Reaper squad may be mounted in a Wave Serpent for +110 pts.

The armour of a Dark Reaper incorporates many specialised systems to further increase their effectiveness as support troops. Their helmet vanes contain a receptor linked directly to their reaper launcher, allowing them to see exactly where their weapon is pointing. To ensure a rigid firing pose, their heavy lower leg armour and boots are fitted with sensitive stabilisers, as well as clamps which secure the Aspect Warrior to the ground.

MUSTERING THE WARHOST

The majority of Craftworld Eldar units are very specialised in their role on the battlefield. To give yourself the best chance of victory in battle, the mix of different units you take into battle must be given careful consideration.

WHERE TO BEGIN?

A good way to start collecting your army is with the compulsory HQ and two Troops units needed to play a Standard mission. A Farseer, with his many and powerful psychic abilities, is one of the best HQ units around, and a strong backbone of Guardians (either Storm squads or Defender squads) is essential to any Eldar force. Alternatively, you could field Dire Avengers instead of Guardian Defender squads. Dire Avengers are better at shooting and have superior armour compared to Guardians. Rangers are another Troops option, but their long rifles mean they are tactically less flexible than Guardians.

On the following pages are a selection of different units from the Craftworld Eldar army, along with some notes on the different tactics you can use when fielding them. For convenience, these units have been divided into three types: 'resilient', 'fast' and 'firepower'. Resilient units can take a lot of damage, fast units move rapidly and firepower units provide massed fire or heavy weapons support.

A Craftworld army can contain almost any mix of these different unit types (following the force organisation chart, of course). However, you will have to decide if you want to collect an even mix of all units, or if you would prefer an army that specialises in one type of attack strategy, with only the odd unit from the other categories.

Eldar Warlocks

Above:
A converted Avatar of the Bloody-handed God.

Left:
Converted Farseer

An Eldar force made from two Guardian Defender squads with heavy weapon platforms (Troops) and a Farseer (HQ).

RESILIENT UNITS

- This type of unit relies upon its high Toughness, thick armour or sheer numbers to withstand damage.

- Resilient units use attrition to win their battles – they must maximise enemy losses whilst sustaining few casualties themselves.

- Resilient units can be tactically flexible, fighting in the centre of the battleline, where they can withstand a lot of enemy fire and attacks, or they should be used as a solid anchor to prevent the enemy sweeping away one flank of your army.

- A unit which falls into this category can be used to attack as well as defend because they are able to advance under fire whilst sustaining only limited losses, thus ensuring that they reach the enemy with enough survivors to do some damage.

- Because they can take so much punishment before being destroyed, resilient units naturally attract a lot of enemy fire and so can be used to divert your opponent's attention away from your more vulnerable or valuable units.

- As tough as they are, you must remember that these units are not indestructible – try not to expose them to more enemy fire than they can handle.

GUARDIANS

STRIKING SCORPIONS

WARP SPIDERS

DIRE AVENGERS

WRAITHLORD

WRAITHGUARD

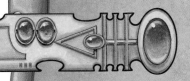

FAST UNITS

SWOOPING HAWKS

JETBIKES

VYPER

• Some fast units, such as Vypers or Jetbikes, use their speed as their main weapon. Others, like Howling Banshees, must get into an assault with the enemy to be most effective.

• These units must keep moving. If they stop, they are not using their greatest asset – speed.

• Fast units are very flexible, and can use their speed to attack where the enemy is not expecting them.

FALCON

FIREPOWER UNITS

• Firepower units can out-shoot most of their enemies. Some of them, such as Fire Dragons and War Walkers, have powerful, high strength weaponry. Others, such as Dark Reapers or Rangers, use their impressive rate of fire to inflict damage.

• Static units (Rangers, Dark Reapers, etc) must stay stationary to shoot and are best deployed where they have a good all-round view of the battlefield.

• Mobile firepower units (Falcons, Fire Prisms, etc) can fire whilst moving, giving them the additional ability to hunt down any enemy who try to hide.

• Firepower units can inflict lots of damage but they can't necessarily take lots of damage!

FIRE DRAGONS

RANGERS

WAR WALKER   **DARK REAPERS**   **FIRE PRISM**

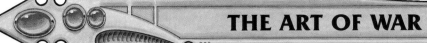

THE ART OF WAR

Here are three strategies for different armies based on the types of troops described.

RESILIENT ARMY

- *The Guardians and Wraithlord target the enemy's toughest units as the opposing army closes in.*

- *The short-ranged but powerful units then spring forward to annihilate the survivors!*

FAST ARMY

- *Swooping Hawks start harassing the enemy.*

- *Vypers hunt down enemy tanks and other tough targets.*

- *A close combat squad in a Falcon or Wave Serpent then smashes into any remaining enemy units.*

FIREPOWER ARMY

- *Static firepower units lay down supporting fire.*

- *The Fire Prism and War Walker use their mobility to hunt down enemy units.*

- *Again, powerful close combat units then spring forward onto the weakened enemy units to finish them off.*

21

EXPANDING YOUR ARMY

TROOPS   **HQ**   **TROOPS**

ELITES

HEAVY SUPPORT

FAST ATTACK

A Farseer and a couple of Guardian squads are a good way to start off your Eldar army, but you'll undoubtedly want to fight large battles which means more units. It's a good idea to think ahead about what to collect next and what sort of units you want your army to have when it's finished.

The Biel-tan army above has had a unit added from each of the other force organisation categories – Elites, Fast Attack and Heavy Support. This gives a well-rounded force which is mobile and has a reasonable shooting capability. Have a look at the different types of army shown in this section and see if any one in particular appeals to you, either

because of the way it looks or the way it plays on the tabletop. As mentioned earlier, you can always collect an army which has a broad mix of firepower, fast and resilient units – you don't have to specialise!

Below is Jonas Ekestam's Maegnár craftworld army. As you can see, much of his army is made up of Aspect Warriors, with just a couple of small Guardian units to fill up his Troops requirements. With the Howling Banshees or Fire Dragons in the Falcon, supported by the Swooping Hawks, jetbikes and Vypers, Jonas' host is quite fast.

Jonas Ekestam's Maegnár craftworld army.

PAINTING ELDAR

Collecting an Eldar army takes time and patience, but the satisfaction of fielding a nicely painted warhost on the battlefield is very rewarding. The many and varied Eldar units offer a great wealth of different painting opportunities.

The main characteristic of painting Eldar models is the contrast between the colour of Eldar helmets and their body armour. You'll notice on this page that the Iyanden Guardian is painted with yellow armour and a blue helmet. Aspect Warriors can also be painted using contrasting colours in just the same way.

ARMOUR

The colour you undercoat your miniatures will play an important part in how your army will look. A Skull White undercoat is perfect for the bright yellow colours of Iyanden, whilst if you are painting a black-armoured Ulthwé craftworld army then of course a Chaos Black undercoat will save you time. Some painters might use a black undercoat for Iyanden, re-coating the raised armour plates in white before painting them yellow. The resulting black lining makes the armour plates really stand out. If you use a white undercoat and then a Chestnut ink wash to shade the yellow afterwards you'll get a similar effect but much more quickly.

THE UNDERSUIT

You'll notice that the plate armour on a lot of Eldar models, in particular the Aspect Warriors, is laid over a flexible undersuit. This is an important element when painting Eldar armour. To keep it simple you could paint over both the armour and the undersuit in a single colour. To add more definition a wash of inks on the armour plates is also a good idea. The alternative is to paint the undersuit and the armour plates in contrasting colours (a darker undersuit usually works best). The Howling Banshee Exarch shown above is a good example of how effective this can be.

HELMETS

A warrior's helmet is a good place to apply a suitable Eldar rune waterslide transfer. The Fire Dragon to the right bears the rune of his Aspect shrine, but other runes are also used as decoration, so feel free to use the transfers in any way you like.

COLOURS

Each craftworld's colours are not a strict uniform. They are often applied with various contrasting colours and on different parts of a vehicle or armour. For example, one squad might have green armour and yellow helmets, while another has green helmets and yellow armour. Some squads may be completely green, or use another secondary colour such as black, blue or white.

WEAPONS

Eldar weapons can be painted in lots of different combinations of colours. Not only can they be painted to contrast or match the warrior's armour, you can also paint them in a variety of metallic colours including Beaten Copper, Brazen Brass or, in the case of this Fire Dragon, Burnished Gold. Use an ink (such as Chestnut Wash) to shade metal and Mithril Silver to highlight it.

Skull White undercoat

Chaos Black undercoat

DIFFERENT TYPES OF BASES

Another way of giving your army a theme is by using a distinctive colour for the bases of your miniatures. This will add character to your army straight away. The most common base colour is Goblin Green, but you can use Snakebite Leather for battles on desert worlds, or Space Wolves Grey for ice worlds and so on. Most hobbyists also like to add texture to the model's base. This can be done with green and/or brown coloured wood dust (called flock), or fine sand. Apply a thin coat of PVA glue to the top of the base and dip it in your chosen basing material. Shake off the excess material and you're left with a nice texture on the base. You can paint the sand any colour you like and highlight it using drybrushing. You can also highlight flock as well and add some small stones and gravel for more texture.

Ice World

Desert World

Jungle World

Scenic Base

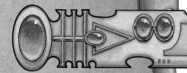

CRAFTWORLD COLOURS

Each craftworld is associated with its own colour or colours. Five of the largest craftworlds are shown on the following pages.

BIEL-TAN CRAFTWORLD

Biel-tan is noted for having a large number of Aspect Warriors, who often incorporate the craftworld's colour scheme into their ritual uniform colours. The main colour of Biel-tan is green, often contrasted with white or light grey.

Above is a selection of Guardian runes, used on the helmets of these Biel-tan units.

Ranger

Ranger

Warlock

Wraithguard

Guardian Defender Squad

Storm Squad Guardian

You can see how Eldar colour schemes can be varied to create diversity within an army.

SAIM-HANN CRAFTWORLD

Saim-hann is most famous for the jetbikes and skimmers of its Wild Rider clans; each of which has its own variation of the Dragon rune which is the craftworld's symbol. Saim-hann's predominant colour is red, often contrasted with white or black.

Defender Squad Guardian

Saim-hann Jetbikes

ULTHWÉ CRAFTWORLD

Ulthwé has more Farseers and Warlocks than other craftworlds. The black of Ulthwé is usually accompanied by a strong contrasting colour such as yellow, orange, red or white.

Wraithguard

Storm Squad Guardian

Warlock

ALAITOC CRAFTWORLD

The Alaitoc army can call upon a large number of Rangers to help it in battle. Alaitoc Guardians and vehicles use deep blue as their main colour, normally in conjunction with yellow or white.

Defender Squad Guardian

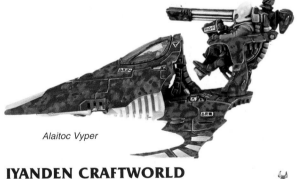

Alaitoc Vyper

IYANDEN CRAFTWORLD

A large number of Wraithlords and Wraithguard make up an Iyanden army. The few Guardians they have wear yellow as their primary colour, with blue or red to contrast.

Iyanden Jetbike

Storm Squad Guardian

Warlock

Some armies use patterns rather than runes to distinguish their squads. The Ulthwé Guardians pictured below show a selection of different helmet patterns you might like to use for your Guardian squads.

As well as runes or patterns, you can use helmet, sash, weapon or gem colours to show which squad a particular model belongs to.

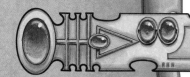

ASPECT WARRIORS

Each Aspect has its own colour scheme. This is not dependent upon the craftworld that the Aspect Warriors originate from, but instead is part of ancient tradition.

ASPECT COLOURS

The traditional colour schemes for the various shrines are:

- *Howling Banshees – White or Bone*

- *Dark Reapers – Black or Midnight Blue*

- *Warp Spiders – Black and Red*

- *Dire Avengers – Deep Blue*

- *Striking Scorpions – Green*

- *Fire Dragons – Red or Orange*

- *Swooping Hawks – Pale Blue or Grey*

- *Shining Spears – White*

Although each Aspect has a unique colour scheme it is quite common for the exact colour and markings to vary from shrine to shrine and even variations within the same shrine or squad are often seen. On some craftworlds the Aspect Warriors incorporate some of the craftworld's own colours, or even have their colour scheme replaced entirely.

Some shrines or squads may even have a colour scheme which bears no resemblance to their Aspect or their craftworld. On this page are just a few of the many different colour schemes and armour patterns used. You should feel free to use whatever colours you find the best or most appropriate.

Dire Avenger Exarch

Warp Spider

Warp Spider Exarch

Striking Scorpion

Striking Scorpion Exarch

Striking Scorpion

Striking Scorpion

Fire Dragon

Fire Dragon Exarch

Swooping Hawk Exarch

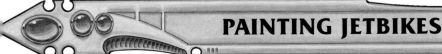

PAINTING JETBIKES

Jetbikes are one of the most easily recognised Eldar units, and it's a rare warhost that doesn't include at least one squadron. Jetbikes offer great scope for a painter, whether you're an expert or just starting out.

When painting Eldar jetbikes think of them as extensions of the rider's armour. Simply paint the jetbike the same colour as the Guardian's uniform. As an alternative, if you are painting lots of different Jetbike squads, as in a Saim-hann army for instance, you could paint the jetbike the same colour as the rider's helmet. With a Biel-tan jetbike you could paint its body white with a green carapace. The carapace itself is excellent for painting special markings and applying large transfers.

Biel-tan jetbike

Saim-hann jetbike

USING SPRAYS

Spray cans of Citadel colour are really useful for painting lots of jetbikes at the same time; you can also use them for painting troops as well. Choose a colour that will cover the majority of the vehicle and undercoat it as normal.

TRANSFERS

Some Eldar runes and designs are quite complex to paint, but there is a wide selection of transfer sheets you can use instead. Looking at the photographs in this book, you will see that transfers can be used in many different places on Eldar warriors and vehicles. They are a quick and simple way of

adding colour and variety to your units and can be used to identify different units in your army. You can also combine transfers together to make new designs or re-paint them different colours, using the design as a kind of template to follow.

Saim-hann carapace designs

VYPERS

Vypers can be painted in the same way as jetbikes. However, it's a good idea to paint the carapace and the clear plastic pilot's canopy separately (Tip: do not varnish the canopy as it will turn cloudy). It's easier to paint the driver of the Vyper without the carapace in place. In fact, painting jetbike riders and crew separately and gluing them into jetbikes afterwards is always a good idea.

PAINTING FALCONS

There is little doubt that the Falcon and its variants are the most elegant-looking vehicles in the Warhammer 40,000 universe. For this reason, it's always a good idea to spend a little bit of extra time painting these centrepiece models.

Painting a large vehicle such as a Falcon is always a rewarding project. This Biel-tan Falcon is painted in Dark Angels Green with contrasting Skull White weaponry. Falcons look good with their top hull painted one colour and the underside painted another. It's also easier to paint the turret separately before the model is glued together. The wide hull is an excellent place for adding detail such as gemstones and transfers.

FIRE PRISM

The Fire Prism is based on the Falcon but is armed with the awesome prism cannon. The huge crystal that constitutes the gun itself could be painted a bright, rich colour to contrast with the rest of the skimmer. This Ulthwé Fire Prism is painted Storm Blue and decorated with fierce lightning flashes painted in Skull White.

A host of Saim-hann vehicles prepares for war.

TOP TIPS

Here are a few 'tricks of the trade' which you might find useful when painting your Eldar army. Many of these techniques may seem a little complicated at first, but with practice they'll become second nature.

ELDAR CHARACTERS

Avatars, Farseers and Warlocks are some of the most detailed models in the Eldar army and it's worth spending some extra time and effort on them. Here are a few tips to help you whether you are a beginner or a veteran.

To paint a cloak or robe, start with a deep base colour, (Shadow Grey is ideal for white robes), and then highlight using the folds in the cloth as a guide. You can use a fairly bright colour along the edges to get a clean, crisp finish, as shown in the photographs below.

These characters also wear armour made of the exotic Eldar substance called wraithbone.

Wraithbone armour

An excellent base colour for wraithbone is Snakebite Leather. Apply a first highlight of Bubonic Brown on top of this, then add Bleached Bone for the lighter shades, finishing off with Skull White. How much of each colour you apply can radically change the look of the bone. By using mostly Bubonic Brown over Snakebite Leather the wraithbone will look dark and ancient. The more Bleached Bone and Skull White you use, the cleaner and more polished it will look.

THORN PATTERN

The thorn pattern is a common Eldar device and is easier to achieve than it first looks. Simply paint a flowing black line and then add small triangles to form the thorns.

Warlock coat

Avatar's loin cloth

Farseer cloak

PAINTING GEMS

Eldar miniatures and vehicles are covered in gemstones and making them look good is simple. You can paint them a flat colour if you want, but there is an easy way to make them really look like gems. First paint the gem in the colour of your choice. Then shade and highlight the gem in reverse, ie, shading at the top and highlighting at the bottom. Finally, add a tiny dot of Skull White at the top. Some painters like to finish off their gemstones with a coat of gloss varnish.

ELDAR SHOWCASE

On this page are some superbly painted Phoenix Lords and Eldar character models. Also featured are winners from the Golden Demon painting competition which is held every year at Games Day.

These three Phoenix Lords were painted by Neil Green. Notice how Neil has used tiny highlights to keep the black armour looking black. Only Baharroth's armour is highlighted to light grey, as befits a Swooping Hawk.

The main feature of both Jain Zar and Maugan Ra is the dramatic bone effect. This is created by using Bestial Brown as the base colour, with layers of progressively lighter colours: Snakebite Leather, Bronzed Flesh and finally Bleached Bone and Skull White painted on top. Interestingly, Neil used a really watery layer of Skull White as a glaze to make the bone look highly polished.

Jain Zar –
The Storm of Silence

Maugan Ra –
The Harvester of Souls

Baharroth –
The Cry of the Wind

Dark Reapers by Ben Jefferson, 1st place Warhammer 40,000 squad, Golden Demon 1996.

Wraithlord by Ben Jefferson, 1st place Warhammer 40,000 vehicle, Golden Demon 1996.

Ben's Wraithlord is beautifully blended from Midnight Blue through Enchanted Blue to Ice Blue. The model is covered with glinting stars and white runes all over its body and loin cloth. Ben has also used a dramatic lightning motif on the machine's head and loin cloth.

Dave used Dark Flesh as the base colour for the head piece and the power claw. This was then highlighted with Bleached Bone painted in a ribbed pattern.

Karandras –
The Shadow Hunter
Painted by Dave Thomas

Fuegan –
The Burning Lance
Painted by Chris Smart

Right: The original Codex Eldar cover that inspired this Exarch conversion.

Eldar Exarch
Converted by Mike McVey.
Painted by Stuart Thomas

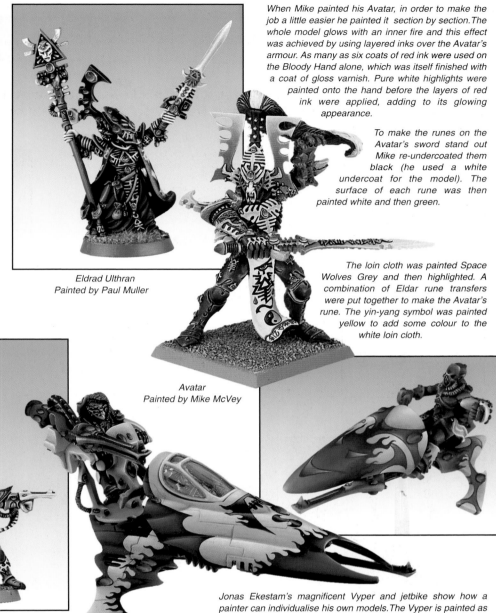

When Mike painted his Avatar, in order to make the job a little easier he painted it section by section. The whole model glows with an inner fire and this effect was achieved by using layered inks over the Avatar's armour. As many as six coats of red ink were used on the Bloody Hand alone, which was itself finished with a coat of gloss varnish. Pure white highlights were painted onto the hand before the layers of red ink were applied, adding to its glowing appearance.

To make the runes on the Avatar's sword stand out Mike re-undercoated them black (he used a white undercoat for the model). The surface of each rune was then painted white and then green.

The loin cloth was painted Space Wolves Grey and then highlighted. A combination of Eldar rune transfers were put together to make the Avatar's rune. The yin-yang symbol was painted yellow to add some colour to the white loin cloth.

Eldrad Ulthran
Painted by Paul Muller

Avatar
Painted by Mike McVey

Asurmen –
Hand of Asuryan
Painted by Paul Muller

Jonas Ekestam's magnificent Vyper and jetbike show how a painter can individualise his own models. The Vyper is painted as part of his Maegnár Craftworld army, whilst the jetbike is more of a display piece. Notice however the crisp painting and the continuing flame motif that characterise Jonas' models.

Wraithguard squad by Valerie Florentin, 3rd place Warhammer 40,000 squad, Golden Demon 1996.

The flame motif is a feature of Valerie's Wraithguard. It appears on the spirit warrior's head and wraithcannon. Notice how the lead Wraithguard is marked with a rune on its loin cloth and a yin-yang symbol on its leg.

THE CRAFTWORLD ELDAR

This section of Codex: Eldar contains a great deal of additional information on the Craftworld Eldar which does not fit into the main army list itself. You will find descriptions and rules for the many exotic weapons and psychic powers of the Eldar, as well as army list entries for certain legendary Eldar characters who you may wish to use in your games of Warhammer 40,000.

Also included is a wealth of information concerning the Eldar race. This includes treatises on their weapons, studies of their culture and language and a rough guide to the subtle interconnections of the various fragments of this once galaxy-spanning race. This is intended for your illumination and entertainment, but maybe it will spark off some ideas for battles or campaigns as well.

Marauth chanted quietly, his voice barely a whisper, the complex spell spilling from his lips in a continuous stream of syllables. As he wove the enchantment, the runes and sigils adorning the portal began to glow; some white, others gold and silver, their inner fire casting flickering shadows across the Farseer's face and his heavy robes. Passing his hand across the gateway, he spoke the final verse, sealing the entrance to the webway for all eternity. He turned and faced the Warlocks around him.

"It is done. The runes of ending have been invoked; our enemies will find no passage through here now," he told them.

The youngest of the Warlocks, Karhaedron, who was barely 300 Eldar years of age, made a respectful gesture to attract Marauth's attention.

"Morfessa, I do not understand the necessity of closing the webway through this portal." The Warlock's young face expressed his sincere concern. "If your divinations are correct and the Dark Kin will find this path to us, then surely is it not better to be prepared with warriors and weapons, and destroy them once and for all?"

"What you say holds much truth," Marauth replied. "As you know, the webway, what remains of it, links us to the ancient places and the other craftworlds. It binds us together and allows us to move swiftly and unseen by the eyes of our enemies. Yet the Dark Kin found their refuge from the Fall within its winding paths and their Dark City spreads like a stain in its depths. They do not know all the ways and means and places within the webway – only the Harlequins could claim that skill, but on occasion they will stumble by chance upon the correct turning or upon one of our portals. The infinity circuit has shown that this will happen, here in this place. However, I have only learnt of the place of their attack, not the time. It would be impossible for us to maintain our guard for an indefinite period. It could be many years before the possible events I witnessed come to pass, and we have not the resources for such vigilance as would be necessary. We require certainty on this, an absolute surety against attack which only the sealing of the portal brings. If even one of the Dark Kin broke through our watchers, they could gain access to the infinity circuit. Such a treasure trove of our ancestors' spirits would be irresistible to them. The carnage they could wreak in our most holy of places is beyond countenance. No, better to be certain."

Karhaedron gave a bow, his eyes lowered out of respect.

"Your wisdom is faultless, as ever, Morfessa. I am most sorry for my lack of understanding and lack of faith," the Warlock apologised.

Marauth smiled warmly, raising the Warlock's chin with a slender finger so that their eyes met.

"Never apologise for asking questions, young seers," the Farseer told his acolytes. "Through asking questions we find answers, and through those answers we gain greater knowledge of ourselves and the universe around us."

Marauth's face hardened, his jaw clenched tightly.

"Our race stopped asking questions once before and our complacency all but destroyed us. That must never happen again."

CRAFTWORLD ELDAR WARGEAR

This section describes the rules for the ancient weapons and equipment used by the Eldar. These rules tend to be more detailed than those in the Warhammer 40,000 rulebook and supersede them if they are different. NOTE: Any items not listed here work exactly as described in the Warhammer 40,000 rulebook.

Banshee Mask: Howling Banshees wear a specially modified helmet, which overloads their foes' nervous systems by using psychosonic amplifiers to turn their battle-cries into powerful energy waves.

A model wearing a Banshee mask always strikes first in hand-to-hand combat in the first round of any assault, regardless of Initiative or other factors. This applies even if they are not in base contact with the enemy. If the combat lasts more than one round, then the mask has no effect in subsequent rounds. If an enemy model also has an ability that allows it to always strike first (such as Dark Eldar Wyches on certain combat drugs) then the attacks are resolved simultaneously.

Biting Blade: The teeth of a Biting Blade tear through flesh and armour, shredding muscle and bone. It is used in an assault and if a model using it wounds an enemy with more than 1 wound, you may roll to wound that opponent again. Keep rolling until you fail to wound. Each wound inflicted must be saved against separately.

Bright Lance: The bright lance is used by the Eldar to destroy heavily armoured targets, using a highly focused beam of laser energy. A bright lance has the profile below. In addition, it treats any armour value higher than 12 as 12.

Rng: 36"   S: 8   AP: 2   Assault 1

Diresword: The Diresword is an extremely potent power weapon, incorporating a spirit stone in its hilt. When the Diresword strikes, the spirit that is within the blade can attempt to destroy the mind of the target.

A Diresword is used in close combat. If a model with more than 1 wound is wounded by a Diresword, it must immediately pass a Leadership test on 2D6 or die automatically. The Diresword ignores normal armour saves.

Distort Cannon: The Distort cannon, or D-cannon, uses the Eldar's advanced knowledge of warp technology to unleash a miniature warp hole onto the battlefield, tearing apart its targets. The D-cannon has the profile given below.

In addition, if it hits a vehicle, use the Ordnance Damage tables to resolve any damage. Note that a battery of more than one D-cannon follows the rules for multiple barrages on page 58 of the Warhammer 40,000 rulebook.

Rng: Guess 24   S:10   AP: 2   Heavy 1 Blast

Eldar Jetbike: The jetbikes of the Craftworld Eldar are designed for extreme manoeuvrability, able to turn on the spot and rapidly change speed. This enables the rider to make swift hit-and-run attacks, emerging from cover to attack before slipping away again. They use all the rules for jetbikes given in the Warhammer 40,000 rulebook. In addition, an unbroken model on an Eldar jetbike may always move 6" in the assault phase, whether they are within 6" of an enemy or not. This move can be in any direction, it does not have to be a charge towards the enemy, even if the jetbike is within 6" of an opposing model.

Eldar Missile Launcher: The Eldar have a grasp of technology which far surpasses that of other races, and it is this knowledge that allows them to make wide use of plasma missiles. Eldar missile launchers may fire krak or plasma missiles, with the profiles given below. A squad that takes casualties from a plasma missile must test for pinning (see page 58 of the Warhammer 40,000 rulebook).

Krak: Rng: 48"   S: 8   AP: 3   Heavy 1
Plasma: Rng: 48"   S: 4   AP: 4   Heavy 1 Blast

Executioner: The Executioner is a long-bladed power weapon capable of slicing an opponent in half with a single blow. It must be wielded in both hands by its user, and so may not be used in conjunction with another close combat weapon or pistol to get +1 Attack. An Executioner adds +2 to the wielder's Strength and ignores normal armour saves.

Firepike: The Firepike is a sophisticated melta weapon, with a distinctive long barrel which can project the deadly melta beam a considerable distance. Like other melta weapons, the Firepike rolls 2D6+Strength for armour penetration against targets that are within half range (9"). It has the following profile:

Rng: 18"   S: 8   AP: 1   Assault 1

Fusion Gun: The fusion gun is a melta-weapon, most commonly carried by the Fire Dragon Aspect Warriors. As a melta weapon, the gun rolls 2D6+Strength for armour penetration when fired at half range (6"). It has the following profile:

Rng: 12"   S: 6   AP: 1   Assault 1

Ghosthelm: A Farseer's Ghosthelm incorporates intricate crystalline psychic circuitry that masks their spirit in the warp, protecting them from the attacks of Daemons and other warp creatures. If the model suffers an attack from the perils of the warp while making a Psychic test (ie, the player rolls a double 6 or double 1) he may ignore the attack on a D6 roll of 4+. In addition, any Daemon that is fighting the model in close combat halves its own Weapon Skill (rounding up).

Hawk's Talon: A Swooping Hawk Exarch often carries a much more powerful version of the lasblasters wielded by his squad, called a Hawk's Talon. This weapon has the following profile:

Rng: 24"   S: 4   AP: 6   Assault 3

Haywire Grenades: The Eldar use Haywire grenades for disabling enemy vehicles. They send out a powerful, short-range magnetic pulse which shorts out electrical wiring and disrupts the energy systems of its target. They may only be used in an assault against vehicles.

A model attacking with these grenades may only make a single attack, whatever their other armaments, Attacks characteristic, or whether they charged. If the attack hits, roll a D6 to determine the effect: 1=no effect, 2-5=glancing hit, 6=penetrating hit. A Haywire grenade may only be used against a Dreadnought if it has already been immobilised or is stunned.

Lasblaster: The lasblaster is a rapid-firing laser weapon, which far surpasses the clumsy lasguns of the Imperium and is used by the Swooping Hawk Aspect Warriors to lay down a hail of fire. It has the following profile:

Rng: 24"   S: 3   AP: 6   Assault 2

**Blood Runs,
Anger Rises,
Death Wakes,
War Calls!**

Battle-chant to Khaine the Bloody-
Handed God

Laser Lance: This is used by Shining Spears Aspect Warriors. They use it to deliver intense short ranged laser blasts as they charge into combat. It is fired in the assault phase when the Shining Spears charge into combat and is worked out just before you move them into combat. The unit's laser lances must be fired at a single unit being charged by the Shining Spears and any casualties count towards the combat resolution for that turn. All of the normal shooting rules apply to this attack, such as rolling to hit, saves for cover and so on. In addition, a model armed with a laser lance counts as having Strength 5 when working out hits in hand-to-hand combat. A laser lance has the following profile:

Rng: n/a   S: 5   AP: 5   Assault 1

Mandiblasters: These are fitted into the helmets of Striking Scorpions Aspect Warriors. Activated by a psychic pick-up in the helmet, it fires a hail of needle-thin shards which act as a conductor for a highly charged laser. A model with a mandiblaster may make a special attack in close combat, worked out at +2 to the model's Initiative. Mandiblasters can be used by models within 2" of an enemy, as well as by models in base contact. They inflict a Strength 4 hit on a D6 roll of 4+. Normal armour saves are allowed. Remove models as you would other close combat casualties. Once these attacks have been resolved, the Striking Scorpions may make any other attacks at their normal Initiative value. Note that since mandiblaster casualties count as close combat casualties, a Striking Scorpion who starts the combat in base contact with the enemy will get their full number of Attacks, even if the model they are in base contact with is removed by mandiblaster fire.

Plasma Grenades: Rather than the crude fragmentation grenades used by other races, the Eldar employ advanced Plasma grenades to stun their enemies when they charge into close combat. These negate the effect of cover in close combat, so that all attacks are worked out in Initiative order.

Powerblades: Powerblades are fitted to the forearm, enabling the wearer to use both hands freely. A well trained warrior can make sweeping strikes with the powerblades as well as their other weapons. A model equipped with powerblades gets +1 Attack. This can be in addition to the +1 Attack for being armed with two other close combat weapons for a total of +2 Attacks. A model using powerblades ignores armour saves.

Prism Cannon: The prism cannon works by focusing a narrow laser beam through a highly complex crystal array and then unleashing it in a devastating burst. It has the following profile:

Rng: 60"   S: 9   AP: 2   Heavy 1 Blast

Pulse Laser: The pulse laser is a highly advanced weapon that fires a stream of powerful laser bolts at its target. It fires D3 shots, rolled each time the pulse laser is fired. It has the following profile:

Rng: 48"   S: 8   AP: 2   Heavy D3

Ranger Long Rifle: The Ranger long rifle is equipped with highly sophisticated sights, allowing the firer to locate weak points in an enemy's armour. The long rifle is treated like a sniper rifle. In addition, if a 6 is rolled for the to hit roll, the shot counts as having AP 1.

Rune Armour: Eldar Farseers and Warlocks are covered by protective runes and sigils that use psychic energy to ward off enemy attacks. A model wearing rune armour has an invulnerable saving throw. If the model is mounted on a jetbike, it may take either a normal 3+ saving throw or its 4+ rune armour save.

Runes of Warding: A Farseer can use runes of warding to divine when an enemy psyker is using his powers and to throw up a psychic shield to protect himself and those nearby. If an enemy psyker attempts to use a psychic power and the psyker or the target is within 6" of the Farseer, the enemy must take the Psychic test on 3D6 and discard the lowest roll. Psykers who do not normally have to take a Psychic test remain unaffected by the runes of warding.

Runes of Witnessing: A Farseer uses runes of witnessing to guide his second sight along the twisting strands of fate, giving him even greater clairvoyance. A Farseer with runes of witnessing rolls 3D6 and discards the highest roll when taking a Psychic test. Note that you must use the lowest two rolls, even if they are a double 1.

Scorpion's Claw: The Scorpion's Claw is an ancient weapon of the Striking Scorpion Exarchs. It takes the form of a powered claw-shaped glove with a shuriken catapult incorporated into its back. The claw may be used both as a power fist and a shuriken catapult, and may be used as both in the same turn.

Shadow Weaver: The Shadow Weaver unleashes a stream of razor-sharp mesh high into the air, which drifts down onto the enemy, slicing through flesh and bone. It is a barrage weapon and a battery of more than one Shadow Weaver follows the rules for multiple barrages on page 58 of the Warhammer 40,000 rulebook. As a barrage weapon, it also uses the rules for pinning (also on page 58). It has the following profile:

Rng: Guess 48", S: 6, AP: –, Heavy 1 Blast

Singing Spear: The Singing Spear is a psychically-charged weapon used by Farseers and Warlocks, which can be thrown at opponents and returns automatically to the user's hand. The Singing Spear has the profile below, and always wounds opponents on a 2+, regardless of their Toughness. If thrown at a vehicle, it has a Strength equal to three times the thrower's Strength and adds +D6 for armour penetration as usual. A Singing Spear may also be used in close combat, but requires two hands to wield and so cannot be used with another close combat weapon or pistol to gain +1 Attack. A model may not throw the Singing Spear and use it in close combat in the same turn.

Rng: 12"   S: Special   AP: n/a   Assault 1

Spirit Stone: Every Craftworld Eldar wears a waystone, to trap their soul when they die and to stop it being consumed by the Chaos god Slaanesh. Waystones containing a soul are known as spirit stones, and can be put to a variety of uses by the Eldar. An Eldar psyker can use the power of a spirit stone to charge themselves with psychic energy. A psyker with spirit stones can use two psychic powers each turn instead of one. However, these must be different powers; a psyker cannot use the same psychic power twice in the same turn.

Starcannon: The starcannon is a highly advanced plasma weapon that uses a sophisticated electromagnetic pulse to guide its lethal bolts to the target. Note that unlike the crude and clumsy plasma weapons of other races, a starcannon does not overheat on a to hit roll of 1. It has the following profile:

Rng: 36"   Str: 6   AP: 2   Heavy 3

Swooping Hawk Grenade Pack: The leg armour of Swooping Hawks incorporates a grenade launcher which can be used to fire a hail of grenades as the Swooping Hawks drop down to attack. If a Swooping Hawk unit uses its Deep Strike ability to deploy, it may use its grenades on the turn it arrives. These have the profile shown below. Place the large Ordnance Blast marker anywhere on the table before the unit lands and roll a Scatter dice. If an arrow is rolled, the marker scatters D6" in the indicated direction. Work out hits and damage as normal. Note that only one marker is placed per Swooping Hawks unit regardless of the number of models in the unit.

Rng: n/a   S: 4   AP: 5   Large Blast

Swooping Hawk Wings: Swooping Hawks have glittering, multicoloured wings. In flight, the 'feathers' of these wings vibrate rapidly, turning them into a blur of vibrant colour. These wings allow the Swooping Hawks to move rapidly across the battlefield, or hover high above it, waiting for the chance to strike. A model with Swooping Hawk wings moves as if equipped with a jump pack. In addition, a squad with Swooping Hawk wings may always use the Deep Strike rules to deploy, even if the mission does not normally allow units to do so.

Vibro Cannon: A vibro cannon uses resonant sonic waves to shake its targets apart and fling troops to the ground. When firing a vibro cannon, pick a target point anywhere within range and line of sight of the vibro cannon, and then roll to hit as normal. If it hits, draw a line between the cannon and its target point, and any unit which the line passes through suffers D6 hits. A vehicle or other target with damage tables hit by a vibro cannon suffers a single glancing hit – do not roll for armour penetration. A unit suffering a casualty from a vibro cannon must take a Pinning test as detailed on page 58 of the Warhammer 40,000 rulebook. If there is more than one vibro cannon in the battery, they all fire a single shot together as described above. Each additional cannon in the battery adds +1 to the Strength of the attack and inflicts a -1 modifier on any Pinning tests taken. A vibro cannon has the following profile:

Rng: 36"   S: 4   AP: –   Heavy 1

Warp Spider Jump Generator: Warp Spiders use a compact but complex jump generator that allows them to teleport short distances via the warp. Although this allows them to move rapidly and avoid any obstacle, the Eldar's exposure to the warp is not without peril. A model with a Warp Spider jump generator may move up to 12" in the movement phase and ignores all terrain during its move. This means that it can pass through woods, over rivers and even through solid walls without penalty. The model may not finish its move in impassable terrain.

The Warp Spiders may make a second jump at the start of the assault phase if you wish. They cannot make a second jump and charge into an assault in the same turn. The jump generator becomes more unpredictable during this second jump and the following rules apply. Nominate the direction the squad is jumping in and move the squad 2D6" in this direction. If you roll a double, one member of the squad has suffered a calamity in the warp and is removed as a casualty (the survivors move the distance rolled).

A unit equipped with Warp Spider jump generators advances and falls back 3D6", ignoring terrain as for its normal move.

Web of Skulls: The Web of Skulls is made from three or more crystalline skulls linked together with lightweight chains, and is thrown like a bolas. It will fly in an arc, smashing through an enemy unit before returning to the thrower. A Web of Skulls is thrown at an enemy unit using the normal shooting rules. If it hits, you may roll to hit again. Keep on rolling to hit the unit until you fail a roll or you have inflicted 4 hits. Make rolls to wound and armour saves for each hit

inflicted as normal. No model in the target unit may be hit more than once. A Web of Skulls may also be used in close combat, and counts as a power weapon. It has the following profile.

Rng: 24"   S: 4   AP: 5   Assault 1

Witchblade: Eldar psykers carry potent force weapons known as witchblades, which may take the form of a spear, sword or some other weapon. A model with a witchblade always inflicts a wound on a roll of a 2+ in hand-to-hand combat (but only inflicts 1 wound with each hit). Armour saving throws are taken normally.

Against vehicles, a witchblade allows the bearer to triple their Strength characteristic when working out armour penetration.

Wraithcannon: The wraithcannon uses the same warp technology found in the larger Distort cannons. It works by opening a small warp space/real space hole, tearing apart the target as it is ripped between dimensions. The wraithcannon always wounds on a roll of 4+ regardless of Toughness, and on a roll to wound of a 6, it inflicts instant death as the target is wholly transported to the warp. Against targets with an armour value, a wraithcannon always inflicts a glancing hit on a roll of 4 and a penetrating hit on a roll of 5 or 6, regardless of the target's armour value. The wraithcannon has the following profile:

Rng: 12"   S: n/a   AP: 1   Assault 1.

CRAFTWORLD ELDAR VEHICLE UPGRADES

Craftworld Eldar vehicles may be fitted with certain extra systems, as outlined in their army list entry. Any vehicle upgrades taken must be represented on the model. No vehicle can be given the same upgrade more than once.

Crystal Targeting Matrix: The crystal targeting matrix allows the crew to rapidly locate their targets whilst on the move. A vehicle with a crystal targeting matrix may shoot during the movement phase rather than the shooting phase – moving, shooting and then completing its move. All the normal restrictions apply to the number of weapons that can be fired due to the total distance the vehicle moves (ie, the distance moved before **and** after shooting).

Holo-field: The vehicle is surrounded by a shimmering holo-field, that distorts its shape and prevents the enemy from targeting its most vulnerable locations. Whenever your opponent rolls on the Damage table for the vehicle, they must roll two dice and apply the lowest result.

Scythes: The vehicle has been fitted with outriggers and blades that allow it to make sweeping attacks on the enemy as it flies past. Any enemy model that rolls a 1 to hit when attacking the vehicle in an assault suffers a Strength 5 hit, normal armour saves allowed.

Spirit Stone: The vehicle incorporates a large spirit stone. The essence contained within it can control the vehicle for short periods of time should the crew be disabled in some way. If the vehicle suffers a 'crew shaken' result, roll a D6. On a 4+ the vehicle

is unaffected. If the vehicle suffers a 'crew stunned' result, roll a D6. On a roll of a 4 or 5 treat this as a 'crew shaken' result. On a roll of a 6 the result is ignored.

Star Engines: The vehicle incorporates a number of secondary engines which can give it a much needed boost. These can be used to move the vehicle 2D6" straight ahead in the shooting phase, instead of firing any weapons. A vehicle may not use its star engines in the same turn that it embarks or disembarks a transported unit. Star Engines and a Crystal Targeting Matrix may not be used in the same turn.

Vectored Engines: The vehicle's engines allow it to turn almost on the spot, allowing the crew to easily steer around intervening terrain. The vehicle may re-roll any failed Difficult Terrain test.

> **"T**he stars themselves once lived and died at our command, and yet you still dare to oppose our will."
>
> Farseer Mirehn Bielann

FARSEER PSYCHIC POWERS

Unless otherwise noted, these work as described in the Psychic Powers section on page 74 of the Warhammer 40,000 rulebook.

Eldritch Storm: The Farseer may summon a corona of crackling energy which strikes out with arcs of lightning and hurls enemies in all directions. A Farseer may create an Eldritch Storm in the shooting phase, instead of firing a weapon. The Eldar player places the large Ordnance Blast template on an enemy unit within 18". Each model touched by the template takes a Strength 3 hit with no AP value. The target unit must also pass a Pinning test as described on page 58 of the Warhammer 40,000 rulebook. Vehicles touched by the template suffer a hit with 2D6 armour penetration (note that it is 2D6, not 2D6+3) and are spun around to face in a random direction

Fortune: The Farseer scries the possible futures to foresee where the enemy will attack, warning his fellow Eldar so that they may avoid enemy fire. This psychic power is used at the start of the Eldar turn. Nominate one Eldar unit with a model within 6" of the Farseer (which may be the Farseer's own unit). This unit may re-roll any failed armour saves or cover saves until the start of the next Eldar turn.

Guide: The Farseer's prophetic powers warn him of the enemy's movements, allowing him to direct the fire of his warriors. This psychic power is used at the start of the Eldar turn. Nominate one Eldar unit with a model within 6" of the Farseer (which may be the Farseer's own unit). This unit may re-roll any missed shooting to hit rolls until the start of the next Eldar turn. If the unit is using a guess range weapon (such as a D-cannon) you may re-roll the Scatter dice if a 'hit' is not scored with the first roll.

Mind War: The Farseer reaches out to attack the mind of an enemy in a desperate mental duel. This psychic power is used in the shooting phase instead of firing a weapon. The Eldar player may choose any enemy model within 18" of the Farseer and within his line of sight as the target for the attack. Both players roll a D6 and add the Leadership of their respective models. For each point the Farseer wins by, the target loses a wound, with no normal armour saves allowed. If the scores are drawn, or the Farseer scores less, the Mind War has no effect.

WARLOCK POWERS

A Warlock's power is available permanently, so he does not need to take a Psychic test to use it. Although they require no Psychic test to use, these powers are treated as psychic powers in all other respects (for example, Destructor could not affect an enemy immune to psychic powers) and the Warlock is treated as a psyker.

Conceal: The Warlock creates a shield of shifting shadows and drifting mists about his squad, concealing them from the enemy. The Warlock's whole squad receives a 5+ cover save. Note that if the squad is also in normal cover, you may only use one cover save.

Destructor: The Warlock gathers up his anger and hatred and unleashes it at the enemy in a roiling blast of raw psychic power. This power is used in the shooting phase instead of firing a weapon. The Destructor is worked out like a normal shooting attack with the following profile:

  Range: Flame   Strength: 5   AP: 4   Assault 1

Embolden: The Warlock instills in his comrades an unshakeable courage, reaching into their minds with visions of mighty heroes and great victories. The Warlock and his squad may re-roll any failed Morale checks or Pinning tests they have to take.

Enhance: The Warlock empowers his fellow warriors with great speed and skill. All models in the Warlock's squad, including the Warlock himself, add +1 to their Weapon Skill and Initiative. The effects of Enhance are not cumulative.

EXARCH POWERS

Exarchs may be given certain warrior powers as detailed in their army list entry. An Exarch can have up to two warrior powers, and these have the effects given below.

Acrobatic: The Exarch's agility allows her to leap over the heads of enemy models to attack anywhere she may choose. Before the Exarch makes her close combat attacks, she may be repositioned in contact with any model of your choice in the enemy unit.

Bounding Leap: The Exarch makes long, graceful jumps, propelling himself towards the enemy. When making an assault move, he may add +D6" to the distance he charges. The Exarch may only charge if he is within 6" of the enemy in the assault phase, but he does not have to engage the first model in his path if his extra charge distance would allow him to reach another model.

Burning Fist: The Exarch summons up all his wrath and hatred and unleashes it in a devastating attack. In close combat the Exarch may re-roll any to wound rolls and ignores normal armour saves.

Crack Shot: The Exarch is a supreme master of all ranged weapons, able to pinpoint his targets with unerring accuracy. The enemy may not make cover saves against shots from the Exarch, and the Exarch may re-roll any failed to wound roll when shooting. Crack Shot may not be used by an Exarch in the same turn as he uses Fast Shot.

Crushing Blow: The Exarch is able to channel his rage and anger and use it to strike with incredible strength. He works out all close combat attacks at +1 Strength.

Defend: The Exarch is adept at self-protection using parries and dodges to avoid enemy blows before striking. He may use this ability in close combat. If he does so, he strikes last, but enemy models must roll a 6 to hit him.

Distract: Using hypnotic gestures and feints, the Exarch confuses and distracts his enemies. The Exarch may use this ability in close combat. One model in base contact with the Exarch, chosen by the Eldar player, is at -1 Attacks. This can reduce a model to 0 Attacks.

Evade: The Exarch is adept at guiding his jetbike to avoid incoming shots and the clumsy blows of the enemy. His saving throw becomes invulnerable.

Fast Shot: The Exarch is adept at laying down a lethal volley of fire from any weapon, firing shot after shot into the enemy. If the Exarch is firing an assault or heavy weapon, add +1 to the number of shots fired (for example, Heavy 3 becomes Heavy 4). If the Exarch is firing a pistol or rapid fire weapon then they never count as moving (ie, they can always fire once up to maximum range or twice up to 12"). Fast Shot may not be used by an Exarch in the same turn as he uses Crack Shot.

Skilful Rider: The Exarch leads his squad unerringly around tree-trunks and branches, down twisting gorges and through rock-strewn passes. He and his squad do not have to roll for difficult terrain.

Stealth: The Exarch is extremely cunning and knows how to use the lie of the land to shield his squad from their enemies as they approach. A squad led by an Exarch with Stealth may infiltrate, if allowed to do so by the mission being played.

Surprise Assault: The Exarch leads his squad on the attack, just when the enemy are least expecting it. When he and his squad charge, they gain +2 Attacks instead of +1.

Sustained Assault: The Exarch keeps his wings in constant motion, darting from one foe to the next in a continuous attack. For each close combat hit he inflicts, he may make an extra attack. If an extra attack hits, then this allows him a further attack. Keep rolling to hit until the Exarch misses. Resolve to wound rolls and armour saves for each hit as normal.

Tank Hunter: The Exarch is well versed in the art of stalking armoured vehicles, able to spot a weak point in the armour almost instantly. He may re-roll any armour penetration rolls he makes for shooting or close combat attacks against targets with an armour value.

War Shout: The Exarch uses her Banshee mask to unleash a terrifying howl of fury and despair, which sweeps over her enemies like a shockwave. When the Exarch uses her mask, the enemy unit she is fighting must pass a Leadership test or reduce its Weapon Skill by -1 for the rest of that assault phase.

Withdraw: The Exarch watches the tides of combat closely, and is able to judge the moment when it will be safe for his squad to withdraw from a fight, ready to attack again. If the squad is in close combat at the end of the assault phase, they may fall back to take them out of the combat if you wish. All the normal rules for a fall back move apply (most importantly, the rules for crossfire) but the enemy squad may only make a 3" move to consolidate. The unit automatically regroups after making its withdrawal.

ELDRAD ULTHRAN, FARSEER OF ULTHWE

	Points	WS	BS	S	T	W	I	A	Ld	Sv
Eldrad	246	5	5	4	4	3	5	1	10	3+ (rune)

An Ulthwé Eldar army of 2,000 points or more may include Eldrad Ulthran. If you decide to take him he counts as one of the HQ choices for the army. Eldrad must be used exactly as described below, and may not be given extra equipment. In addition, he may only be used in a battle where both players have agreed to the use of special characters.

Wargear: Shuriken pistol, Staff of Ulthamar, runes of warding, runes of witnessing, Ghosthelm, spirit stones, rune armour.

Psychic Powers: Eldritch Storm, Fortune, Guide and Mind War.

SPECIAL RULES

Staff of Ulthamar: The Staff of Ulthamar is a potent artefact made from the purest wraithbone. Eldrad can channel his immense psychic powers through the staff, increasing his abilities or using it as a powerful weapon. The staff can be used in two ways, but it may only be used in one way in any single turn. Firstly, it can be used to allow Eldrad to use another psychic power. This can be a psychic power he has already used that turn. Secondly it can be used in an assault, in which case it always wounds on a roll of 2+ and ignores armour saves.

Divination: Eldrad Ulthran is possibly the most accurate and powerful Farseer of the Eldar, and his powers of precognition and prophecy are legendary. After both sides have deployed at the start of a game, the Eldar player may reposition D3 units in his army. No unit can be repositioned outside its normal deployment zone, and may only be moved up to 6" from its original position. In addition, when using reserves you may add +1 to a single reserves roll each turn (declare before rolling the dice).

Independent Character: Unless accompanied by a bodyguard, Eldrad Ulthran is an independent character and follows all the Independent Character special rules as given in the Warhammer 40,000 rulebook.

Bodyguard: Eldrad Ulthran may be accompanied by a bodyguard of Warlocks. See the separate entry in the army list.

"Eldrad is the greatest among us. He is the sun which eclipses the light of our stars. He is Ulthwé and the fate of our kind rests in his hands. His eyes are the keenest, no detail goes unnoticed. Four thousand runes can he cast, guiding our path through torment and war, death and salvation. He is the pathfinder, the seeker, the true guide. Even your race has trembled before his might, though you may not have known it. It was he who guided us to the Ork known as Ghazghkull, and commanded us to steer his path to your world of Armageddon. Ten thousand Eldar lives would have been lost if he had not done so. What sacrifice is a million humans for such a cause?

He knows your affairs better than you do yourself. He warned that weakling seer you call Emperor of the treachery of Horus and the strife which would engulf us, just as it engulfed the rest of the galaxy, but your arrogance deafened you to his words. Your stupidity almost destroyed the galaxy, yet you never knew how close the forces of light were to our ultimate defeat. He saw the Great Devourer and warned our kin on Iyanden, even before they had neared our galaxy.

To him all futures are laid out, just as your crude implements of torture are laid out on the cold metal of that shelf. You say we are random and capricious, we say you are vulgar and idiotic. Some of you call us your enemies. All races are our enemy in time. Some of you call us your allies. You are not allies, any more than a butcher's knife is his ally. You are tools, nothing more. To be used and expended to protect our race, that is your fate.

Your kind think you are so magnificent, yet even now, at the nadir of our power, we can manipulate you, turn you to our ends, as easily as you might pull a trigger and fire a gun. Our time will come again, Eldrad has promised us. Once more you upstart mon-keigh [subject spits] shall kneel before our power! This time we will not be so lenient! We will exterminate you, every world, every vessel, every one of you! Eldrad has seen the stars stained red with your blood, and it pleases him!

You think us weak, but we will be your doom, children of Earth."

Interrogation of captured Eldar Ranger Prisoner no. 28264. Prisoner Awaiting Termination.

	Points	WS	BS	S	T	W	I	A	Ld	Sv
IYANNA ARIENAL, SPIRITSEER OF IYANDEN										
Iyanna	75	4	4	3	3	2	4	1	9	Special

An Eldar army from the craftworld of Iyanden may include Iyanna Arienal. Arienal may be included in a single Wraithguard squad in the army, instead of a Warlock from a Farseer's bodyguard. She must be used exactly as described below, and may not be given extra equipment. In addition, Arienal may only be used in a battle where both players have agreed to the use of special characters.

Wargear: Shuriken pistol, Spear of Teuthlas, Armour of Vaul.

Warlock Powers: Destructor, Enhance.

SPECIAL RULES

Spear of Teuthlas: This is an ancient singing spear which dates back to the founding of the Iyanden Craftworld. The Spear of Teuthlas follows the rules for a singing spear but has a range of 18" instead of 12".

Armour of Vaul: The Armour of Vaul is a strange and truly ancient artefact. It uses the wearer's mental powers to throw up a virtually impenetrable screen of psychic energy. Instead of making a normal armour save, Arienal must pass an unmodified Leadership test in order to save against any hits she suffers. If the test is failed she loses a wound as normal. The Armour of Vaul can make saves against attacks that ignore normal armour saves, just like an invulnerable saving throw.

Spiritseer: Arienal is skilled at conversing with the spirits of the dead, and can communicate with them over greater distances than an ordinary psyker. Any Wraithguard unit with a model within 12" of Arienal does not have to test for its Wraithsight, just as if a Warlock were accompanying the unit.

Arienal ran a slender finger along the tracing lines of wraithbone that were wreathed around the infinity circuit channel. At her touch, the wraithbone began to pulse with psychic energy, a warm glow spreading outwards along the web of the psychotropic material. Singing softly, offering her prayers to those whose bodies had been destroyed, Arienal took a waystone from a pouch at her belt and placed it in the niche at the centre of the channel.

The Spiritseer took a deep breath, bringing calm to her mind. This was the most difficult part of the ceremony, the part that so few seers were able to bear. Taking a small ceremonial blade from its sheath within her sleeve, Arienal cut the flesh on the palm of her right hand. Clenching her fist over the waystone, she let five droplets of blood spill onto the psychic gem. The blood soaked into the waystone as if its shiny surface was porous.

Arienal waited for a moment and then suddenly the infinity circuit was alive with energy. A blaze of light shot along one of the conduits into the waystone, making it glow with inner power. She felt the spirit inside enriched by her own life essence; the psychic contact between her and the dead had been made with the offering of blood. Her head filled with visions, vistas of death, a barren plain of existence from beyond the veil, and her mouth was dry with the taste of grave-dust.

Arienal knew this spirit well. His name was Althenian, an Exarch of the Fire Dragons who had fallen at the Battle of Two Hundred Pyres. Even in death, Althenian was as eager to serve the craftworld as he had been in life. With due reverence, Arienal took the living spirit stone from the matrix and cradled it in her hands, a single tear running down her cheek. Even now, after so many centuries, she was loath to disturb those in eternal rest, to snare a spirit back from paradise. But war was close at hand and the fighters of Iyanden were few. Such sensibilities had to take second place to survival. With a heartfelt sigh, Arienal stood and started walking to the chamber where the shells of the Wraithguard awaited their spirits.

NUADHU 'FIREHEART'

Come brothers, follow me, we hunt across the skies!

Come, chosen of Khaine, and see how our prey, the gangly humans flee! There is no place for them to hide under the pale face of Lileath the moon, nor under the sun, the face of Asuryan.

Feel the rush of the wind against your skin and hear her keening cry in your ears. Listen to her call well, for are we not the Wild Riders, the children of the storm?

Enjoy the hunt brothers, let sword swing and blood spill. Feel the beat of your heart in your chest and know that you yet live.

Fear not the death brothers, for she is old and slow, and will never catch the Windrider host. It is our enemies who are afraid, for each kiss of our weapons brings the sweet oblivion that they crave by opposing us.

Follow me, brothers, battle awaits!

NUADHU 'FIREHEART', WILD RIDER OF SAIM HANN									
					Armour				
	Points	WS	BS	S	Front	Side	Rear	I	A
Nuadhu	75	5	4	4	11	11	10	6	3

An Eldar army from the craftworld of Saim Hann may include Nuadhu. He may be added to a Jetbike squadron or taken as a Fast Attack choice on his own. He must be used exactly as described below, and may not be given additional equipment. Nuadhu may only be used in a battle where both players have agreed to the use of special characters.

Wargear: Power weapon, Alean – the Steed of Khaine.

SPECIAL RULES

Alean – the Steed of Khaine: Nuadhu rides upon a Vyper that has been built to allow him to fight from its back, much as Khaine rode to war on the legendary steed Alean. Nuadhu and the Vyper are considered a single model; he cannot leave his fighting platform. Think of Nuadhu as a unique type of Vyper, with the special rules detailed below:

Type: Fast, Skimmer, Open-topped.

Weapons: Shuriken cannon.

Close Combat: Nuadhu fights in close combat in the same way as a Dreadnought, using the profile above. He may make a 6" assault move in the assault phase, providing he moved no more than 12" in the movement phase, otherwise he may only fight if assaulted. He carries a power weapon, so no armour saves are allowed for wounds inflicted by him in close combat. Note that because he is treated as a Dreadnought he never falls back, even if defeated in close combat.

Hit and Run: Nuadhu can make hit-and-run attacks just like an Eldar jetbike. This means that he can always move 6" in the assault phase, as long as he did not move more than 12" in the movement phase.

Reckless: Nuadhu pays no heed to danger, gladly flying into the heaviest enemy fire. Nuadhu ignores all 'crew stunned' and 'crew shaken' results on the Damage tables.

Chief of the Wild Riders: Nuadhu's Wild Riders will gladly follow him into the fiercest fighting. Any unit led by Nuadhu will never fall back and cannot be pinned.

Independent Character: Nuadhu is an independent character and follows all the Independent Character special rules as given in the Warhammer 40,000 rulebook. He may only join Guardian Jetbike squadrons. As he has an armour value, not a Toughness, if he is with a squad, you must distribute hits before rolls to wound/penetrate armour are made (see page 52 of the Warhammer 40,000 rulebook).

PHOENIX LORDS

An Eldar army may include one or more Phoenix Lords. Each Phoenix Lord counts as one of the HQ choices for the army. They must be used exactly as described below, and may not be given any additional equipment. Furthermore, Phoenix Lords may only be used in a battle where both players have agreed to the use of special characters.

	Points	WS	BS	S	T	W	I	A	Ld	Sv
PHOENIX LORD										
Phoenix Lord	See below	6	5	4	4	3	7	3	10	3+

SPECIAL RULES

Independent Character: A Phoenix Lord is an independent character and therefore follows the Independent Character special rules as given in the Warhammer 40,000 rulebook.

Fleet of Foot: Despite having a save of 3+, Phoenix Lords are an exception to the normal rule and may still benefit from the Fleet of Foot special rule as detailed on page 4 of this book.

Fearless: Phoenix Lords have travelled the galaxy for millennia and are utterly fearless. A Phoenix Lord will never fall back or be pinned – even by attacks that do not normally allow a Morale check to be taken. If the Phoenix Lord is leading a squad of their Aspect, then the whole squad becomes fearless.

Maugan Ra

THE HARVESTER OF SOULS . . . . 130 pts

Aspect: Dark Reapers.

Wargear: The Maugetar (shuriken cannon with built-in Executioner).

Warrior Powers: Crack Shot, Fast Shot & Crushing Blow.

Like the craftworld of Ulthwé, Altansar craftworld stayed close to the Eye of Terror after the Fall. Then, many millennia ago, a brief expansion of the Eye of Terror caught Altansar in its grip. Despite battling fiercely against the pull of the massive warp storm, the Altansar Eldar were doomed, slowly dragged into the warp over five centuries. It is claimed that the only survivor was the Phoenix Lord Maugan Ra, first of the Dark Reapers. Carrying the Maugetar (lit. Harvester) and wearing the morbidly decorated armour of his now destroyed temple, the Harvester of Souls reaps a high toll of blood from those who oppose the Eldar.

Jain Zar

THE STORM OF SILENCE . . . . . . 143 pts

Aspect: Howling Banshees.

Wargear: Banshee Mask, Executioner, the Silent Death (as Web of Skulls with Strength 5).

Warrior Powers: Acrobatic, War Shout & Bounding Leap.

The most swift and ferocious of the Asuryan was Jain Zar (trans. the Storm of Silence), first chosen of Asurmen. Like her mentor, Jain Zar travelled widely across the webway, teaching her skills to many, and so it is that the Howling Banshees can be found on all but the remotest craftworld.

For centuries at a time she may be beyond the knowledge of mortals, but she always returns, and her shrines keep constant vigil for her. Her chosen weapons are the Jainas Mor (lit. Silent Death) and the deadly Zhai Morenn (Blade of Destruction) which she wields with devastating speed and skill.

Karandras

THE SHADOW HUNTER . . . . . . 165 pts

Aspect: Striking Scorpions.

Wargear: Biting Blade, Scorpion's Claw, Scorpion's Bite (as mandiblaster, but makes attacks at Strength 5).

Warrior Powers: Crushing Blow, Stealth & Surprise Assault.

Legends of Karandras the Shadow Hunter tell of one of the most mysterious of all the Phoenix Lords. No one knows where he originally came from, but it is known that he was not the first Exarch of the Striking Scorpions.

It is said that before him came one called Arhra, the Father of Scorpions, and that Arhra turned to the darkness and waged war upon his own kin. Eventually he fled into the webway to become the Fallen Phoenix and whether he lives still is unknown, but many have their suspicions that he is still alive.

Baharroth

THE CRY OF THE WIND . . . . . . 160 pts

Aspect: Swooping Hawks.

Wargear: Swooping Hawk wings, Swooping Hawk grenade pack, Plasma grenades, Hawk's Talon, power weapon.

Warrior Powers: Bounding Leap, Sustained Assault, Withdraw.

As Asurmen is the first Exarch, so his finest pupil is believed to have been Baharroth (trans. the Cry of the Wind). He was the Winged Phoenix, who taught the skills of what would become the Swooping Hawk Aspect.

Like all the Phoenix Lords, Baharroth has been re-born many times, and it is claimed that he will meet his end fighting alongside his fellow warriors at the Rhana Dandra, the final battle against Chaos.

Fuegan

The Burning Lance . . . . 149 pts

Aspect: Fire Dragons.

Wargear: Firepike, Melta-bombs, Fire Axe (power weapon, +1 Strength).

Warrior Powers: Burning Fist, Tank Hunter, Fast Shot.

Fuegan (trans. the Burning Lance) is the founder of the Fire Dragon Shrine, teaching many Exarchs the art of death through flame and fire. Fuegan refused to flee when the Shrine of Asur was destroyed by the Fallen Phoenix, and he was thought lost for many centuries, until he reappeared to fight alongside the Eldar at the Haranshemash (trans. World of Blood and Tears).

It is said that Fuegan will call together the Phoenix Lords for the Rhana Dandra, and that he will be last to die in that final conflict.

Asurmen

The Hand of Asuryan . . . . 137 pts

Aspect: Dire Avengers.

Wargear: Twin-linked shuriken catapults, Sword of Asur (Diresword, re-roll misses in close combat).

Warrior Powers: Defend, Distract, Battle Fate (Asurmen's save is invulnerable).

The Eldar believe that the first Exarch was Asurmen, founder of the Shrine of Asur, which was to grow into the most widespread of the Aspects – the Dire Avengers. It is claimed that he was first killed whilst fighting the minions of the Great Enemy, and that his many reincarnations have continued this fight. Asurmen's ritual armour is said to incorporate two mighty vambraces containing shuriken catapults, and it is believed that his Diresword was the first ever created and contains the sprit of his brother Tethesis, who was slain by a Daemon.

After the great cataclysm known as the Fall, the Eldar craftworlds were scattered across the galaxy. The Eldar path was begun as a means of controlling the raging emotions and desire for perfection that is the lure all Eldar must resist. The Path of the Warrior was founded on many craftworlds, and grew into the Aspects that survive to this day. These original Aspect temples were created by mighty Eldar fighters, the original Exarchs of the Bloody-handed God. The Eldar believe that these ancient figures survive to this day, continuing their fight for the Eldar cause whenever and wherever they appear. Many Eldar also believe that these Phoenix Lords, or Asurya as they are known to the Eldar, are no longer truly alive, that their armour is animated by the spirits of Exarchs who have been absorbed into the consciousness of the Phoenix Lord. If this is true, then the Phoenix Lords would indeed be mighty warriors, with many thousands of years of experience.

Extracts from Inquisitor Czevak's "Ancient Wanderers – the Phoenix Lords of the Eldar examined."

"War is my master; death my mistress."

Maugan Ra

SOULSTONE
Relates to secretive, denial, hiding or preservation key runes.

THE ELDAR OF THE STARS
Rune for Craftworlds and their inhabitants. Connotations of imprisonment, eternity, rigidity and self-denial. Stylised in present inactive tense.

SALVATION
Lit: One who has passed from the shadows into the light.

ELDAR OF HISTORY
This rune represents the Exodites. Similarities to pre-Fall runes for regression and escape. Stylised in the ancient past inactive tense. Phrase above main rune means isolated, lost, divorced and disenchanted.

FREEDOM
Other meanings include victory or transcendence. Similar to runes for hope and future.

THE SOUL-LESS ONES
Formed from runes connected with hunger, so could mean ones with no hunger, devoured ones or even the ever-hungry ones.

SOLITAIRE
Soulless or living dead. However, also a rune of hope or rescue.

THE GOD OF LAUGHTER
'Those who travel' or 'The Fearless'. Stylised in the future active tense, suggests things to come rather than things which have come to pass. Also based upon an inverse rune which means the 'Great Enemy'.

WORLD SPIRIT
Abandoned or forgotten, also connections with runes for eternally lost or beyond retrieval.

OUTCASTS
'Those who wander'. Ancient meaning is shadow or the lost.

THE DARK KIN
Anarchy and violence, corruption and torture, pain and misery. Forms of this rune mean soulthieves, thirsting ones, the predators or forever damned.

SOUL-DRINKER
Also scavengers, looters, parasites, vermin, dying, diabolical, demoniac and decaying. One of the most hideous rune-concepts in Eldar language.

Sirs... The artefact above [illustrated by my Uncle Langstri Mung] was taken from an ancient Eldar site on the third moon of Paravax by my great, great-grandfather Eleusis Mung. Since then our family has endeavoured to understand its full import. We have done much research into the mysterious Eldar language and believe it to be some form of physical representation of the beliefs of the disparate Eldar races.

Your Servant, Obelius Mung

INTROSPECTIONS UPON PERFECTION
by Kysaduras the Anchorite

In the spring storms of youth, it is common amongst our people to question the validity, and indeed sanity, of our ways, most especially the pursuit of perfection in but one field of endeavour at a time: the Path, as it has been called since our ancestors created it. While the young are intellectually capable of studying the tragic lessons of the Fall and the Great Enemy unleashed by our ancestors, their view of the universe is too narrow to truly see the lessons intrinsic in the terrible events which destroyed our home worlds and drove forth the survivors to wander the stars. It is true of all that in youth there is great bravery and great foolishness in equal measure, an abiding belief that no obstacle is too great to overcome, no foe too mighty to defeat, no problem so complex that it cannot be solved. Conversely it may be said that those who survive the galaxy's tumult for long enough come to believe that all obstacles, foes and problems may not be resolved, only allayed for a brief sliver of history, which in turn is but an instant in the slow dance of the universe.

Thus it is that young and old clash incessantly over the necessity of the Path. The young rail against the restrictions it imposes upon them. Much as our doomed forebears did, they wish to taste every sensation, every emotion within their new-found world as soon as possible. They do not fear the Great Enemy that was created by the desires of our ancestors, for their whole conception of her evil is gleaned from distant tales and legends, and that which brought fear in the nursery is spurned and ridiculed in adolescence. Only with time can they begin to feel the terrible thirst which we gave her and begin to understand that she is a mirror image, a reflection of our worst excesses given life by the debauchery and depravity which preceded the Fall.

Long ago our race realised that the only way to elude the Great Enemy was to shatter the reflection, to live a life of denial and focus upon but one aspect of life, pursuing it unto perfection. This is anathema to the young, just as it is to the Great Enemy. The young do not desire the discipline of the Path, but rather their curiosity drives them to try every fruit from the tree. Thus it is that so many take the Path of Wandering or the Path of Damnation in their first years of adulthood, and so the great tragedy of our kind is played out again and again as the number of our people shrink from generation to generation.

Of all the things I learnt during my sojourn upon that vast and aptly named 'craftworld', none was more horrifying than the secret of the infinity circuit (a poor translation of the Eldar term for the device, but one which must suffice). I had been amongst these strange alien creatures for some months, and was slowly learning something of their customs. The Eldar are an enigmatic race, who will rarely give a direct answer to a direct question. To learn anything of their ways I was forced to slowly piece together information from observation and careful consideration of the truth hidden in the largely allegorical answers my guide and host would deign to give me.

So it was that I had for some time been trying to fathom the Eldar's attitude to that which must needs come to us all, namely the restful sleep of death. Of all my questions, it was those upon this subject which my host seemed the most reluctant to answer. All intelligent creatures must struggle to come to terms with their own mortality, and I know of no-one who has not had to conquer his fear of dying at some time. Even the bravest of warriors must overcome this fear; indeed, it is by such confrontation that they prove themselves truly brave. Yet, for such an aloof and seemingly wise race, the Eldar appear to have a fear of dying which exceeds that of any people I have ever met. Yet slowly I was able to piece together the truth, and came to understand why the Eldar view death with such horror, and how they avoid it through the medium of the infinity circuit.

It is common knowledge that each Eldar bears upon his breast a highly polished gemstone. Some consider these an affectation or mere decorative bauble. Nothing could be further from the truth! These devices, whose name best translates as 'spirit stones', are actually psycho-receptive crystals attuned solely to the mind of their owner, and which are designed to capture the very souls of the Eldar at the moment of death. Why exactly the Eldar should go to such lengths to capture this psychic energy I was never able to find out; all I could ascertain was that the Eldar appear to have a belief that should their soul not be captured in this way then it would be lost to a strange shadow-realm where it would, quite literally, suffer a fate worse than death. What fate could be worse than the half-death of imprisonment in the cold crystal of a spirit stone is hard to imagine, yet the fact remains that the Eldar prefer an infinity trapped in this way, and that the alternative is looked upon with a dread unmatched by any I have ever seen.

But enough pointless hypothesising - the fact is that at death the soul of an Eldar is captured by the spirit stone he wears on his breast. The majority of such 'inhabited' spirit-stones are taken to a place known as the Dome of the Crystal Seers, or at least, such was the case in the craftworld upon which I resided, and I have no reason to doubt that it is the same elsewhere. Here they rest and, one hopes, find some sort of peace. Sometimes, however, a spirit stone is grafted to the robotic body of a Wraithguard or Eldar Dreadnought, imbuing its artificial form with a living intellect. The horror of such a fate is difficult to imagine, dooming the recipient, as it does, to an eternal shadow-life trapped in a shell of cold unfeeling steel. These spirit-warriors defend their craftworld and are much revered by their mortal counter-parts... yet I cannot help but think that such honour is little reward for so great a sacrifice.

Extract from "My Time Amongst the Eldar, or How I Visited Iyanden Craftworld and Lived!" by Ieldan Soecr

ADDENDUM

Extract from Imperium training document file Eld/465./version 2c

The information contained herein is intended as a field reference guide only.

For more detailed information and reports on Eldar weapons and known technology see files 729585 – 944475.

The Eldar are famed for their shuriken weapons. These range in size from the shuriken pistol up to the shuriken cannon, and all work on a similar principle. The ammunition is stored as a solid core of plasti-crystal material that is forced up from the magazine by a magnetic repulsor. A series of rapid high-energy impulses originate at the rear of the weapon then move it forward at a terrific speed. These impulses detach a monomolecular slice of the ammunition core and hurl it from the weapon's barrel, while the ammunition core is kept in the line of the firing impulse by the magnetic repulsor.

This allows the weapon to fire up to a hundred rounds of ammunition in a burst of one or two seconds, and each ammunition core is good for ten or more bursts of fire before it needs replacing. The downside of this firing mechanism is its lack of rifling on the barrel, which drastically reduces its accuracy, keeping the weapon's effective range below that of standard solid ammunition weapons of similar size.

Eldar laser weapons appear to function in a similar way to our own, using highly focused light beams to cause traumatic temperature change on impact with the target. However, they have far more efficient power generation and transmission systems, using artificially grown crystals to filter and refine the laser bursts to their optimum power and potency. This has given rise to such weapons as the bright lance, scatter laser, lasblaster and prism cannon.

The bright lance's highly charged energy bolts are much more accurate than those of an Imperial issue lascannon, making armour under a certain thickness redundant. The scatter laser and lasblaster use crystalline power cells to store up a charge of laser energy and then fire them in a concentrated burst, much like our own multi-lasers but in a far more energy-efficient manner. The prism cannon uses a two-stage firing process, with a medium magnitude laser charge fed into a crystal prism which greatly amplifies the shot in a fraction of a second, whilst dispersing the energy burst to target a wider area.

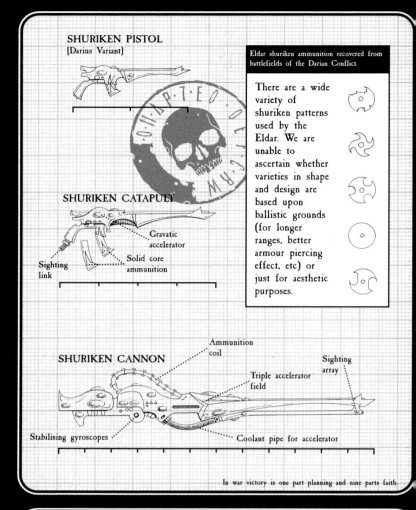

SHURIKEN PISTOL
[Darius Variant]

Eldar shuriken ammunition recovered from battlefields of the Darian Conflict.

There are a wide variety of shuriken patterns used by the Eldar. We are unable to ascertain whether varieties in shape and design are based upon ballistic grounds (for longer ranges, better armour piercing effect, etc) or just for aesthetic purposes.

SHURIKEN CATAPULT
— Gravatic accelerator
Sighting link
Solid core ammunition

SHURIKEN CANNON
Ammunition coil
Triple accelerator field
Sighting array
Stabilising gyroscopes
Coolant pipe for accelerator

In war victory is one part planning and nine parts faith.

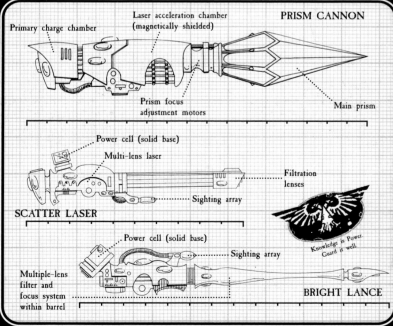

PRISM CANNON
Primary charge chamber
Laser acceleration chamber (magnetically shielded)
Prism focus adjustment motors
Main prism

Power cell (solid base)
Multi-lens laser
Filtration lenses
Sighting array
SCATTER LASER

Knowledge is Power. Guard it well.

Power cell (solid base)
Sighting array
Multiple-lens filter and focus system within barrel
BRIGHT LANCE

DISTORT CANNON

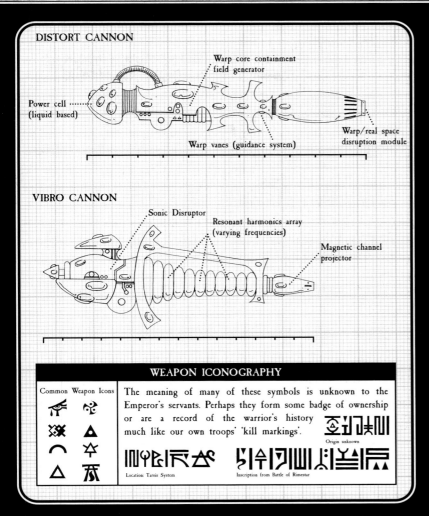

Warp core containment field generator

Power cell (liquid based)

Warp vanes (guidance system)

Warp/real space disruption module

VIBRO CANNON

Sonic Disruptor

Resonant harmonics array (varying frequencies)

Magnetic channel projector

WEAPON ICONOGRAPHY

Common Weapon Icons	

The meaning of many of these symbols is unknown to the Emperor's servants. Perhaps they form some badge of ownership or are a record of the warrior's history much like our own troops' 'kill markings'.

Origin unknown

Location: Tarsis System

Inscription from Battle of Rimestar

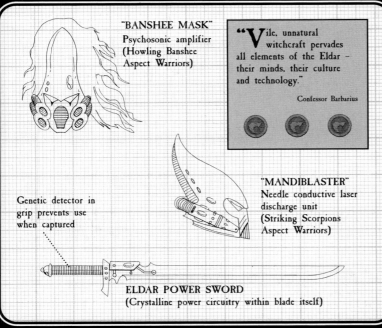

"BANSHEE MASK"
Psychosonic amplifier (Howling Banshee Aspect Warriors)

> **"V**ile, unnatural witchcraft pervades all elements of the Eldar – their minds, their culture and technology."
>
> Confessor Barbarius

Genetic detector in grip prevents use when captured

"MANDIBLASTER"
Needle conductive laser discharge unit (Striking Scorpions Aspect Warriors)

ELDAR POWER SWORD
(Crystalline power circuitry within blade itself)

Although shuriken and laser technology form the bulk of the Eldar arsenal, they employ a number of large support weapons that use very sophisticated technologies. Most common of these are the Distort cannon, Shadow Weaver and vibro cannon.

The Distort cannon uses a warp engine to collapse an area of real space, effectively creating a miniature warp hole. If the target is not wholly swept into warp space, it is most usually torn to pieces by the complex gravitational forces employed. Luckily, the D-cannon is relatively short-ranged and inaccurate by Eldar standards.

The Shadow Weaver creates a dense monofilament mesh from an as yet unidentified organo-polymer compound. This compound is kept in a liquid state within a magnetic reservoir, and when released through the thousands of microscopic firing ducts is woven into a web-like cloud by spinning gravity clamps. The clouds are forced high into the air before they drift down, making them ideal for disrupting an attack and causing the Eldar's enemies to seek shelter.

The vibro cannon contains a sonic field generator that creates a rapidly scaling wave of sonic energy ranging from ultrasound to hypersonic frequencies. This wave is directed along a magnetic tunnel, and when it hits, the resulting resonant frequencies shake the target apart. This is most evident when two or more vibro-cannons cross their sonic beams, causing tremendously powerful disparate energy waves which can damage even the toughest target.

Like all wargear of the Eldar, their close combat weapons are highly advanced and well crafted. As with most races, they make wide use of mono-molecular edged blades and disruption powerfields. However, there are also numerous other devices designed to overcome the foe in hand-to-hand fighting. Perhaps the most fearsome is the Banshee Mask, which contains complex circuitry designed to amplify the Howling Banshee's war cries into a psychic shockwave which obliterates the nervous system and scrambles the brain's neural pathways, rendering the victim incapable for a few seconds, or causing death in extreme cases.

The Striking Scorpions' so-called mandiblaster is another example of exotic wargear. Triggered by a psychic pick-up in the helmet, the mandiblaster fires a hail of needles at the foe which act as a conductor for a short-ranged but powerful laser 'sting'. Comparisons have been drawn between the mandiblaster and the Tormentor Helm of certain Eldar pirates, which utilises a similar mechanism.

To gain greater understanding of the Eldar mind, one must begin to understand their language. It is a highly evolved system of communication, developed over many tens of thousands of years. The Eldar have many tongues and dialects, some specific to craftworlds and castes within their social structure, others more widespread. It must be noted that words can appear in many different sources and languages, and yet have widely different meanings in each occurrence. This is because many Eldar words carry a range of connotations and references to the Eldar myth cycles. Thus, the name given to their most prevalent war engine, Faolchú, is most readily translated as 'Falcon'. However, Faolchú is not just any bird of prey, but is in fact the legendary Falcon of Eldar myth. It is a word that is replete with implications of vengeance and retribution, of justice and the slaying of wrongdoers. In this way Faolchú appears in many texts and is wrongly translated as meaning the warcraft of the same name, rather than the concept of revenge. Similarly, murehk, the name given to the shuriken pistol, means 'Sting of the wasp', another mythical creature which was said to plague the gods throughout the war in Heaven.

The term normally translated as human, "mon-keigh", can actually be found in stories dating thousands of years before the first contact between humans and Eldar, and refers to a race of sub-intelligent beasts that lived in the twilight realm of Koldo. These beasts invaded the Eldar lands and subjugated them for many years. The mon-keigh of legend were cannibalistic, misshapen monstrosities, eventually cleansed from the galaxy by the hero Elronhir. It can thus be surmised that the word mon-keigh refers to any non-Eldar species the Eldar deem inferior, in need of extermination.

It is almost impossible for an outsider to understand anything but the most basic attributes of the Eldar language, as many of its references draw directly upon the Eldar psyche, mythical peoples and places, and long-lost times and events. For example, rhiantha means, at the fundamental level, 'starlight'. However, a full translation would read more like 'the starlight which shines upon the waters of Rhidhol during the winter'. Without knowing where Rhidhol is, or even if it is a real or mythical place, the full meaning is impossible to ascertain.

Things become even more convoluted when these words are placed within a sentence - 'Elthir corannir rhiantha en' is translated literally as 'the Eldar maiden who weeps tears for the warrior-folk in the starlight which shines upon the waters of Rhidhol during the winter'. In our own rather basic terms, the phrase would translate as 'widow' or 'mourner', but in the Eldar tongue it is a much deeper expression of grief and loss, with implications of eternal woe and heartache.

Personal names are of two types - titles of rank and purpose or inherited names from the ancient tales. The first of these may also include references to the individual's prowess and temperament. Many Eldar have a ceremonial name and a common name. Warriors often have a third, a battle name. Times and circumstances when different names are used is a matter of old customs - to use a common name at the wrong moment would be a great insult, while to refer to a person by their ceremonial name at the wrong time is seen as fussy and aloof.

Ceremonial names often grow longer with age, as achievements and abilities grow. The learnéd Eldar who was kind enough to educate me in these matters was called Alai-Eltanomorreiasalonethatil - Alai (of the Alaitoc craftworld) - Eltano (his birth name) - mor (the wise) - reia (rising star) - salo (the teacher) - nethatil (family name ending). During common conversation he is called Elthil (simply a contraction of his birth-name and family name). The second type of name is most often used by the warriors, and is handed down from generation to generation, from the time of legends. The Exarchs have the names of great warriors from myth and each successive Exarch who wears the sacred armour takes on that name, forgetting their past life.

The Eldar written language is similarly complex. Each rune is not a simple letter form like our own written Gothic, but is a symbol of a concept, much as many words in the language are more than just description. Even more strangely, many of these word-concepts, while being pronounced the same, have a subtly different meaning when committed to writing. It is an area that I am barely conversant with, and do not truly understand myself. It has taken myself and my one hundred and six predecessors this long to even grasp the most basic ideas behind the spoken tongue, and it will be many more centuries of study by myself and my successors before we can ever claim to truly understand these enigmatic aliens.

Report of Lexicos Aldus Mari, last diplomatic envoy to the Alaitoc Craftworld, relations ceased due to outbreak of Beelze Conflict [453.M36].

Inquisition Authorisation:
Essential use only
Subject: Eldar language
File Status: Pending completion
Future project authorisation:
Subject to Inquisitiorial control

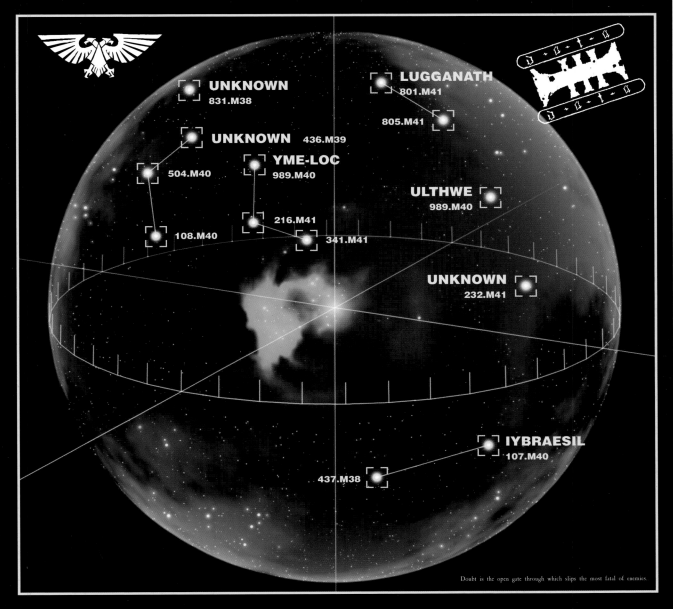

UNKNOWN
831.M38

LUGGANATH
801.M41

805.M41

UNKNOWN 436.M39

YME-LOC
989.M40

504.M40

ULTHWE
989.M40

108.M40

216.M41

341.M41

UNKNOWN
232.M41

IYBRAESIL
107.M40

437.M38

Doubt is the open gate through which slips the most fatal of enemies.

STUDY OF THE DISPLACEMENT OF THE ELDAR IN SEGMENTUM OBSCURUS, WESTERN QUADRANT

By the hand of Lord Captain Morley, attaché to the Fleet Insturum of Alien Studies, docking complex Heracles, Cypra Mundi.

By the very nature of space, it is only with extreme difficulty that one can locate and record the positions of enemy installations. When one is dealing with the slippery Eldar, the problem is increased. Firstly, they are ill-disposed to the rightful presence of Imperial ships in their vicinity. They jealously guard their meagre territories and are exceptionally loath for outsiders to discover their presence. Secondly, the deceitful Eldar are highly adept at evasion and stealth. Their technology is geared towards deception, misdirection and eluding detection, so that if one were unlucky enough to stumble upon them, it is entirely possible that one would be unaware of the fact. Their ability to move their vessels more rapidly from system to system than any other known race also makes hunting them down difficult in the extreme. Despite these factors, we can collate a rough picture of their whereabouts in the area of this study.

Most Eldar live upon gigantic vessels known as craftworlds, which travel the void endlessly. It is supposed, and highly likely, that these enormous constructions are capable of warp travel, although there are no confirmed accounts of such an occurrence. Every craftworld is a self-contained environment, with no need for planetary colonies or outposts. Each operates independently from the others. Though they often aid each other in times of conflict, it is also not unknown for the self-interested Eldar to fight among themselves (this is however a rare occurrence). Due to the sheer spatial distances between most craftworlds, it seems unlikely that the fragmented Eldar have much contact with each other. The schematic above shows the approximate locations of several craftworlds (and years of their discovery) which have been detected in the study region. Annotations (including names) are derived from information released by the Inquisition.

CODEX™

ELDAR®

"Trust not in their appearance, for the Eldar are as utterly alien to good, honest men as the vile Tyranids and savage Orks. They are capricious and fickle, attacking without cause or warning. There is no understanding them for there is nothing to understand – they are a random force in the universe."

Imperial Commander Abriel Hume.

The craftworlds of the Eldar are scattered across the stars, massive drifting starships that are home to the last survivors of a race that once ruled the universe. Now the Eldar battle against the other races of the galaxy, even resorting to piracy in their fight to avoid extinction.

Inside you will find:

Army List: A full army list for the Eldar of the craftworlds, with details of their potent war machines, the specialised Aspect Warriors and the powerful Warlock and Farseer psykers.

Hobby Section: A full colour guide to collecting and painting an Eldar army, including information on the different craftworlds encountered by mankind, and the forces that fight for them in battle.

The Eldar: Numerous pieces of background information regarding the enigmatic Eldar, along with rules for famous warriors such as the Phoenix Lords, Eldrad Ulthran and Nuadhu 'Fireheart' – Wild Rider Chief of the Saim Hann.

This is a supplement for Warhammer 40,000. You must possess a copy of Warhammer 40,000 to be able to use the contents of this book.

ISBN 1 869893 39 5

PRODUCT CODE **60 03 01 04 001**

UK
Games Workshop Ltd
Willow Road,
Lenton,
Nottingham,
NG7 2WS.

AUSTRALIA
Games Workshop
23 Liverpool Street,
Ingleburn,
NSW 2565.

US
Games Workshop Inc
6721 Baymeadow Drive,
Glen Burnie,
Maryland,
21060-6401.

CANADA
Games Workshop
2679 Bristol Circle
Units 2&3
Oakville, Ontario
Canada
L6H 6Z8

JAPAN
Games Workshop
Willow Road,
Lenton,
Nottingham,
NG7 2WS.

PRINTED IN THE UK

CITADEL® MINIATURES

5 011921 968923